Delmar Publishers
is proud to support
FFA activities

Ecology of Fish and Wildlife

DeVere Burton

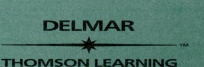

ITP

DELMAR

THOMSON LEARNING

Africa • Australia • Canada • Denmark • Japan • Mexico • New Zealand • Philippines
Puerto Rico • Singapore • Spain • United Kingdom • United States

NOTICE TO THE READER

Publisher does not warrant or guarantee any of the products described herein or perform any independent analysis in connection with any of the product information contained herein. Publisher does not assume, and expressly disclaims, any obligation to obtain and include information other than that provided to it by the manufacturer.

The reader is expressly warned to consider and adopt all safety precautions that might be indicated by the activities herein and to avoid all potential hazards. By following the instructions contained herein, the reader willingly assumes all risks in connection with such instructions.

The Publisher makes no representation or warranties of any kind, including but not limited to, the warranties of fitness for particular purpose or merchantability, nor are any such representations implied with respect to the material set forth herein, and the publisher takes no responsibility with respect to such material. The publisher shall not be liable for any special, consequential, or exemplary damages resulting, in whole or part, from the readers' use of, or reliance upon, this material.

Cover Art: Pete Palumbo   Cover Design: Kristina Almquist

Delmar Staff
Publisher: Tim O'Leary   Production Manager: Wendy Troeger
Administrative Editor: Cathy Esperti   Senior Project Editor: Andrea Edwards Myers

Printed in the United States of America
4 5 6 7 8 9 10 XXX 01 00

For more information, contact Delmar, 3 Columbia Circle, PO Box 15015, Albany, NY 12212-0515; or find us on the World Wide Web at http://www.delmar.com

International Division List

Asia
Thomson Learning
60 Albert Street, #15-01
Albert Complex
Singapore 189969
Tel: 65 336 6411
Fax: 65 336 7411

Japan:
Thomson Learning
Palaceside Building 5F
1-1-1 Hitotsubashi, Chiyoda-ku
Tokyo 100 0003 Japan
Tel: 813 5218 6544
Fax: 813 5218 6551

Australia/New Zealand:
Nelson/Thomson Learning
102 Dodds Street
South Melbourne, Victoria 3205
Australia
Tel: 61 39 685 4111
Fax: 61 39 685 4199

UK/Europe/Middle East
Thomson Learning
Berkshire House
168-173 High Holborn
London
WC1V 7AA United Kingdom
Tel: 44 171 497 1422
Fax: 44 171 497 1426

Latin America:
Thomson Learning
Seneca, 53
Colonia Polanco
11560 Mexico D.F. Mexico
Tel: 525-281-2906
Fax: 525-281-2656

Canada:
Nelson/Thomson Learning
1120 Birchmount Road
Scarborough, Ontario
Canada M1K 5G4
Tel: 416-752-9100
Fax: 416-752-8102

For permission to use material from this text or product contact us by Tel (800) 730-2214; Fax (800) 730-2215; www.thomsonrights.com

Library of Congress Cataloging-in-Publication Data:
Burton, L. DeVere.
        Ecology of fish and wildlife / DeVere Burton.
             p. cm.
        Includes index.
        ISBN 0-8273-6065-7 (textbook)   1. Animal ecology—Juvenile literature.   2. Wildlife conservation—Juvenile
             literature. [1. Animal ecology. 2. Ecology. 3. Wildlife conservation.] I. Title.
        QH541.B87   1995   94-44242
                                                   CIP
                                                   AC

Contents

iv

SECTION V: CONSERVATION AND MANAGEMENT   341

Preface

The fish and wildlife populations of North America are among our most treasured natural resources. . . and one of the most vulnerable. They are treasured not only because of the deep appreciation many people have for wild creatures, but also for the part they play in our ecosystem. They are vulnerable because certain activities can upset delicate relationships between wild animals and their living environments.

This textbook is about fish and wildlife ecology. It defines relationships between the wild creatures and their habitats, and with each other. It also discusses human activities that impact fish and wildlife populations. It is a pragmatic approach to managing our fish and wildlife resources in the world that we have inherited from past generations, and that our own generation has helped to create.

A basic premise of this textbook is that decisions regarding the management of fish and wildlife populations should be based on reliable research subjected to the peer review process. Broad areas of disagreement exist among those who conduct research on ecology issues. It seems that some management decisions are affected as much by politics as they are by science.

A serious attempt has been made to present both sides of major ecology issues in this text. It is the contention of the author that we should seek "middle ground" in resolving difficult environmental issues, and that it is wise to avoid radical positions on either side of an issue. It is also important to consider the long-term effects that some management strategies are likely to have on humans. It is probably unrealistic to expect that people will abruptly cease to use resources that have historically provided income and security for their families

This textbook is divided into five sections:

SECTION I   — Ecology Basics
SECTION II — Ecology of Mammals
SECTION III — Ecology of Birds
SECTION IV — Ecology of Fishes, Reptiles & Amphibians
SECTION V   — Conservation and Management

The first section deals with our understanding of how the earth and the various ecosystems function. Sections II, III & IV present specific information about the fish and animals that are found in the North American ecosystem. Section V addresses the issues of conserving and managing fish and wildlife populations.

Included in each chapter are features entitled Objectives, Terms For Understanding, Animal Profiles, Career Options, Looking Back, Review and Learning Activities. Each chapter is filled with photographs and illustrations that will aid students as they seek understanding of the concepts that are presented.

The School-To-Work movement that is sweeping through our education system requires integration of academic and vocational education. This textbook is intended to facilitate integration of science in curricula dealing with management of wildlife and natural resources.

About the Author

L. DeVere Burton, author of *Ecology of Fish and Wildlife*, is State Supervisor, Agricultural Science & Technology in the Idaho State Division of Vocational Education. As this text is published, he is serving a term as President of the National Association of Supervisors of Agriculture Education.

The author was a high school agriculture teacher for 15 years, and has been involved as a professional educator in agricultural education since 1967. He has experienced teaching assignments in large and small schools, and in single and multiple teacher departments. He has taught at four different high schools, and at a major land grant university. He has been involved in program supervision since 1987. All of these experiences have contributed to his philosophy that "education must be fun and exciting for those who learn and for those who teach."

His first experience with agricultural education occurred when his high school agriculture teacher recruited him to a ninth grade agriculture class at Star Valley High School in Afton, Wyoming. By the end of his second year as an FFA member, he had discovered his career in agricultural education. He has never looked back, or regretted his choice.

A wide range of agricultural experiences have prepared the author for his career as an educator in agriculture. He was raised on a farm in Western Wyoming that produced hay, grain, milk, eggs, and a few hogs and sheep. His high school jobs included testing milk for butterfat content, and caring for the hogs on a combination beef, swine, and trout ranch. During college he worked as a maintenance/warehouse worker in a feed mill, manager of a dairy farm, sawmill worker, logger, finish carpenter, and animal research assistant. He has worked in the food processing, metal fabrication, and concrete industries, and he owned and managed a purebred sheep and row crop farm for several years.

Dr. Burton earned his B.S. degree in Agricultural Education from Utah State University in 1967. He was awarded his M.S. degree in Animal Science from Brigham Young University in 1972. His Ph.D. degree was earned at Iowa State University in 1987 where he was also an instructor in the Agricultural Engineering Department.

This is the second textbook that Dr. Burton has authored. His first text is entitled *Agriscience & Technology*. It was written in a serious attempt to strengthen the science content, and to expand the breadth of the curriculum in agricultural education.

Acknowledgments

This book is dedicated to students and teachers who appreciate the outdoors, and the wild animals that live there. It is also dedicated to the people who work in the agriculture and natural resource systems of North America. These are the individuals whose management decisions directly affect our wild animal resources and our citizens.

Many people are involved in the creation of a textbook such as this one. Gratitude is expressed to family members, friends, and colleagues whose patience and encouragement have contributed to the completion of this work. Special thanks to the editors and reviewers for their many hours evaluating the manuscript and gathering materials to support the publication of this book.

Thanks to all who have contributed ideas, pictures, materials, and technical expertise:

Idaho Department of Fish and Game
Boise, Idaho

U.S. Fish and Wildlife Service
Washington, D.C.

David Milton
Fish and Wildlife Instructor
Boise, Idaho

Robert Pratt
Agriculture Teacher
Nezperce, Idaho

Birds of Prey National Monument
Boise, Idaho

Weldon Sleight, Ph.D.
Utah Agriculture Experiment Station
Logan, Utah

Gary Newenswander
Utah Agriculture Experiment Station
Logan, Utah

The Idaho Statesman
Boise, Idaho

Starkey Experimental Range and Forest
La Grande, Oregon

University of Idaho
Moscow, Idaho

Boise State University
Boise, Idaho

United States Department of Agriculture
Washington, D.C.

United States Department of the Interior
Washington, D.C.

Museum of Natural History
Washington, D.C.

Life Magazine

National Geographic Magazine

Project Wild

Using Our Natural Resources
1983 Yearbook of Agriculture

Biology, The Living World, by Peeter Alexander, Mary Jean Bahret, Judith Chaves, Gary Courts, and Naomi Skolky D'Alessio, copyright 1989 by Prentice-Hall.

Wild Animals of North America, by Wayne Barrett and Edward Park, copyright 1960 by the National Geographic Society.

Field Guide To North American Fishes, Whales & Dolphins, by Herbert T. Boschung, Jr., James D. Williams, Daniel W. Gotshall, David K. Caldwell, and Melba C. Caldwell, copyright 1983 by Chanticleer Press Inc.

Mammals, Third Edition, by William H. Burt and Richard P. Grossenheider, copyright 1980 by Houghton Mifflin Company.

Mammals, John A Burton and Jim Channell, copyright 1991 by Smithmark Publishers inc.

Agriscience & Technology, by L. DeVere Burton, copyright 1992 by Delmar Publishers.

Environmental Science For Agriculture And The Life Sciences, by William G. Camp and Roy L. Donahue, copyright 1994 by Delmar Publishers.

Reptiles And Amphibians, by Roger Conant and Joseph T. Collins, copyright 1991 by Houghton Mifflin Company.

Managing Our Natural Resources, Second Edition, by William G. Camp and Roy L. Donahue, copyright 1991 by Delmar Publishers.

The Kingfisher Illustrated Encyclopedia Of Animals, by Michael Chinery, copyright 1992 by Kingfisher Books.

Agriscience Fundamentals & Applications, by Elmer L. Cooper, copyright 1990 by Delmar Publishers.

Grzimek's Animal Life Encyclopedia, by H. Grzimek and C. Bernhard, copyright 1975 by Van Nostrand Reinhold Company.

Magnificent Voyagers Waterfowl Of North America, by Albert H. Hochbaum, copyright 1988 by Thomasson-Grant Inc.

Environmental Science: The Way The World Works, by Bernard J. Nebel, copyright 1990 by Prentice-Hall.

A Field Guide To Freshwater Fishes, by Lawrence M. Page and Brooks M. Burr, copyright 1991 by Houghton Mifflin Company.

The Encyclopedia Of Birds, by Christopher M. Perrins and Alex L. A. Middleton, copyright 1985 by Equinox Ltd.

Birds Of America, by T. Gilbert Pearson, copyright 1936 by Doubleday and Company, Inc.

Environmental Science: How The World Works And Your Place In It, by Jane L. Person, copyright 1989 by J. M. LeBel Enterprises Inc.

Eastern Birds, by Roger Tory Peterson, copyright 1980 by Houghton Mifflin Company.

The Audubon Society Field Guide To North American Birds, by Miklos D. F. Udvardy, copyright 1977 by Chanticleer Press.

Mammals, by Herbert S. Zim and Donald F. Hoffmeister, copyright 1987 by Western Publishing Company, Inc.

REVIEWERS

Lynne Cook
Georgia

Roy Crawford
Texas

Emily Gurlich
Washington

David Jensen
South Carolina

John Joyner
Georgia

Gerald McDonald
Texas

Bob Vandenbusch
Wisconsin

Dave Wyrick
Michigan

SECTION I

Ecology Basics

Ecology is the branch of biology that describes relationships between living organisms and the environments in which they live.

1 Principles of Ecology

THE BRANCH of biology that describes relationships between organisms and the environments in which they live is called **ecology**. Different organisms each relate differently to their environments. They eat different foods, seek different kinds of shelter, and require different environments in which to raise their young. They are also used as food by a variety of other organisms. All of these relationships are described in the science of ecology. It is concerned with all of the activities that affect the environment and the living organisms that depend upon it for food and shelter.

OBJECTIVES

After completing this chapter, you should be able to:

■ explain the law of conservation of matter
■ suggest ways that waste materials can be properly disposed of to reduce or eliminate damage to the environment
■ relate pollution of the environment to the law of conservation of matter
■ examine the positive and negative effects of waste control methods that are used by modern society
■ define the role of energy in the science of ecology
■ explain the first and second laws of energy
■ describe the major events that occur in natural cycles such as the carbon, nitrogen and water cycles
■ investigate the importance of water to living organisms
■ explain how a food chain is organized
■ distinguish among food chains, food webs, and food pyramids.

CONSERVATION OF MATTER

A basic law of physics is the **law of conservation of matter**: Matter can be changed from one form to another, but it cannot be created or destroyed by ordinary physical or chemical processes. This law holds true in the study of ecology.

The law of conservation of matter applies to everything that exists. Most organisms use only those materials that make up their food supply, and little waste material is generated. The waste that is generated is capable of being recycled through natural processes. The human population generates waste materials that are not so easily disposed of. Some of the materials that humans create persist in the environment almost indefinitely, and some of these waste materials are harmful or toxic to other organisms.

Waste materials have become a major problem in many of the industrialized nations of the world. By-products of industrial processes, solid wastes from our population centers, pesticide residues from farms, gardens, and yards, petroleum leaks and spills are only a few of the waste materials that pollute the environment, see Figure 1-1. These materials do not just go away when we are finished with them.

The law of conservation of matter applies also where waste materials are concerned. We may change the form of waste materials to make them more compatible with the environment, but we cannot destroy them. With this in mind, we must properly dispose of all waste materials to prevent serious environmental problems, see Figure 1-2. We can help solve this problem by using fewer materials that are disposable, and by recycling waste materials.

Industrial wastes have been a serious problem for many years, see Figure 1-3. They include a variety of harmful chemicals, poisonous metallic compounds, and acids and other caustic materials that are left over from manufacturing processes. Many of our **surface waters** (rivers, streams, ponds, and lakes) are seriously polluted by industrial wastes, see Figure 1-4. In many cases fish

Figure 1-1: Surface water pollution is a major problem and causes serious damage to the environment.

Figure 1-2: Waste material must be properly disposed of to prevent serious environmental damage.

and other **aquatic species** of plants and animals that live in surface water environments have been poisoned.

Many industries have made a serious effort to reduce or eliminate water and air pollution by using water treatment plants and installing special equipment in smokestacks. New environmental laws have mandated improvements in the ways that **pollutants** (dangerous waste materials) are handled. It is important to research new ways to eliminate all forms of pollution to the environment regardless of its source.

Figure 1-3: Industrial air pollution has been a problem for years and many industries have made a serious effort to reduce it.

Figure 1-4: Many of our natural resources are seriously polluted by waste.

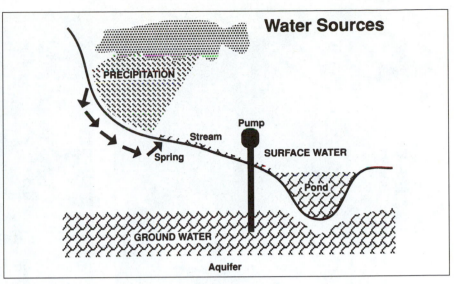

Figure 1-5: Water that is used to produce agricultural products is obtained from several sources. Many areas depend upon surface water and ground water for use in irrigation. *From Burton, Agriscience and Technology, copyright 1992 by Delmar Publishers.*

Most of the pollutants that affect surface water are also a problem in **groundwater**. This is the water that is located underground. It is stored in large underground reservoirs where it occupies the space between soil particles such as sand, gravel, and rocks, see Figure 1-5. Groundwater provides a water supply to man-made wells and naturally flowing springs.

Many waste materials can be changed to reduce or eliminate pollution. **Solid wastes** include most of the materials that are gathered by trash collectors for disposal. Much of this trash is buried in landfills. In recent years, we have learned that highly toxic liquids, see Figure 1-6 often ooze out of the landfills and pollute the water and soil in the local area. Material buried in landfills tends to remain for long periods of time, and it does not break down as

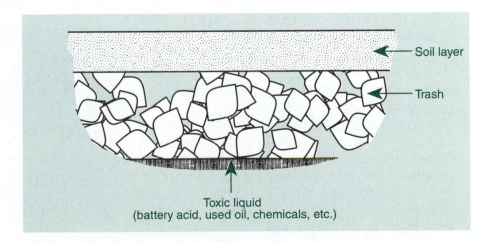

Figure 1-6: The layers of a landfill.

Figure 1-7: Plastic is one of the many products that we currently recycle. *Photo courtesy of Waste Management, Inc.*

Figure 1-8: Recycling metals is considered to be a good alternative to burying it, however, it is not a perfect solution to the problem.

quickly as many people expected. Much of it will still be there centuries from now in much the same form as when it was buried.

Some communities burn the combustible portion of their solid waste as a source of energy. Metals, glass, and plastic materials are often separated out to be recycled, see Figures 1-7, 1-8. Although this approach is considered to be a good alternative to burying garbage, it is not a perfect solution to the problem.

Career Option

ENVIRONMENTAL SCIENTIST

An environmental scientist conducts research related to pollution of the environment. He or she gathers and analyzes data to determine how the environment is affected by different approaches to controlling pollution.

They also use experimental results to determine pollution standards for government regulations, and they propose improved practices for managing pollution problems.

A career as an environmental scientist requires an advanced professional degree in chemistry or biology with emphasis on environmental science, wildlife and/or natural resources. High school preparation should include a strong curriculum in science and mathematics with a broad experience base in agriculture and natural resources.

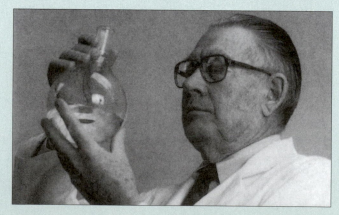

Photo courtesy of Utah Agricultural Experiment Station.

Care must be taken to prevent air pollution from the by-products of burning. The burning process releases large amounts of carbon monoxide and sulphur dioxide into the air along with ash and other gases that can be harmful to the health of the people, animals, and plants that live in the surrounding area. New technologies have been developed that are effective in trapping pollutants, but many of these processes are expensive and their use is limited.

Pesticides are chemicals that are used to control insects or weeds. When a pesticide is used only to kill insects, it is called an **insecticide**, see Figure 1-9. A pesticide that is used to control plants is called an **herbicide**. A **rodenticide** is used to poison rodents. Large amounts of pesticides are used each year on lawns, gardens, golf courses, and farms to control unwanted species of plants, insects, and rodents. All of these materials can be dangerous to the environment when they are not properly used.

Empty pesticide containers are dangerous when proper disposal methods are not used. Most pesticides are sold in metal or plastic containers. Those who use these materials should carefully triple-rinse empty containers with water and dispose of the container and the rinse water according to the directions on the label. Most empty pesticide containers are considered to be **hazardous materials**. Such materials are treated as threats to the environment, and their disposal is controlled by law.

Petroleum is an oily, flammable liquid that occurs naturally in large underground deposits. We call it crude oil, and from this basic material a large variety of products are manufactured. Gasoline and diesel fuels are the best-known and most widely used products obtained from petroleum.

When petroleum or petroleum products are spilled or leaked into the environment, they are often hazardous to the organisms that live there, see Figure 1-10. Thousands of aquatic animals, particularly fish and waterfowl, die each year due to spills of crude oil into the surface waters.

Figure 1-9: Chemicals that control weeds and pests can be dangerous to the environment. It is important that they are used properly. *Photo courtesy of Utah Agricultural Experiment Station.*

Figure 1-10: A live, oiled goldeneye injured by a petroleum spill. *Photo courtesy of U.S. Fish and Wildlife Service.*

Ecology Profile

EXXON VALDEZ OIL SPILL

One of the most damaging environmental accidents ever to occur took place on March 24, 1989 off the coast of Alaska. A large oil tanker ran aground, damaging the ship and spilling 11 million gallons of crude oil into the ocean at Prince William Sound. Despite a desperate effort to contain the spill, it became widely dispersed. It damaged a large area along the coast and killed massive numbers of birds, marine mammals, and fish. The ship's captain and owners were prosecuted for negligence that led to the spill, but we may never know the full extent of the damage to this ecologically sensitive area.

Petroleum products are also damaging to the environment when they leak or spill on land. Leaking underground fuel tanks have polluted groundwater, contaminating drinking water and destroying plant and animal life. Poisonous fumes from petroleum spills are hazardous to the health of humans as well as other animals, and they create potential fire hazards.

New environmental laws have been passed in recent years requiring the inspection of underground tanks to detect leaks. In some cases, leaks have been found after many years of environmental pollution, see Figure 1-11. Spills of this type are often difficult to find, and expensive to clean up. Some states have even required that underground fuel tanks be removed.

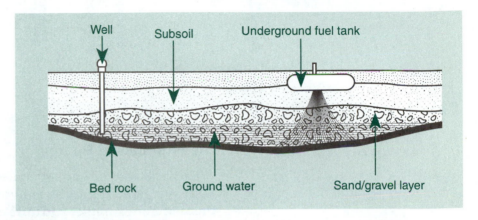

Figure 1-11: Underground fuel tank leakage.

LAWS OF ENERGY

Energy is the ability to do work or to cause changes to occur. It is the power or force that allows animals to move or the tides to flow. Energy flows through systems from areas where it is concentrated to areas where it becomes dispersed or unorganized. For example, food is the source of energy for animals. As it is digested, it gives off heat to warm the body of an animal, and it provides power to the muscles making movement possible. Even the ability of the brain to think depends on a supply of energy.

Energy also flows through entire ecosystems, see Figure 1-12. About two-thirds of the solar energy that passes from the sun to the earth is trapped by the land, water, plants, and atmosphere.

Solar energy is captured by plants in the form of sunlight and stored as molecules of sugar and starch. When a plant is eaten by an animal, the energy is released from the plant cells allowing the animal to do work, or it is stored in the body of the animal in the form of fats, proteins, or carbohydrates until it is needed. When plants or animals die, the energy may be stored in the form of fossil fuels (coal or oil deposits) that store energy for long periods of time, or given up in the form of heat as the remains decompose. Sometimes energy is transferred to other animals when the remains of dead plants and animals are eaten and digested for food.

Energy cannot be recycled, but can be stored for later use.

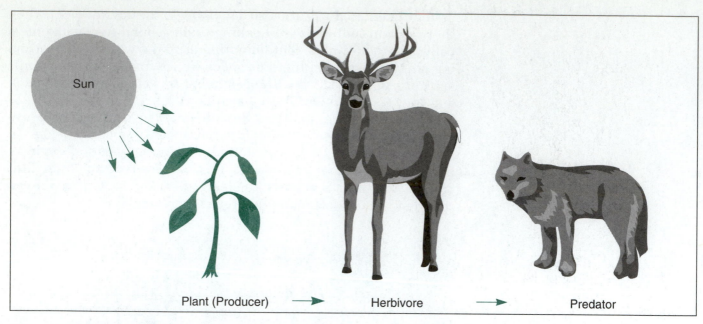

Figure 1-12: Energy flow in an ecosystem.

The First Law of Energy

The **first law of energy** states that energy cannot be created or destroyed, but it can be converted from one form of energy to another. For example, **radiant energy** which comes from the sun is converted to **chemical energy** (sugars and starches) by plant

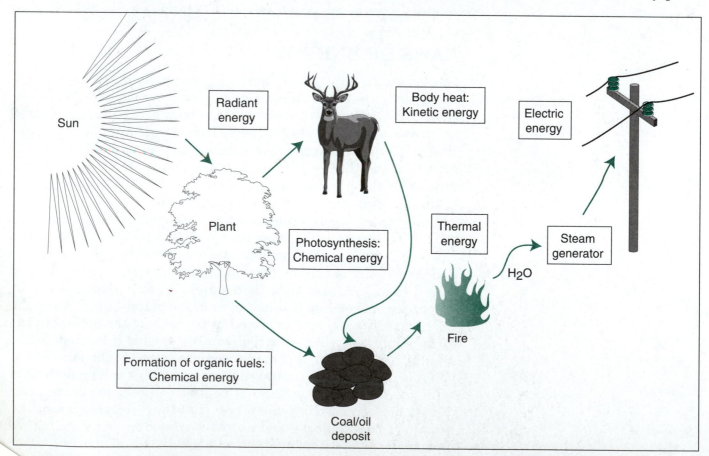

Figure 1-13: First law of energy.

leaves during the process of photosynthesis. Chemical energy from a plant that is eaten by an animal is converted into body heat and **kinetic energy**, the energy associated with motion and movement.

Large deposits of coal or crude oil have been formed from decayed plant materials. During the formation of these materials, energy is stored as chemical energy. **Thermal energy** is created in the form of heat when these fuels are burned. Thermal energy can be converted to **electrical energy** by heating water to operate a steam engine that uses kinetic energy to drive a generator, see Figure 1-13.

The Second Law of Energy

The **second law of energy** states that each time energy is converted from one form to another, some energy is lost in the form of heat, see Figure 1-14. The heat that is lost is not destroyed, but simply becomes unavailable for later use. Every single change in the form of energy results in the loss of heat.

It is possible for some of the heat that is lost from living plants and animals to be trapped when the energy in the food is converted from one form to another. The process of digestion releases heat into the body cavity where it is used to maintain body temperature in warm-blooded birds and mammals, see Figure 1-15.

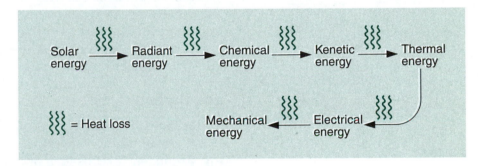

Figure 1-14: Second law of energy.

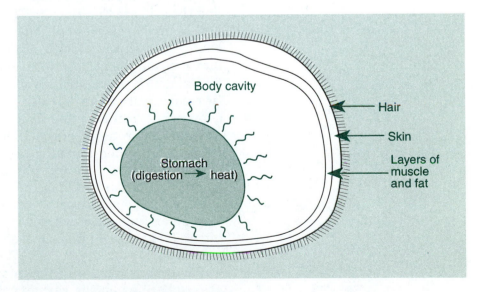

Figure 1-15: Maintaining body heat.

All living things respond to the laws of energy and all ecosystems on earth depend upon energy sources. Much of the controversy surrounding the energy conservation issue deals with this important principle: **we cannot continue to use more energy than the earth can replace.**

NATURAL CYCLES

Only a few of the known elements that are found in the upper crust of the earth, in water, or in the atmosphere are abundant in the tissues of living organisms. The most plentiful of these elements are carbon, hydrogen, oxygen, and nitrogen. These four elements account for 96 percent of the material that is found in living organisms. Over thirty other elements are known to make up the other 4 percent of living tissue.

These elements that are so important in the formation of living organisms are used over and over again. An atom of carbon may exist as part of a sugar molecule in a plant that is later eaten by an animal. It may then become part of the muscle of the animal. When the animal dies, the same atom of carbon may be passed into the soil or into the atmosphere when the tissue of the animal decomposes. Finally it may be taken up by another plant to form new plant tissue. In this manner, elements cycle from living organisms to nonliving materials and back, again and again. This circular flow of elements from living organisms to nonliving matter is known as an **elemental cycle**.

Cycles exist for all of the elements that make up living tissue. We will consider only the carbon and nitrogen cycles at this time.

The Carbon Cycle

Carbon is the most abundant element found in living organisms. It makes up the framework of the molecules that are found in living tissue. In the absence of water nearly half of the dry matter found in the bodies of animals or humans consists of carbon. Carbon moves readily between living organisms, the atmosphere and the soil, see Figure 1-16. The respiration process of both plants and animals releases carbon dioxide (CO_2) into the atmosphere. Using light in the process of photosynthesis, plants take carbon dioxide from the atmosphere and turn it into sugars that they use to make new tissue such as roots, stems, and leaves. When plant tissue decays, carbon dioxide is released back to the atmosphere as a gas or converted over a long period of time to **fossil fuels** such as natural gas, crude oil, or coal. Sometimes plant materials are eaten by animals. When animals die, their bodies decompose releasing carbon dioxide to the atmosphere as a gas, or the carbon from their bodies may be converted over long periods of time to fossil fuels.

People mine or extract fossil fuels from the surface of the earth for use as fuels and other purposes. When these materials are burned, the combustion process releases carbon dioxide to the atmosphere.

The oceans absorb large amounts of carbon dioxide from the atmosphere when the carbon content of the atmosphere is high, and they release carbon dioxide to the atmosphere when atmos-

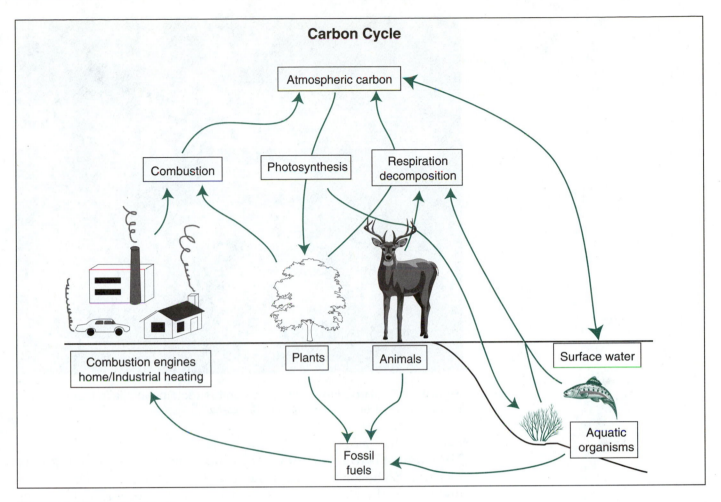

Figure 1-16: Carbon cycle.

pheric carbon dioxide decreases. Until recent years, the carbon content of the atmosphere has remained nearly the same due to this action by the oceans.

During the past one hundred years, the large amounts of fossil fuels that have been burned have increased the levels of carbon dioxide in the atmosphere. We are burning these fuels faster than the oceans can absorb the extra atmospheric carbon. We do not know the effect of our massive CO_2 inputs to the atmosphere, but one effect is thought to be global warming. Long-term climatic change could have drastic consequences for natural and agricultural ecosystems.

The Nitrogen Cycle

Nitrogen is the most abundant element in the atmosphere. It makes up about 80 percent of the air supply. In its elemental form (N_2) it is a colorless, odorless gas that cannot be used by plants or animals. It must be combined with oxygen or other elements before it is available as a nutrient for living organisms. Plants and animals use nitrogen compounds to form protein and other important molecules like DNA and vitamins.

Nitrogen fixation is a process by which nitrogen gas is converted to nitrates. This can occur in several different ways.

Figure 1-17: Nitrogen-fixing nodules on the roots of an alfalfa plant. *Photo courtesy of Dr. J. Burton, The Nitragin Company.*

Nitrogen-fixing bacteria are able to convert nitrogen gas to nitrates. Some forms of these bacteria live in the soil. Others live in nodules on the roots of clover, beans, peas, and other legumes, see Figure 1-17. These types of plants are able to make their own nitrogen fertilizer. Some types of blue green algae and fungi are also capable of nitrogen fixation.

Several industrial processes also convert nitrogen gas to nitrates. One of these processes converts nitrogen gas to ammonia as a by-product of steel production. Ammonia can also be obtained directly from natural gas. The ammonia is then converted to a form of nitrate which can be used for fertilizer, or added directly to soil as a fertilizer.

Nitrogen fixation occurs naturally in the atmosphere when lightning strikes. The electrical current that passes through the atmospheric nitrogen converts some of the nitrogen gas to nitrogen compounds that can be used by plants. Nitrates are also released from animal wastes, and from plants and animals that die and decay.

At the same time that nitrates are being produced from nitrogen gas, other nitrates are breaking down to release nitrogen gas back to the atmosphere. This process is called **denitrification.** It occurs when some forms of bacteria come into contact with nitrates. A similar process occurs when nitrates are carried by runoff water into surface water which constantly exchanges nitrogen with the atmosphere.

The circular flow of nitrogen from free nitrogen gas in the atmosphere to nitrates in the soil and back to atmospheric nitrogen is known as the **nitrogen cycle**, see Figure 1-18.

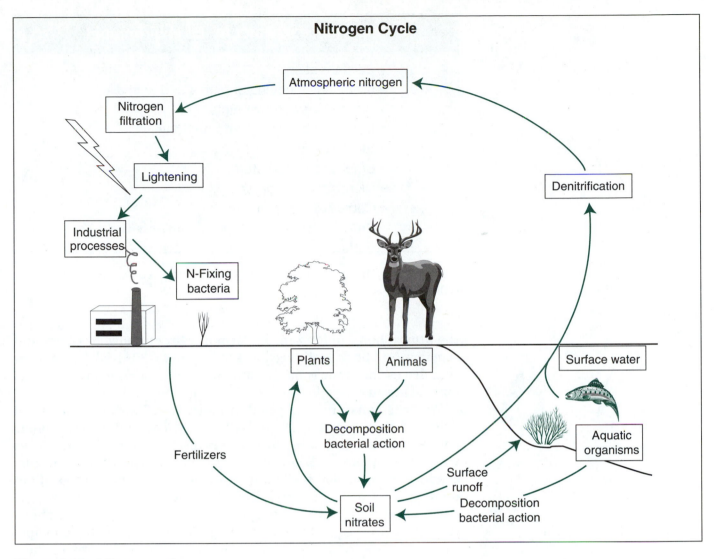

Nitrogen Cycle

Figure 1-18: Nitrogen cycle.

The Water Cycle

Water is one of the most important resources in the environment, see Figure 1-19. It provides a living environment to many species of organisms. It supports the growth of plant life.

Water is a required nutrient for living things. It makes up approximately 70 percent of the weight of living plants and animals, see Figure 1-20. Water is used to control the temperature of organisms and as a solvent for nutrients. It also performs many other functions related to maintaining life. Water in living tissues is also a part of the water cycle. It also dissolves nutrients and carries them to the tissues that need them. It stores heat and helps maintain a more constant temperature in the environment. Water action helps to create soil by breaking down rock into small particles. It also cleans the environment by diluting contaminants and flushing them away from vulnerable areas. Evaporation of water cools the surfaces of leaves and skin. Water even provides protection to some species of organisms, see Figure 1-21. Living things require large amounts of water to survive.

The **water cycle** occurs when water moves from the oceans to

Importance of water to the environment

- Soil formation.
- Living environment for plants and animals.
- Supports growth of plants.
- Dissolves and transports nutrients.
- Stores heat.
- Stabilizes temperature.
- Cleans the equipment.
- Cools living organisms.
- Provides protection from natural enemies.

Figure 1-19: This is a chart showing the importance of water in the environment.

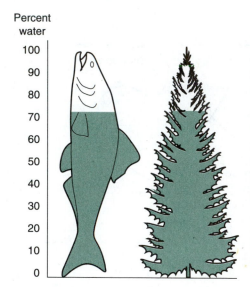

Figure 1-20: Water is an important component of living tissue.

the atmosphere, to land in the form of rain or snow, to rivers and streams, and back to the oceans, see Figures 1-22, 1-23. The energy that drives this cycle comes from two sources: solar energy and the force of gravity.

Water is constantly recycled. A molecule of water can be used over and over again as it moves through the water cycle. Solar energy that is trapped by the ocean is a source of heat that causes water to evaporate. Additional water enters the atmosphere by evaporating from soil and plant surfaces, especially in areas of hot temperatures and high precipitation.

Plants give up large amounts of water to the atmosphere through a process call **transpiration**. This is a controlled evaporation process by which plants lose water through pores in their leaf surfaces. These pores open wide when leaf surfaces are hot,

Figure 1-21: Beaver pond, auxiliary, main dams and lodge. *Photo courtesy of Leonard Lee Rue IV.*

Figure 1-22: A watershed is an area in which rainwater and melting snow are absorbed to emerge as springs of water or artesian wells at lower elevations. *Photo courtesy of Utah Agricultural Experiment Station.*

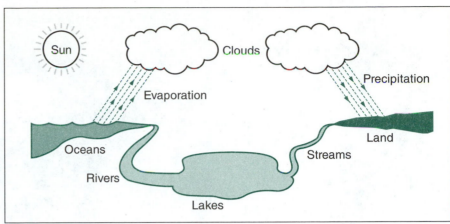

Figure 1-23: The water cycle.

and they close down when leaf surfaces are cool. Transpiration also creates a negative pressure that moves water and nutrients into the plant from the soil, see Figure 1-24.

Mammals control the temperature of their bodies by giving up water through their skin pores in the form of sweat. Cooling takes place when the sweat evaporates from the surface of the skin. Sweating in mammals occurs in much the same way that leaves transpire. Additional moisture is released to the atmosphere from the moist inner surfaces of the lungs, see Figure 1-25.

Moisture from all of these sources builds up in the atmosphere forming clouds. Clouds release stored water to the earth surface in the form of rain or snow. Gravity draws the water back into the earth, and causes it to flow from high to low elevations.

Large amounts of the earth's water supply are stored for long periods of time. Storage occurs in aquifers beneath the surface of the earth, in glaciers and polar ice caps, in the atmosphere, and in

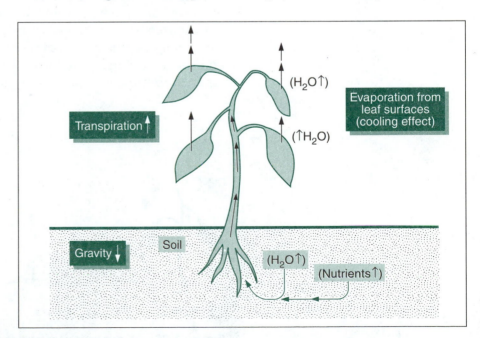

Figure 1-24: Effects of transpiration.

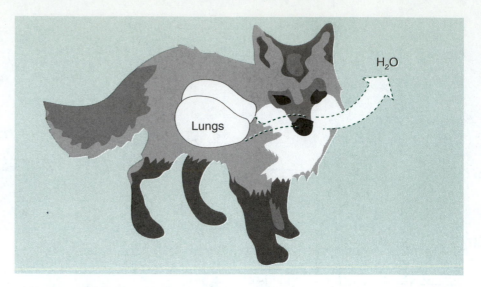

Figure 1-25: Cooling by evaporation.

deep lakes and oceans. Sometimes this water is stored for thousands of years before it completes a single cycle.

FOOD CHAINS

A **food chain** is made up of a sequence of living organisms that eat and are eaten by other organisms that live in the community. Each member of the chain feeds upon lower-ranking members of that chain. The general organization of a food chain moves from organisms knows as **producers** (usually considered to be food plants) to **herbivores** (plant eating organisms). These plant eating organisms are also called **primary consumers**. **Secondary consumers**, or **carnivore**s, are meat-eating organisms that eat primary consumers. A typical food chain begins with a plant as a food source and ends with a large predator. For example, a food chain may begin with meadow plants. Field mice eat plant roots and seeds, and are prey for raptors such as hawks and eagles and other predators such as foxes and coyotes, see Figure 1-26.

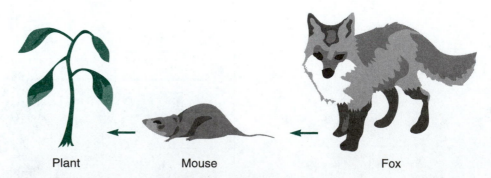

Plant   Mouse   Fox

Figure 1-26: An example of a food chain.

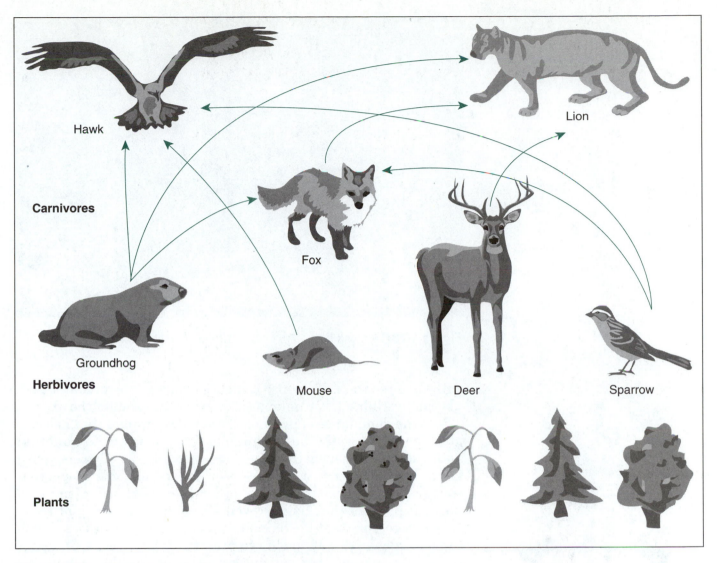

Figure 1-27: An example of a food web.

Most food chains become quite complicated because many predators will eat nearly any animal that they can catch and kill. Each food chain is interwoven with other food chains to create a **food web,** see Figure 1-27. A **food pyramid** arranges organisms in a ranking order according to their dominance in a food web.

The most versatile predators usually occupy the highest rank in a food web, see Figure 1-28. These mammals or birds usually have few natural enemies, and they are capable of preying on a large variety of other species. They maintain their positions at the top of the food pyramid unless a stronger predator migrates into the area that is capable of competing more favorably for the existing food supply. Sometimes a new predatory species even preys upon the species that previously occupied the highest rank in the food pyramid.

When changes occur in the kinds of organisms that occupy an ecosystem, they affect nearly every other species in the ecosystem. The movement of humans into a new area has the effect of displacing the predators that occupy the top ranks in a food web. Humans assume these positions by preying on the herbivores in competition

Figure 1-28: A black bear. *Photo courtesy of U.S. Fish and Wildlife Service.*

with the predators. They also control the size of the predator population by killing these animals or driving them out of the area.

It is the dominance of humans in the food pyramid that has created the controversy that surrounds the movement to maintain and restore natural environments, see Figure 1-29. Extreme positions have been taken on both sides of the issue. Some people contend that human dominance of other species of organisms is a natural process that has evolved since the beginning of time. Others recognize that the human species is the only species that is capable of changing its habits to preserve other species of organisms. They contend that humans have the moral obligation to protect all other forms of life, see Figure 1-30.

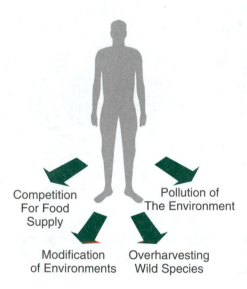

Competition For Food Supply

Pollution of The Environment

Modification of Environments

Overharvesting Wild Species

Figure 1-29: Human dominance.

Figure 1-30: Our natural resources should be appreciated and cared for.

LOOKING BACK

Ecology is the science that describes relationships between living organisms and their environments. It includes all of the activities that affect the environment and the organisms that live in it. The law of conservation of matter states that the form of matter can be changed, but matter cannot be created or destroyed by ordinary physical or chemical processes. Humans are capable of changing matter into a variety of materials that are not found in nature. When they are thrown away, they become pollutants. Pollution of the environment by waste materials is one of our greatest problems as a society. Environmental scientists spend their careers solving such problems and studying ways to maintain or restore the environment.

Energy is the force that drives all systems in nature. It originates from the sun and is lost to outer space in the form of heat. The first law of energy states that energy cannot be created or destroyed, but it can be converted from one form to another. The second law of energy states that each time energy changes from one form to another, some of it is lost in the form of heat.

The circular flow of an element from living organisms to nonliving matter is known as an elemental cycle. The carbon cycle and the nitrogen cycle are two of the most important elemental cycles. The water cycle is the circular flow of water from the oceans to the atmosphere, to the land as precipitation, to rivers and streams, and back to the oceans.

A food chain consists of a sequence of living organisms in which each member of the chain feeds upon lower-ranking members of that chain. Overlapping food chains form food webs. The organization of a food pyramid progresses from organisms that produce food (plants) to animals that eat plants (herbivores), to animals that eat other animals (carnivores). Human dominance of the food pyramid has resulted in controversy over the role of humans in conserving the environment and the species that occupy it.

REVIEW QUESTIONS

1. State the law of conservation of matter and explain what it means.
2. Suggest some ways that waste materials might be disposed of to maintain or improve environmental quality.
3. Explain how pollution of the environment is related to the law of conservation of matter.
4. List several current methods of disposing of waste materials, and discuss the positive and negative effects that each method has on the quality of the environment.
5. Define what energy is and discuss the role that energy plays in nature.
6. Define and explain the first and second laws of energy.
7. Illustrate the carbon cycle, the nitrogen cycle, and the water cycle, and explain the importance of each cycle to living organisms and to the environment.
8. List some factors that make water important to all living organisms.

9. Describe the organization of a food chain.
10. Explain how a food chain is different from a food web.
11. Describe how a food pyramid is organized.

LEARNING ACTIVITIES

1. Develop an environmental research project to study the waste disposal system that is used in your area. Identify actual and potential pollution problems that are associated with each of the ways that the different waste materials are disposed. Invite local sanitation authorities to visit your class to discuss local pollution problems. Develop possible solutions for each of the local waste disposal problems that are identified.

2. Take a field trip to an unpopulated area or nature center that is near your school. Have class members prepare and give reports on the habits of each of the organisms that are found there. Investigate and illustrate the ranking of each organism in the food pyramid. Suggest possible ways that the food chain might be organized, and illustrate how the food chains overlap to create a food web.

2 Understanding Relationships Between Ecology and Agriculture

KEY TERMS

biological succession
biosphere
climax community
community
competitive advantage
competitive exclusion
  principle
ecologist
ecosphere
ecosystem
habitat
misuse
niche
organism
pioneer
population
primary succession
range of tolerance
secondary succession

AGRICULTURE is the practice of raising plants and animals in a controlled environment. Competition from unwanted plants and animals is usually controlled. Fences define the borders of this artificial environment, keeping domestic animals in and restricting the movements of wild animals. Tillage machines and agricultural chemicals are often used to restrict or eliminate unwanted plants, animals, and diseases in the environment. Such practices disrupt the natural ecosystem that existed before agriculture was introduced in the area.

The practice of agriculture does not destroy ecology, but it does change ecosystems. Some of the original species of organisms find it difficult or impossible to survive in areas where agriculture is practiced. Other species of animals and birds enjoy an abundance of food and shelter and find it easier to survive in agricultural areas.

OBJECTIVES

After completing this chapter, you should be able to:

- explain how scientists organize living organisms for the study of ecology
- suggest ways that an unbalanced ecosystem might be brought into balance
- describe impacts that modern agriculture has had on the ecosystems of North America
- identify relationships between farming practices and the environments of different species of fish and wildlife
- distinguish between primary and secondary biological succession
- predict the biological succession of organisms in a particular environment.

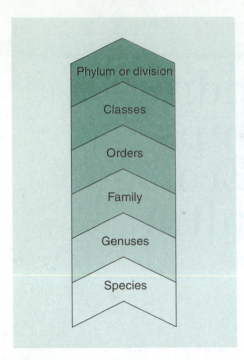

Figure 2-1: A system that biologist use to organize living things according to their relationships to one another is called Taxonomy.

FUNDAMENTALS OF ECOLOGY

As students of ecology, it is important to understand the language and organization of this branch of science. All living organisms that are known to man are classified into naturally related groups in a systematic way. These are known as taxonomic groups and include the following divisions: species, genus, family, order, class, and phylum (animals) or division (plants), see Figure 2-1. Organisms of the same species are closely related and exhibit many of the same characteristics. Organisms of the same order exhibit some of the same characteristics, but are not considered to be closely related.

In a similar manner, scientists who study ecology organize living things into different classifications. An **organism** is an individual living plant or animal. A group of similar organisms that is found in a defined area is known as a **population** of plants or animals. A **community** includes all of the populations of organisms that live within a defined area such as a woodland, marsh, or cornfield.

An **ecosystem** is made up of the community of living organisms plus all of the nonliving features of the environment such as water, soil, rocks, and buildings. Each of the components of an ecological system or ecosystem has an impact on the other components of the ecosystem.

Relationships also exist between two or more ecosystems. For example, soil erosion in Canada has an effect on the ecosystem from which the soil was lost, but it also affects survival rates of aquatic organisms in the Columbia River system and in the saltwater ecosystem where the silt is deposited in the Pacific Ocean. When all of the ecosystems of the earth are considered as a whole, we call it a **biosphere** or **ecosphere**, see Figure 2-2.

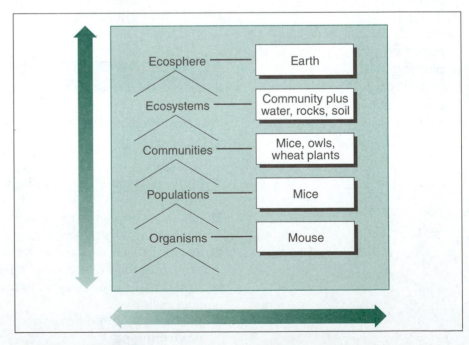

Figure 2-2: Ecologists organize living organisms and their environments into groups within an area to make it easier to define and understand their relationships.

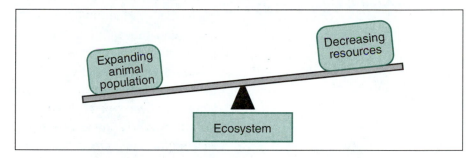

Figure 2-3: A balanced ecosystem exists when living organisms and non-living resources are maintained at constant levels.

When an ecosystem supports all of the living community within it, and nonliving resources are maintained at constant levels, it is said to be balanced, see Figure 2-3. A balanced ecosystem is rare, however, because the slightest change in any component of the ecosystem will affect all of the other components. If a population of organisms increases more rapidly than a resource is replaced in the system, it will eventually deplete the resource and cause the death of the organisms. Such an ecosystem is unbalanced, see Figure 2-4.

Figure 2-4: An unbalanced ecosystem exists when the community of living things uses up nonliving resources faster than they can be replaced.

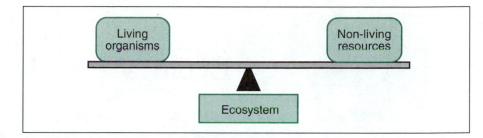

Career Option

ECOLOGIST

A scientist who studies relationships between living organisms and their environments is known as an **ecologist**. A person in this career will need a strong science background, and he or she should plan on additional graduate work after college. Ecology is a relatively new employment field that is continually evolving. With the emphasis on protecting the environment that is expected in the years ahead, a person who is educated as an ecologist can expect a high demand for his or her skills.

Photo courtesy of Cliff Coles, Photographic Services, Oregon.

Ecology Profile

BIOSPHERE II

Biosphere II was an experimental man-made environment located in the Arizona desert. It was originally sealed to prevent movement of air or water into or out of the facility. It was an experimental prototype of an environment to sustain human life on the moon or another planet. It contained many scientific features to produce food and oxygen for people and carbon dioxide for plants. Water was recycled by plants and through evaporation.

The original plan was for the people who lived in Biosphere II to remain sealed inside the structure for two years. Shortly after the structure was sealed, however, one person was removed due to an injury. Controversy has followed the return of this person to the system following medical care. It has been alleged that additional supplies were taken into the system at that time. A number of scientists contend that this invalidated the original scientific study.

After several months of operation, oxygen levels in the closed environment became too low and the people living in Biosphere II began to suffer from lack of sufficient oxygen. Eventually new air was added to the environment to raise the oxygen back up to a level that was considered safe.

During the remainder of the experiment, the people raised food for themselves and their animals using only the resources that were available within the biosphere. This experiment was a good learning experience that illustrated the difficulty of creating a stable environment. This is man's first real attempt to create an artificial ecosphere.

Photo courtesy of Space Biospheres Ventures, Arizona.

Figure 2-5: Intensive cropping system. *Photo courtesy of USDA Soil Conservation Service.*

AGRICULTURAL IMPACTS ON ECOSYSTEMS

Early in the twentieth century, most Americans were farmers. They lived in the country and spent most of their time producing food and clothing for their own families. Today the agricultural industry has changed. New intensive farming methods have been introduced, and less than 2 percent of our citizens now live and work on farms.

Nearly all of the land that can sustain profitable crop production is now used for growing crops, see Figure 2-5. The native plants that used to grow abundantly on this land are mostly gone, and the **habitats** or environments of many birds, animals, and other

Figure 2-6: The healthy deer shown in this figure have plenty of food. *Photo courtesy of L.G. Kesterloo, Virginia Commission of Game and Inland Fisheries.*

Figure 2-7: The Whooping Crane is an endangered species. *Photo courtesy of Utah Agricultural Experiment Station.*

organisms lived, have changed. Even the kinds of species that live in agricultural areas have changed as farming methods have been modified.

Agricultural practices have resulted in both positive and negative effects on living organisms. Some of the changes that have occurred have been favorable to wild birds and animals because they have resulted in a more abundant supply of food or shelter, see Figure 2-6. The white-tail deer, for example, is an animal that has benefited from modern farming practices. Many agricultural areas have large populations of resident deer that feed on cultivated plants.

Some species of birds and animals have declined because farming practices have reduced or poisoned their food supplies, eliminated their shelter, or destroyed their breeding grounds, see Figure 2-7. The whooping crane and several other species of wetlands birds are examples of organisms that have suffered reduced populations due in part to modern agricultural practices and to human encroachment upon their habitat, see Figure 2-8.

The use of agricultural chemicals is often cited as a practice that has resulted in extensive damage to the environment. Some chemicals, such as DDT, are known to have side effects that were not known when they were approved for use. Damage to the environment due to **misuse** of agricultural chemicals sometimes occurs, just as it occurs when industrial, lawn, or garden chemicals are used improperly. Public criticism should be directed against the

Figure 2-8: Wetland areas are a natural habitat to many organisms.

Figure 2-9: A pest control specialist sees to it that when it is necessary to use chemicals that they are used safely and wisely. *From Burton, Agriscience and Technology, copyright 1992 by Delmar Publishers.*

misuse of all chemicals in agriculture and every other segment of human society.

Chemicals that are applied for their intended purposes according to the recommendations of the manufacturer can usually be safely used, see Figure 2-9. Agricultural chemicals that are in use today are subject to extensive scientific testing to assure that they do not pose a threat to the environment.

Incorrect disposal of unused chemicals and chemical containers poses a threat to the environment. Those who use agricultural and industrial chemicals must assume responsibility for applying them properly, and for appropriately disposing of the waste that is associated with chemical use, see Figure 2-10. Most of the laws and regulations that apply to disposal of hazardous materials

Figure 2-10: Laws and regulations for the disposal of hazardous waste materials should be strictly obeyed.

Figure 2-11: The Grizzly Bear is an endangered species. *Photo courtesy of Stacy Riggert.*

apply equally to agricultural chemicals and they should be strictly obeyed.

BIOLOGICAL SUCCESSION

Animals and plants that were unable to adapt to the changes that occurred in their habitats as farms were established are no longer present or now exist as smaller populations, see Figure 2-11. They have been replaced by different species that are capable of surviving in the new environments that were created by farming the land, see Figure 2-12. The change that occurs as one kind of living organism replaces another organism in an environment is called **biological succession** or ecological succession.

Two forms of biological succession are known to exist. **Primary** succession occurs where organisms did not exist before. Such an area might be found on a lava flow that has cooled and hardened, see Figures 2-13, 2-14.

Secondary succession occurs when an ecosystem is damaged or partly destroyed and remnants of the former community still exist. The changed environment will support only those organisms that are naturally found in an earlier stage of biological succession. Examples of secondary succession can be found in areas such as Yellowstone Park where large tracts of old-growth forests were destroyed by fire, see Figure 2-15.

Figure 2-12: The coyote is a highly adaptable animal. *Courtesy of U.S. Fish and Wildlife Service. Photo by R.H. Barrett.*

Ecology Profile

PRIMARY BIOLOGICAL SUCCESSION IN A LAVA FLOW

The material that remains after hot, molten lava has cooled is sterile and without life. The surface of such a flow consists of hardened lava rock and cinders. Over time, simple plants such as lichens and fungi begin to grow on the hard surface. The first plants to grow in an environment are called **pioneers.** The lichens begin to decay when they are acted upon by small fungus organisms that feed off the plants. Tiny soil particles carried by the wind become trapped in the lichen growth, and a soil base begins to develop on the hard lava surface.

This makes it possible for other simple plants to germinate and grow. Hardy plants, such as thistles and weeds, are the first complex plants to grow in this environment. They are followed by shrubs, trees and other plants that are adapted to the conditions found in the area. Hundreds or even thousands of years may be required before sufficient soil exists to allow complex plants to survive. The plants that occupy an environment when succession of species is complete and plant populations become stable are called **climax communities**.

Figure 2-13: Lava flow that has cooled and hardened. *Photo courtesy of Shannon Kolbe.*

Biological succession is a process that goes on all the time. An environment is seldom stable and unchanging. Every time a living environment changes in some way, the creatures that seek shelter or food from it are also affected. Some organisms find it difficult to adjust to change in their habitat. The change may even reduce

Figure 2-14: An example of Primary succession. *Photo courtesy of James O. Sneddon.*

Figure 2-15: An example of Secondary succession. *Photo courtesy of Cliff Coles, Photographic Services, Oregon.*

their ability to survive. Other organisms may benefit from a change in their environment. Sometimes a change makes it easier for an organism to compete with other organisms that live in a particular environment.

A **niche** is a description of the role that an organism fulfills in an environment. It includes its position in the food web, where it lives, and when it is active. Several animals appear to be competitors in a single habitat, but if they occupy different niches, competition between them may be minimal.

A **competitive advantage** exists when one organism is better able to survive in an environment than another. A species that enjoys a competitive advantage will increase in numbers as time passes, and the populations of organisms without a competitive advantage will decrease. Eventually, the weakest competitor is lost from the environment, see Figure 2-16.

The loss of an organism from a specific niche or habitat is an example of the **competitive exclusion principle**, which is the hypothesis that two or more species cannot coexist on a single resource that is scarce relative to the demand for it. The Bighorn

Ecology Profile

SECONDARY SUCCESSION AFTER A FOREST FIRE

Forest fires are sometimes responsible for the destruction of old-growth forests and the habitat that they provide to the animals, birds, and other organisms that live there. The heat that is generated by a forest fire often sterilizes the soil. It kills many of the organisms that are in the soil and frequently moves the stage of succession backward.

The 1988 fires that burned large tracts of the pine forest in Yellowstone Park were destructive because there was an abundance of dead plant material on the forest floor. This created extreme heat causing the forest canopy to burn.

After a destructive fire occurs, hardy grasses, thistles and other pioneer plants are some of the first forms of vegetation to be found in the area. In some instances there are still living trees in the area that have been protected by their thick bark. In other instances, trees are reseeded due to heat from the fire. Some kinds of pine cones depend on the heat from a fire to open them up and release the seeds.

Secondary succession following a fire often begins with cheatgrass and crabgrass growth followed by tall grasses and food plants. Pine trees eventually invade the grasslands, establishing pine forests. Hardwood trees are usually the last species to come into an area. They grow up through the pine forest, and eventually become the dominant species.

Photo courtesy of United States Department of Agriculture, Forest Service.

Figure 2-16: An organism that can exercise competitive advantage over another usually takes over the niche in the environment and excludes the competing species.

sheep is a good example of an organism that declined because it could not adjust to a changing environment, see Figure 2-17. In this instance, the wild burro, which is not a native species has established itself in habitats formerly occupied by Bighorn sheep. The burros are more competitive grazers than the Bighorn sheep and they have a competitive advantage when food supplies are inadequate for both populations.

Figure 2-17: Bighorn sheep. *Photo courtesy of U.S. Fish and Wildlife Service.*

Figure 2-18: The comfort zone for the penguin is extreme cold, whereas the comfort zone for the flamingo is extreme warmth. *Photo courtesy of Leonard Lee Rue IV.*

The ability of an organism to survive changes in its environment depends on its **range of tolerance** for change. All organisms have comfort zones in which living conditions are matched with their survival needs, see Figure 2-18. As the range of conditions to which an organism can adjust becomes greater, its ability to survive increases.

The range of tolerance of an organism for its environment is largely determined by its inherited ability to adjust to new environmental conditions. A bird that is able to gather or digest only one kind of food will be unable to survive if that food supply is lost from the environment. Some organisms depend totally on a single kind of shelter; others cannot tolerate pollution in the air and water, see Figure 2-19. In addition, the temperature range in the environment often determines whether an organism is capable of surviving. A narrow tolerance range makes it very difficult for an organism to survive environmental changes.

Figure 2-19: Pollution in rivers, streams, and ponds can jeopardize populations of many fish species. Through concerted efforts to clean up waters and through restocking programs we can avoid this problem. *Photo courtesy of U.S. Fish and Wildlife Service.*

LOOKING BACK

Ecology is the branch of science that describes relationships between living organisms and their environments. Organisms are organized into populations, communities, and ecosystems that overlap and react with each other. Modern agricultural practices have resulted in modified ecosystems and populations of organisms in cultivated areas. Environments are constantly changing, and biological succession of living organisms occurs in an orderly and predictable manner. Organisms that have a competitive advantage over other members of the community, or a wide range of tolerance to changes in the environment, tend to survive while other organisms do not.

REVIEW QUESTIONS

1. Explain how scientists organize living organisms to aid their study of ecology.
2. What differences exist between balanced and unbalanced ecosystems?
3. What impacts do modern farming methods have on ecosystems in North America?
4. What benefits and risks are associated with pesticide use on farms, lawns, roadsides, and gardens?
5. How are primary and secondary biological succession different? Give examples of each type of succession.
6. Predict how the science of ecology will be of benefit to the world in the twenty-first century.

LEARNING ACTIVITIES

1. Plan and set up an aquarium in which a balance between plant and animal life is considered. Observe the ecosystem carefully to determine whether adequate oxygen and carbon dioxide is produced to sustain the plants and animals that were placed in the system.
2. Conduct a public awareness campaign to encourage all users of agricultural and industrial chemicals to dispose of waste materials in a legal and safe manner.

3 Biomes of North America

KEY TERMS

biome
canopy
coniferous forest biome
conifers
continental shelf
deciduous forest
desert biome
estuary
forest floor
freshwater biome
grassland biome
herb layer
intertidal zone
lentic habitat
limnologist
lotic habitat
marine biologist
marine biome
neritic zone
oceanic zone
plankton
phytoplankton
salinity
shrub layer
strata
temperate forest biome
terrestrial biome
thermal stratification
tundra biome
turbid
understory
wetlands
zooplankton

AN ECOSYSTEM is made up of plants, animals, and microorganisms that interact with each other, and with the non-living materials that surround them. Sometimes the boundaries of an ecosystem are easily identified because they consist of physical barriers such as mountains or water. Most of the time, however, ecosystems are not discrete communities, and they blend together where they meet. The boundaries of an ecosystem resemble a permeable membrane that allows organisms and materials to flow both in and out of the ecosystem, see Figure 3-1. The effects of an event that occurs inside an ecosystem usually spills over into neighboring ecosystems, as well.

OBJECTIVES

After completing this chapter, you should be able to

- explain how events in one ecosystem affect events in a neighboring ecosystem
- describe the relationship between an ecosystem and a biome
- list the distinguishing characteristics of a freshwater biome
- identify similarities and differences between freshwater and marine biomes
- discuss ways in which wetland habitats function to cleanse the environment
- name the terrestrial biomes that are found in North America and describe their similarities and differences
- design a map of the North American continent that illustrates the locations of the major biomes found there.

Ecosystems that are located at similar latitudes and elevations often have many of the same characteristics. The temperature and the amount of precipitation usually does not vary significantly. A

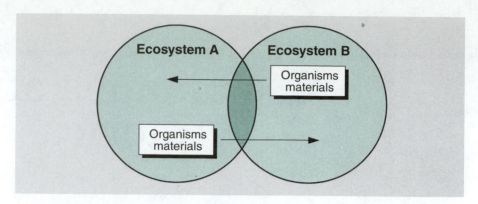

Figure 3-1: The boundaries between ecosystems frequently overlap and they are porus enough to allow organisms and materials to pass through in both directions.

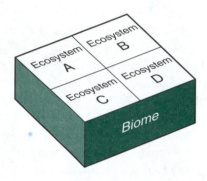

Figure 3-2: A biome consists of a group of similar ecosystems.

biome is a group of ecosystems within a region that have similar types of vegetation and similar climatic conditions, see Figure 3-2.

The climate of a region is in part determined by its latitude or distance from the equator. The temperature declines as the distance to the equator increases. Climate is also determined by altitude, see Figure 3-3. Conditions are different according to altitude within the same mountain range. The temperature decreases as the altitude increases. Ecosystems support different kinds of organisms as we move north or south from the equator and as we move from low to high elevations.

FRESHWATER BIOMES

A **freshwater biome** is composed of plants, animals, and microscopic forms of life that are adapted to living in or near water that

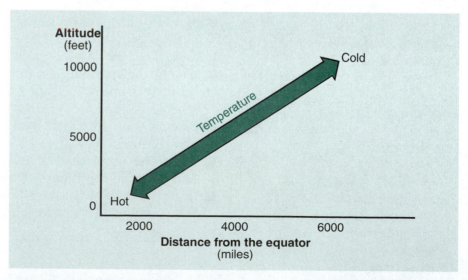

Figure 3-3: Climate is determined by altitude and by distance from the equator. The temperature becomes cooler as you travel away from the equator and as you gain altitude.

Figure 3-4: An example of freshwater biome. *Photo courtesy of Brian Yacur.*

Figure 3-5: A freshwater ecosystem may have many different kinds of plants and animals depending upon environmental conditions. *Photo courtesy of Bill Camp.*

is not salty, see Figure 3-4. The term "freshwater" can be misleading, however, because some freshwater environments depend upon water that may be stagnant, muddy, or heavily polluted. Such water may not appear to be "fresh" at all.

There are many different environmental conditions that are found in freshwater ecosystems, and each set of conditions creates a unique living environment, see Figure 3-5. The water in a particular area may be hot, cold, or loaded with dissolved materials. In each instance, different kinds of plants and animals have become adapted to the water conditions that are present. In some cases organisms have become so specialized that they may be unable to survive in any other known location.

Ecology Profile

SOUTHERN CAVEFISH
(TYPHLICHTHYS SUBTERRANEUS)

The Southern cavefish has become adapted to living in the darkness of a cave environment. It has no eyes, and its skin color is pink-white. It has two rows of sensory feelers located on the caudal fin to help it navigate. It lives in caves east of the Mississippi River from Indiana on the north to Georgia and Alabama on the south.

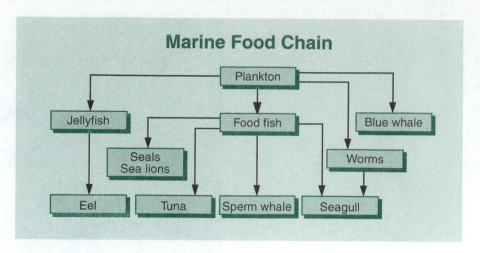

Figure 3-6: Plankton are the food source either directly or indirectly for nearly all marine animals.

Among the organisms that live in freshwater biomes are **plankton.** These are microscopic plants called *phytoplankton* and animals called **zooplankton** found floating on the surface of the waters. They serve as food producers in the food chain, and they are eaten by fish and other aquatic animals, see Figure 3-6.

The organisms that are capable of living in a freshwater habitat are limited by conditions such as water temperature, light intensity, concentration of dissolved material and the flow rate of a stream, see Figure 3-7. Water temperature is one factor that determines which species of organisms can survive in the habitat. Some organisms have very little tolerance for changes in water temperature, and they die when the temperature becomes too hot or too cold for them. Some organisms can live in water that is quite hot while others survive best in cold water, but few species can survive in water temperatures that fluctuate rapidly more than a few degrees. Some organisms can survive over a broad water temperature range as long as the temperature changes gradually.

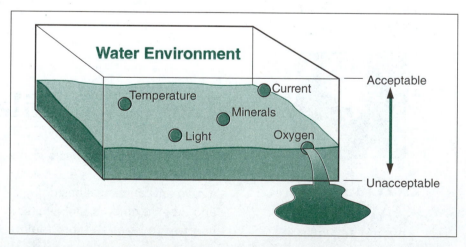

Figure 3-7: The most limiting factor in the water environment determines how acceptable the environment will be. Like holes in a water tank, the lowest hole determines the capacity of the tank.

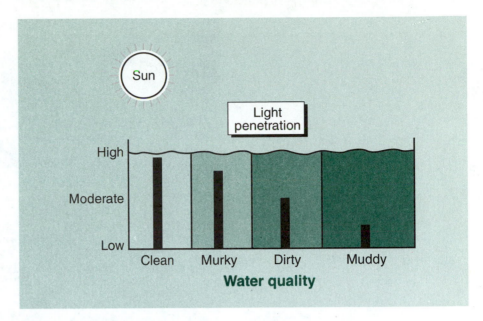

Figure 3-8: Light penetration in an aquatic environment.

A second condition that affects freshwater organisms is the amount of light that can penetrate the water. This determines the amount of photosynthesis that can occur and the kinds of plant life that can exist there. When water is **turbid** or cloudy with suspended particles of silt, photosynthesis is limited in water plants. This occurs because suspended particles block the sunlight and prevent light from reaching water plants, see Figure 3-8.

Another factor that limits the kinds of organisms that can live in an aquatic environment is the concentration of dissolved minerals in the water. When high levels of nitrates and phosphates are present dense blooms of blue green algae and other plants may occur in the surface water. At the same time, fish and other aquatic animals may die if the concentration of these materials in the water becomes too high. They may also die if the concentration of dissolved oxygen in the water is too low. Fish and other aquatic animals that need a rich supply of oxygen may die, or be replaced by other fish that can survive in a low oxygen environment.

The current or flow of the water in a river or stream may be the limiting environmental factor for some organisms, see Figure 3-9. The intensity of the current contributes to the amount of force which an organism must be able to withstand to prevent being carried downstream. It also affects the availability of food.

A fifth factor that affects the organisms in a freshwater habitat is the amount of oxygen that is dissolved in the water. Cold water that is flowing rapidly tends to carry more dissolved oxygen than warm water that is moving slowly through a pond or swamp, see Figure 3-10. Flowing water tends to dissolve oxygen readily, and cold water temperatures help to stabilize oxygen in the water.

These five criteria interact to create the conditions found within freshwater ecosystems. If any one of them becomes unfavorable, an entire population of organisms may be eliminated or replaced by other species better suited to the new conditions.

Figure 3-9: Water current in a stream may be the limiting environmental factor for some organisms. *Photo courtesy of Wendy Troeger.*

Two different kinds of habitats are found in freshwater biomes. A **lotic habitat** exists where water flows freely in streams and rivers. Free flowing water tends to support little plant growth. It is typical of a food web where food for fish and other aquatic animals is carried in by the water from distant sources.

In contrast a **lentic habitat** is one in which water stands for long periods of time, as in swamps, ponds, and lakes. Lentic environments contain areas of differing light, and lack a strong current. The water temperature also differs. Deep water tends to be cold, and the water near the surface is usually warmer. This is called **thermal stratification.** Some common plants and animals that live in these types of areas are cattails, lilies, herons, and frogs, see Figure 3-11.

A term that is currently attracting public interest is **wetlands**.

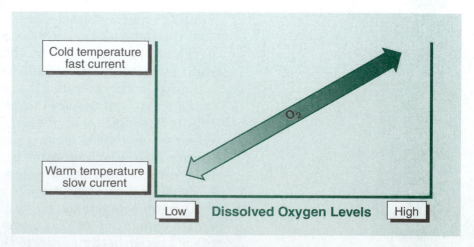

Figure 3-10: Disolved O_2 in an aquatic environment.

Figure 3-11: Herons thrive well in a lentic habitat. *Photo courtesy of U.S. Fish and Wildlife Service.*

A term that is currently attracting public interest is **wetlands**. Wetlands are defined as land areas that are flooded during all or part of the year. A wetland may consist of dry land for much of the time and still be classified as a wetland if it is subject to flooding during some season of the year.

It is clear that the area of wetlands in North America has declined with the advance of civilization. We have drained many acres of marshes and swamps to prepare the land for raising crops. The USDA has reported that up to 90 percent of the original prairie wetlands located in Iowa, Minnesota, eastern North Dakota and southeastern South Dakota have been converted to farms. Similar trends are evident in other areas. A strong movement initiated by advocates for waterfowl, and aided by federal legislation is attempting to reverse this trend in the United States.

Organized groups of sportsmen such as Ducks Unlimited have taken a lead role in the effort to restore wetlands. They have purchased large tracts of land with donated funds and returned them to marshlands. Through such efforts they hope to create favorable conditions for migrating waterfowl such as ducks and geese, thereby increasing the populations of these birds, see Figure 3-12. Other organizations that are involved in restoration of wetlands include state and federal agencies, conservation groups such as the Nature Conservancy and ecological researchers.

Some attempts to increase the amount of wetlands have created controversy between landowners and the government agencies that are responsible for determining and enforcing wetland policy. Laws and regulations that encourage restoration of wetlands include: Executive Order 11990; Protection of Wetlands, 1977; The Federal Water Bank Program; State Water Bank programs; and Watershed Protection and Flood Prevention Act.

Figure 3-12: There has been a big effort to restore and maintain wetlands. *Photo courtesy of U.S. Fish and Wildlife Service. Photo by Bob Ballou.*

The general sentiment today is that wetlands are difficult to restore. It is best not to disturb them in the first place. Some landowners believe that the law reaches too far by designating some lands as wetlands that historically did not flood. They define as wetlands some of the land that floods only seasonally as a result of land modification projects that were implemented on the property. Such changes include the diversion of water into new channels and irrigation practices that result in ponding of water.

Regulations to modify the ways in which wetlands can be used are currently being developed by federal and state government agencies. All efforts to promote or reduce government control of wetlands will affect the freshwater ecosystems of North America.

MARINE BIOME

The world's largest biome is the **marine biome**. It consists of the oceans, bays, and estuaries, and makes up approximately 71 percent of the surface area of the earth, see Figure 3-13. A distinguishing characteristic of the marine biome is salinity, or the salt concentration of the water; oceans have a salt concentration ranging from 3 to 3.7 percent. **Salinity** is an important factor that affects the ability of an organism to survive in a marine environment. It varies quite a bit in surface waters, especially near land where rivers enter, but tends to be quite stable in deep ocean waters. The marine biome is also quite diverse. It covers the hot southern zones of the North American coast to the cold arctic regions of the continent's northern reaches.

Figure 3-13: The world's largest biome is the marine biome. *Photo courtesy of Shannon Kolbe.*

The ability of organisms to live in the marine biome depends on several conditions that are necessary to sustain life in the ocean, see Figure 3-14. Water temperature, for example, varies considerably in the marine biome. Deep ocean waters tend to be colder than surface waters, and arctic waters are colder than ocean waters located near the equator. Ocean currents such as the Gulf Stream, which is warm, and the Humboldt current, which is cold, distribute ocean temperatures differently.

The amount of light that penetrates the water's surface influences the amount of plant life that exists there. The presence of food-producing plants attracts fish and other aquatic organisms to waters where food is plentiful. A supply of nutrients for plants to use in photosynthesis is also important in producing food for aquatic animals. Certain areas of the ocean have strong ocean cur-

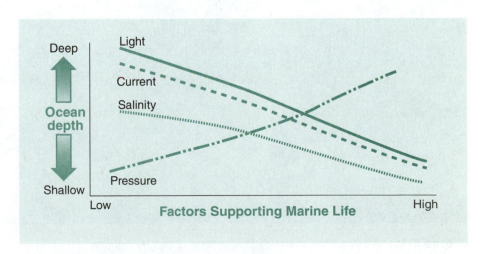

Figure 3-14: Factors supporting marine life.

rents that circulate deep ocean waters to the surface. These currents are known as upwelling currents. These areas are important to the commercial fishing industry because deep waters tend to be high in nutrients, and fish concentrate in these areas to feed. Ocean currents are important in moving nutrients from deep ocean waters to surface waters where photosynthesis can take place. Plankton are found in abundance where sunlight and nutrients are available, and these organisms play an important role as food producers in the aquatic food chain.

Water pressure becomes greater as the water gets deeper. Water pressure occurs because of the action of gravity on water molecules. As more water molecules are stacked above an organism, they exert greater pressure on its body. A good example of this is the pressure that a person feels on his/her eardrums during a dive in deep water. The deeper you dive, the greater the pressure becomes.

Some organisms can withstand the pressure of deep ocean water, but most aquatic organisms tend to seek and remain in surface waters where food is more plentiful and where water pressures are not extreme.

Scientists have identified three important zones in ocean environments. The **intertidal zone** is located near the shore, see Figure 3-15. When the tide is low, this zone is an exposed beach that is above the water. During the high tide period, it is covered with water. The width of this zone is greatest when the slope of the ocean floor is gradual. It is narrow when the slope of the ocean floor is steep.

Figure 3-15: An intertidal zone. *Photo courtesy of Wendy Troeger.*

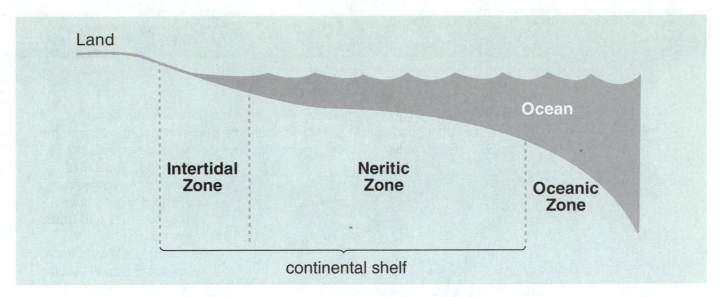

Figure 3-16: The continental shelf.

both in and out of water. They must be able to cope with frequent changes in the temperature of their environment. They must also be capable of retaining moisture when the tide is out, and of tolerating salinity when the tide is in.

The intertidal zone is subject to strong water currents that can wash plants and animals out to sea. Organisms that live in this zone must be capable of withstanding these currents by attaching themselves to rocks or burrowing into the sand. The types of animals that can survive in tidal zones depends in a large part on the type of material composing the ocean floor.

The **continental shelf** is land that is submerged under the surface of the ocean, see Figure 3-16. It slopes gradually away from the shore toward deeper water. The area beyond the intertidal zone that extends to the outer edge of the continental shelf is called the **neritic zone**. The most favorable living environment in the ocean is found here. Plankton and other food-producing organisms are abundant in this zone due to the abundance of nutrients and light. For most aquatic organisms, the neritic zone is the most productive area in the ocean.

The **oceanic zone** begins at the outer edge of the continental shelves and includes the deep ocean waters between them. The surface waters in this zone are productive waters. The presence of light and nutrients supports the production of plankton and other food-producing organisms to a depth of about 200 yards. Below that depth, the waters become dark and the temperatures become cold. Some production does occur to depths of 2,000 yards, but beyond those depths, the living environment is generally cold and less productive. Deep-ocean volcanic vents, explored in the last ten years, provide warmth and energy for strange communities of worms, mollusks and shrimp-like crustaceans in some parts of the ocean floor (e.g., off Seattle).

Mixing of fresh and saltwater occurs where rivers and streams

Career Option

AQUATIC BIOLOGIST

An education in aquatic biology prepares a person for a number of different occupations that study relationships between aquatic organisms such as fish, plankton, clams, and snails and the water environments in which they live. All of these careers require college degrees, and typically advanced graduate training.

An aquatic biologist who specializes in saltwater aquatic life is known as a **marine biologist**. A person who chooses a similar career specializing in freshwater aquatic life is a **limnologist**.

Figure 3-17: This river opening is very close to sea level. At high tide, the sea flows in and the estuary becomes salty, almost like sea water. At low tide, the water flows out and the water becomes fresh almost like fresh water.
From Camp, Environmental Science, copyright 1994 by Delmar Publishers, photo courtesy of Bill Camp.

Mixing of fresh and saltwater occurs where rivers and streams enter the ocean. Such areas are often marshy, and shallow, with abundant supplies of food-producing plants. An area such as this is called an **estuary.** The salinity of the water varies depending on the strength of the current as well as seasonal rainfall, tides, and evaporation in shallows. These areas are prime habitat for oysters and young fish of many kinds. An estuary has characteristics of both freshwater and marine biomes, see Figure 3-17.

TERRESTRIAL BIOMES

A **terrestrial biome** is a large ecosystem consisting of plants, animals, and other living organisms that live on land. Several distinct biomes are found on the North American continent, see Figure 3-18. Among these are the desert tundra, grasslands, temperate forests, and coniferous forests. Each of these biomes is distinctly different from the others, and each provides a set of environmental conditions that support different kinds of living organisms.

Desert Biome

The **desert biome** is an environment that is very dry. Precipitation is usually less than 10 inches per year, only enough to allow specially-adapted plant and animal species to live. Daytime temperatures are usually very hot and the nights are often quite

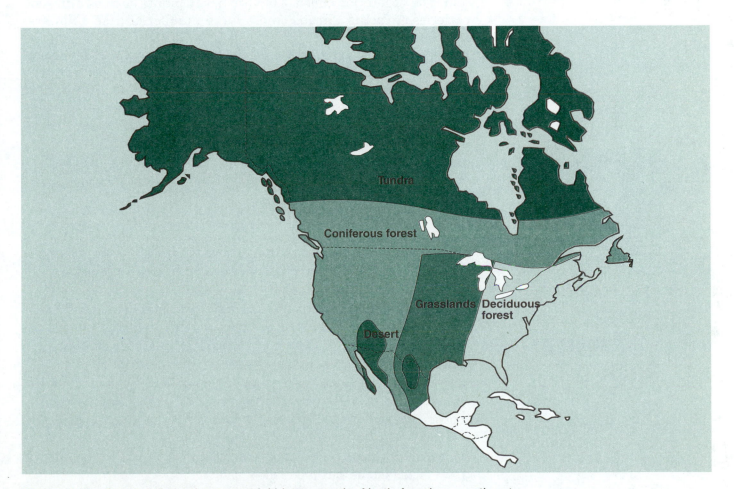

Figure 3-18: Several distinctive terrestrial biomes on the North American continent.

Figure 3-19: A desert biome. *Photo courtesy of Shannon Kolbe.*

cold. The vegetation consists of very hardy plants such as different varieties of cactus, shrubs, and a large number of desert flowers, see Figure 3-19.

Some plants survive in the desert because they have developed deep taproots that can draw moisture from deep in the soil. Others have very short growing seasons; they germinate and grow to seed stage using the moisture of the melting winter snow or from a single rain storm. The desert literally does "blossom like a rose" before the hot, arid climate robs the soil of its moisture.

Ecology Profile

THE GREAT BASIN — A DESERT PROFILE

Jim Bridger was a famed mountain man during early explorations of the American west. He was a trapper and trader who spent much of his time alone or among the Native American tribes that lived in the region. Brigham Young once inquired about The Great Basin as a possible location for a settlement and Jim Bridger replied that it was unfit for human habitation. He further stated that crops could not be grown there, and he was reported to have offered $1,000 for the first bushel of corn grown in the valley of the Great Salt Lake.

Before Brigham Young and the Mormon pioneers brought irrigation to this desert region, it was a land that fit the description of Jim Bridger: a desert unsuitable for human settlement. Today, irrigation has made it possible to convert large desert tracts into productive farmland.

Figure 3-20: Roadrunner. *Courtesy of U.S. Fish and Wildlife Service. Photo courtesy of J.C. Leupold.*

Desert conditions are found on nearly 35 percent of the land area of the earth. In North America, desert lands make up much of the Great Basin region located in Utah and Nevada, the southwestern United States, and northern Mexico.

The animals that inhabit the desert biome are highly specialized to survive with a limited supply of water, see Figure 3-20. Some of them get all the water they need from the plants that they eat. Others lap up the drops of dew that condense on the leaves of plants during the cool nights. Still others get needed moisture from the body fluids of the animals upon which they prey. In some cases, desert animals and birds travel long distances to drink at the few water holes that are found in these arid regions.

Among the animals that are found in the desert are these: small rodents such as rats and mice, birds, lizards, snakes, insects and birds of prey. In some instances larger animals such as jackrabbits (hares), coyotes and pronghorn are found within traveling distance of water.

Tundra Biome

The **tundra biome** is located in the frozen northern regions of the continent, extending from timberline on its southern limit to the areas of permanent snow and ice cover in the north, see Figure 3-21. Temperatures are well below freezing for most of the year, and the soil remains permanently frozen (permafrost) underneath the surface. Although less than 10 inches of precipitation falls annually in these areas, evaporation rates are low and the water cannot penetrate the frozen soil, so much of the area remains wet and spongy during the summer growing season.

Figure 3-21: Alaskan range and tundra. *Photo courtesy of Len Rue, Jr.*

Climate and productivity are very seasonal in the tundra biome. Although the summer growing season is short, the sun shines most of the time at these northern latitudes and the tundra plants must mature and produce seed before the winter season returns. Water ponds up in the low spots, and the abundant aquatic and semi-aquatic insects support migrating waterfowl and other birds in the summer. Caribou migrate into the tundra regions during the summer to graze on the grasses, mosses, lichens, sedges and shrubs that grow there. As the growing season ends, the migratory birds and animals move back to the south.

The musk ox is a permanent resident in the tundra. Most of the precipitation in these areas comes in the form of rain during the fall months. Since snowfall is quite limited, the musk ox is able to graze year round on the short vegetation. This animal has adapted well to the difficult climatic conditions of the tundra environment. It is protected from the cold by a long, thick haircoat and a layer of body fat. Its hooves are useful for defense against predators, and they are used to remove snow from the food supply.

Rodents such as lemmings become very abundant during some years and nearly disappear at other times, see Figure 3-22. They provide food for predators such as the lynx, artic fox, ermine, birds of prey, and bears that move in and out of the tundra region. This natural cycle of the lemmings from abundance to scarcity controls the numbers of predators.

Grassland Biome

The **grassland biome** in North America is often referred to as the prairie and is characterized by a lack of trees. It is located in the middle of the continent and includes the prairies of Canada and

Figure 3-22: Lemmings.

the midwestern region of the United States, and the grasslands that extend south into Mexico. Historically, the grassland areas formed the largest biome in North America, see Figures 3-23, 3-24. The term grasslands is misleading in that many plants besides grasses are found there. Such plants include sedges, forbs and a number of other types of plants. Much of the grasslands in North American have now been converted into farmland where corn, wheat, and soybeans are the main crops.

The native species that covered the grassland region before the land was cultivated were well adapted to the region's climate and frequency of natural fires. Both bunch grasses that grow in tufts and sod grasses with matted roots have deep root systems that make good use of soil moisture. Periodic burning is characteristic of this environment and deep roots also protect these plants against fire damage. Fire may completely consume the foliage of these plants, but new shoots arise from the roots when favorable growing conditions return.

Figure 3-23: This tallgrass prairie is located in the Flint Hills area, Riley County, Kansas. *From Camp, Managing Our Natural Resources, copyright 1991 by Delmar Publishers, photo courtesy of Dr. Jay McKendrick, University of Alaska, Fairbanks.*

Figure 3-24: This is an example of a shortgrass prairie. It is located in the Custer State Park in South Dakota. *From Camp, Managing Our Natural Resources, copyright 1991 by Delmar Publishers, photo courtesy of Dr. Jay McKendrick, University of Alaska, Fairbanks.*

The climate of the grassland biome is continental, meaning it has moderate to hot summers and cold, freezing weather in the winter. Grasslands are located in areas where severe droughts occur from time to time. These drought conditions contribute to the lack of trees in temperate grasslands. Lightning fires are also extremely damaging to trees not specifically adapted to burning and keeps them from becoming established.

It was in the prairie regions that the great herds of bison once ranged. Today they have been replaced in this area by herds of domestic cattle and sheep. Only in Yellowstone National Park are bison found in abundance. Other wild animals that are commonly found in the grasslands include prairie dogs, mice, snakes, rabbits, pronghorn, coyotes (prairie wolves), and many kinds of birds and insects.

Temperate Forest Biome

The **temperate forest biome** is identified by broadleaf trees such as oak, maple, cherry, ash, hickory, and beech that shed their leaves in the fall. The temperate forest biome in North America begins south of the coniferous forests of Canada and Maine, and extends southward along the east coast and westward until gradually replaced by grasslands. In its natural state, this entire area was a **deciduous forest.** As civilization moved west, many woodland tracts were cleared and replaced by farms.

Precipitation in this habitat is generally in excess of 30 inches per year. Four distinct seasons are observed in these regions, with a bright-colored display of maple and ash leaves after the first frosts in the fall. This biome tends to be less homogeneous in its plant population than some other biomes because its climate is less homogeneous.

Several levels or layers of vegetation called **strata** are found in a deciduous forest. The tall trees form the **canopy** or ceiling of leaves at the highest levels. Smaller trees fill in the area beneath the canopy. These shorter trees make up the **understory** of the forest. Short bushes make up the **shrub layer** in the zone beneath the understory. The shortest plants, including ferns and grasses, and other flowering plants are collectively called the **herb laye**r. The **forest floor** is composed of a layer of decaying plant materials that act as a mulch in preserving the soil moisture.

Mammals and birds of many kinds are native to this environment. Squirrels and many species of birds prefer to live in the canopy of the forest while other species prefer the understory. White-tailed deer, opossums, skunks, foxes, birds of prey, snakes, squirrels and mice are all common residents of broadleaf forests.

The passenger pigeon which was a native species in this environment became extinct due primarily to hunting pressure, and to the loss of critical nesting habitat. Wild turkeys and black bear are species that have suffered as deciduous forest habitat has been lost to timbering and land development. Today the wild turkey has been introduced into many areas of the country where habitat is available, and the species is making a dramatic comeback in numbers.

Coniferous Forest Biome

The **coniferous forest biome** is an evergreen forest of pine, spruce, fir, and hemlock. It forms a broad northern belt across the continent, extending from the grasslands and temperate forests on the south to the tundra regions on the north, and from the Northeast coastal region to the Pacific Northwest.

The vegetation in this biome consists mostly of trees known as **conifers** that produce seed in cones. The foliage of conifer trees is dense, and light intensity on the forest floor is inadequate to support the growth of most plants. A heavy carpet of dead needles covers the forest floor, and very few shrubs, grasses, or other plants are found there.

Precipitation in the region is mostly in the form of snow, and it generally ranges from 15 to 40 inches per year. The winters tend to be long and cold, and summer temperatures are moderate with cool nights. The needle-shaped leaves of conifers are well-adapted to cold temperatures, and to conserving moisture during dry times. The shape of the trees and the flexibility of the branches allows heavy snow loads to fall to the ground without breaking the limbs.

The coniferous forest biome is the home to many birds, insects and mammals. Large mammals such as elk, moose, mule deer and caribou often graze in the meadows and wetlands that are scat-

Figure 3-25: Forests that are harvested under good management practices will continue to prosper along with the wildlife that depend upon them. *Photo courtesy of Shannon Kolbe.*

tered throughout the coniferous forest. Predatory species that are found in these regions include black bears, grizzly bears, wolverines, lynx, timber wolves, foxes, mink, hawks, and owls. Squirrels, porcupines, hares, mice, and a variety of birds also live in coniferous forests.

Coniferous forests provide much of the lumber that is used for construction, see Figure 3-25. These forests are an important renewable resource, and forests that are harvested under good management practices can continue to provide a healthy environment for the living organisms that depend upon them.

LOOKING BACK

An ecosystem is seldom isolated from neighboring ecosystems because its boundaries usually allow organisms and materials to flow in and out of the environment. Ecosystems that exist in areas with similar vegetation and climate are grouped together into biomes. A freshwater biome exists when water that makes up the living environment is not salty. Ocean water makes up the marine environment. The terrestrial biomes of North America include all environments that are land-based. Water habitats are sometimes called wetlands. Federal regulations are in place that require restoration and preservation of wetlands.

REVIEW QUESTIONS

1. Explain how a problem in one ecosystem affects neighboring ecosystems that share its boundaries.
2. Assess the role of climate in the formation of biomes.
3. Identify the characteristics of a biome, and define the relationship of a biome to an ecosystem.
4. Create a chart that lists the distinguishing characteristics of each of the biomes in North America. It should also list the plants and animals that are found there, and describe the climates of the various biomes.
5. Identify ways that freshwater and marine biomes are similar, and contrast their differences.
6. Describe what plankton are, and appraise their importance in the aquatic food chain.
7. Describe why wetland habitat restoration is important.

LEARNING ACTIVITIES

1. Take a field trip to an area near your school and instruct the students to perform the following tasks:
   - describe the major characteristics of the environment
   - list the kinds of plant life that are observed
   - list the animal species that are known to inhabit the area
   - identify the insect species that are observed
   - develop a chart that shows how each organism might fit into the food web
   - discuss how human activity in the area might benefit or harm the organisms and the environment in which they live.

   Note: If a field trip is not possible, a good video showing the essential elements of the environment is a reasonable alternative.

2. Identify experts in your community who can provide factual information about the plants and animals that are found in your area. Invite one or more of these people to visit the class to discuss local environmental issues. Opposing points of view on controversial issues should be explored.

4 Our Wildlife Resources

WILD ANIMALS are some of the greatest treasures on planet earth. The native Americans who were here when Columbus first arrived had depended on wild animals for hundreds of years to supply much of their clothing, shelter, and food. The early colonists could not have survived in the New World without the wild animals from which they, too, obtained food, shelter, and clothing, see Figure 4-1.

OBJECTIVES

After completing this chapter, you should be able to

- describe how wild animals have contributed to the survival and comfort of humans
- distinguish between responsible and abusive stewardship of the land and of the environment
- explain how the U.S. Endangered Species Act provides protection to organisms that are in danger of becoming extinct
- discuss the controversial issues that have resulted from implementing the U.S. Endangered Species Act
- list several species of wild animals and birds that are protected by the U.S. Endangered Species Act, and describe what steps have been taken to improve their chances of survival
- identify environmental factors that contribute to extinction of organisms
- describe how a high degree of specialization in a species makes it more vulnerable to extinction
- suggest ways that nonadaptive behavior and low biotic potential contribute to extinction of some species of organisms
- defend the role of dependable research as a management tool for protected wildlife species.

Figure 4-1: Humans have depended on wild animals for food, clothing, and shelter for hundreds of years.

A NATIONAL TREASURE

The kinds and numbers of wild animals in North American ecosystems have changed with the movement of civilization across the continent. Some species have been lost due to natural changes in the environment; some have declined due to human competition, neglect, or abuse. Other kinds of birds and animals have benefited from the changed environments created by humans and they have become more abundant, see Figure 4-2.

Figure 4-2: Pheasants have been able to adapt to their changing environment. *Photo courtesy of U.S. Fish and Wildlife Service.*

Figure 4-3: It is important to our environment that we exercise responsible and careful management over natural resources and property. *Photo courtesy Michael Dzaman.*

Farmers, outdoor sportspersons, **naturalists** (people who study nature), and others who benefit from the use and ownership of land must exercise good **stewardship** by properly caring for the land, see Figure 4-3. This includes conserving and protecting the soil resources and the overall environments of the plants and animals that live there, see Figure 4-4. Landowners, managers, and users

Figure 4-4: The United States Department of Agriculture is responsible for the management of forest lands which are owned by the federal government. *Photo courtesy of Stacy Riggert.*

are responsible not only to their own generation, but to all future generations for care and management of our land and water resources. While most farmers and other property owners make serious efforts to protect the ecology of the area, some people abuse the right to control property by failing to manage it in a reasonable and responsible manner. Abuse of property and wildlife resources has made it necessary to pass laws that protect wild animals and the environments that they live in.

The U.S. Endangered Species Act

Congress passed legislation in 1969 that protected animal species that were declining in numbers. The act was expanded in 1973 to require the U.S. Fish and Wildlife Service to identify species of animals and plants that might become **extinct** due to the death of the entire population. The act identifies two classes into which those species that are found to be at risk may be placed. Those in immediate danger of extinction are classed as **endangered species**. These are the plants and animals that have small numbers in their population. In many cases, the population is also becoming smaller throughout most or all of the range that is occupied by the species, see Figure 4-5. Species that are in less danger of extinction, but which are at risk of becoming endangered, are classed as **threatened species.** These are species that can reasonably be expected to survive if immediate steps are taken to protect the environment in which they live.

The act protects both the species and its **habitat**, see Figure 4-6. Habitat is defined as the environment in which an **organism** or living creature makes its home and from which it obtains its food. Species that are protected under this legislation cannot be hunted or killed without heavy legal penalties being assessed. The act also protects the organism's habitat from development or other disruptive uses by people. Restricted use of the land area where protected species live is necessary to prevent the delicate balances of nature from being destroyed.

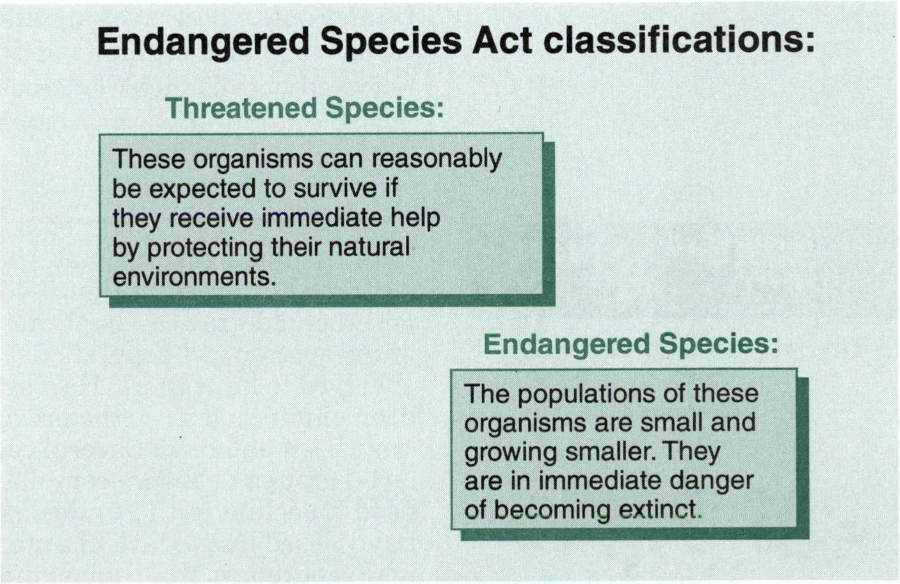

Endangered Species Act classifications:

Threatened Species:

These organisms can reasonably be expected to survive if they receive immediate help by protecting their natural environments.

Endangered Species:

The populations of these organisms are small and growing smaller. They are in immediate danger of becoming extinct.

Figure 4-5: Endangered Species Act classifications.

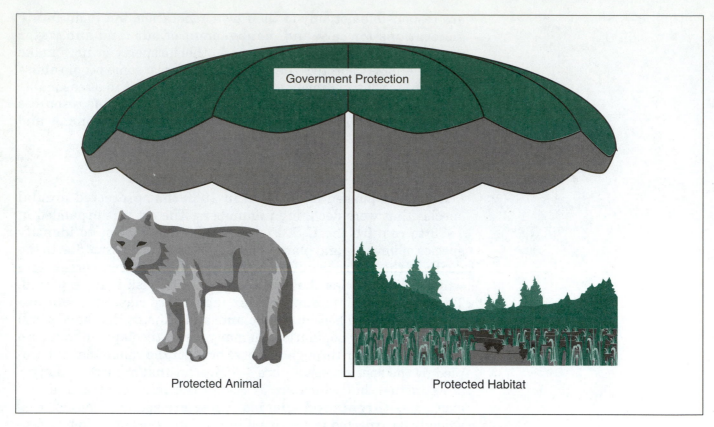

Figure 4-6: The Endangered Species Act protects both the species and its habitat.

CONTROVERSY AND PROTECTIONISM

Protection of endangered or threatened species has resulted in controversy over the provisions of the Endangered Species Act. This is because the act requires non-use or restricted use of natural resources in the areas where protected species live. The economic cost to humans of enforcing the Endangered Species Act in a particular area can be very high.

The U.S. Fish and Wildlife Service often comes under pressure from people whose lives are affected by this act. These people want to prevent enforcement of the law or to change the way it is enforced because jobs, businesses, and sometimes the welfare of entire communities may be adversely affected if the law is fully implemented.

THREATENED SPECIES PROFILE

The Northern Spotted Owl

In the Pacific Northwest, the spotted owl has been identified as a threatened species (Figure 4-7). Its favored habitat is mostly composed of old-growth forest, however recent scientific studies have found spotted owl populations in young forests, and they have been observed to hunt there. Harvest of timber in these areas has nearly been eliminated since the provisions of the Endangered Species Act have been enforced. Several court actions have been filed by interested groups to assure compliance with the act, and the result has been a decline in the timber industry of the region. Lumber mills have closed due to lack of a steady supply of logs, and many people who worked in the timber industry lost their jobs. Protecting a

Figure 4-7: The Northern spotted owl. *Courtesy of U.S. Fish and Wildlife Service. Photo by Randy Wilk.*

threatened or endangered species is important, however, it is just as important that the issues on both sides are clearly and accurately presented and supported.

Many controversies of a similar nature exist in different areas of the United States where species of birds and animals have been classified as threatened or endangered. In each case, conflict occurs between people who rely financially on the resources that make up the limited habitat of the listed species, and environmentalists who insist that there can be no compromise on the law that protects endangered and threatened species.

THREATENED SPECIES PROFILE

The Loggerhead Sea Turtle

The loggerhead sea turtle is a threatened species that begins its life on the east coast of Florida (Figure 4-8). Females lay about one hundred eggs each year in the nesting area before returning to the sea. However, only one in ten thousand of the baby turtles that hatch survives to adulthood. Many of the turtles are caught in shrimp nets where they drown. Shrimp fisherman are reluctant to modify their nets to protect the turtles because their catch of shrimp tends to be reduced. The issue is an economic one. Can shrimp fish-

Figure 4-8: The loggerhead sea turtle. *Photo courtesy of U.S. Fish and Wildlife Service.*

ermen afford to lose 25 percent of their catch to save fifty thousand turtles per year? The law also restricts development of the nesting areas along the shore.

Critics of the Endangered Species Act are concerned that the law does not take into account the human costs associated with its enforcement. Many of these people object to the law because local economies are often weakened and jobs are lost when it is strictly enforced. They believe that there should be a balance between human rights and the rights of the protected species, see Figure 4-9.

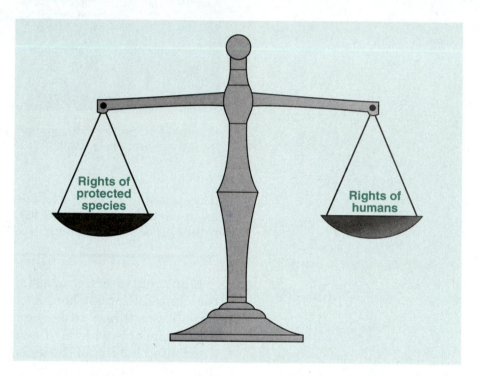

Rights of protected species

Rights of humans

Figure 4-9: Many of the people whose jobs have been affected by The Endangered Species Act would like to see a balance established between the rights of humans and the rights of protected species.

ENDANGERED SPECIES PROFILE

The Red Wolf

The red wolf is an endangered species whose native range extends from the southeastern United States to central Texas (Figure 4-10). Like other species of wolves, it is a predator. It feeds mostly on small animals and birds, but packs of wolves may also attack and kill large animals including domestic livestock. Like the Grey wolf, Red wolves have been hunted and killed because their feeding habits put them in competition with humans. They were nearly eliminated from many livestock ranges during the

Figure 4-10: The red wolf. *Courtesy of U.S. Fish and Wildlife Service. Photo by Steve Maslowski.*

past 100 years. As an endangered species, it is illegal to kill these animals. The recovery of this species is generally opposed by livestock producers and by some hunting advocates.

Another controversial issue is the difficulty in distinguishing between species that are victims of natural selection and those that are victims to competition from humans and their management of the environment. Some protected populations continue to decline even when no human interference is evident. In some instances, no amount of human intervention can prevent the species from becoming extinct. This is not a new phenomenon. Natural selection and extinction are processes that have been happening for as long as life has existed on this planet. The problem has become much worse as the world population of humans has expanded into many areas. Much of the wild animal domain has now been converted to farms and other human uses. The black market trade in protected wildlife has become a major problem for some animal species. Ivory, fur, pets and animal parts are sold illegally. On a world scale, trade in wildlife has been reported to rank second behind drug trafficking among illegal trade activities. Such activities lead to extinction of a species. How do we know when we have done all that we can do, and when should we let nature run its course. Is it even ethical to interfere with the natural selection process?

ENDANGERED SPECIES PROFILE

The Northern Swift Fox

The historical range of the northern swift fox extends across the northern plains of the United States and into Canada. It is protected as an endangered species in Canada (Figure 4-11).

Figure 4-11: The Northern swift fox.

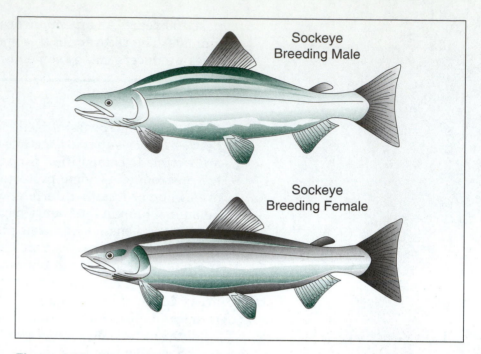

Figure 4-12: The sockeye salmon.

One endangered species that has generated recent controversy is the sockeye salmon, see Figure 4-12. The sockeye spawns in Redfish Lake in Idaho and enters the Columbia River system at one to two years of age. They migrate past eight dams as they swim down the river to the Pacific Ocean. The dams create a series of barriers that restrict the movement of the fish, and they interfere with the ability of the fish to migrate safely.

Efforts to draw down the reservoirs behind the dams have resulted in controversy over ownership of the water rights. The water in the reservoirs is legally managed for production of electrical power, commercial shipping, and for irrigation of crops. The reduced water levels that are necessary to eliminate slack water during the annual migration of the sockeye is expected to restrict shipping and commerce on the Columbia River and to deplete the supply of stored water needed for producing electricity and crops.

EXTINCTION AND ITS CAUSES

Extinction of a species of organisms is not something to be taken lightly. Diversity of species is considered to be an indicator of a healthy environment. Humans have become a dominant species because they are able to adapt to nearly any environment. They also act as predators toward many of the animals and birds with which they share the environment. They frequently disrupt the habitats upon which other species depend.

Destruction or modification of habitat is the greatest single cause of extinction. When organisms lose their food supply, they

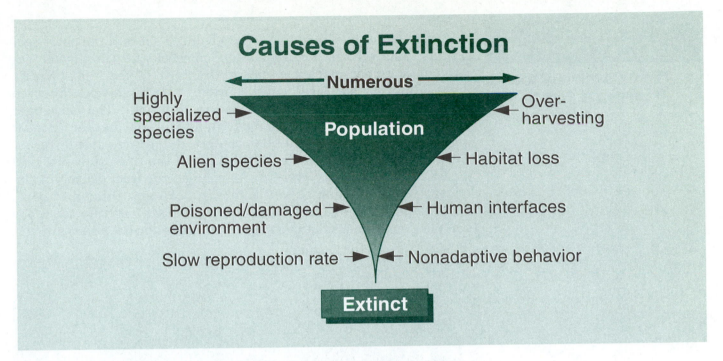

Figure 4-13: There are many factors that can lead to the extinction of a species.

soon starve to death. When their shelter is damaged or destroyed, they can more easily be eliminated by natural enemies or by unfavorable weather conditions. Even modest changes in weather conditions are dangerous to an organism that has lost the shelter to which it is accustomed. Natural disasters such as severe storms or extreme temperatures are among the greatest threats to an already threatened or endangered species of organisms.

Many organisms are unable to adapt quickly to changes in environmental conditions or to heavy losses from predators. They decline in numbers as their environments are modified, or as predators increase. A single cause can result in the extinction of some plants or animals, but a species is more likely to become extinct due to a combination of factors that impact it in a negative way, see Figure 4-13.

An important factor that sometimes leads to extinction of an established species is the introduction of an **alien species** into the ecosystem. The new species may compete with the native species for food and shelter, or prey upon it as a source of food. When this happens, the balance in the ecosystem is upset, and the weaker species tends to decline while the newly introduced species increases in numbers. Alien species are sometimes sources of new diseases. Entire populations of organisms have been lost because of such diseases.

Still another major factor that contributes to extinction is overhunting of a species by humans. In some instances this has been done commercially, but in other instances sport hunting of a vulnerable species has contributed to its extinction.

EXTINCT SPECIES PROFILE

The Extinction of the Passenger Pigeon

A combination of negative factors led to the loss of the passenger pigeon as a species (Figure 4-14). This species of wild pigeon was the most numerous bird species in the world in the 1800s. Nearly two billion of these birds were reportedly in a single flock observed by ornothologist Alexander Wilson in Kentucky. The passenger pigeon was intensely hunted for food, and a large market for meat developed in eastern population centers, made accessible by recent expansion of the railroads. Destruction of the trees that provided nesting habitat, and harvesting of young birds from densely-populated nesting areas played a role in reducing their numbers.

The passenger pigeon lived in large flocks and this tendency made them vulnerable to disease and predation. In addition to all

Figure 4-14: The passenger pigeon. *Courtesy of U.S. Fish and Wildlife Service. Photo courtesy of Luther Goldman.*

of these problems, a pair of passenger pigeons only produced one egg and raised a single offspring during each nesting period. The last known passenger pigeon died in a zoo in Cincinnati in 1914. The lesson to be learned from the passenger pigeon is that even the most numerous species can become extinct when their habitat is destroyed and when unrestricted hunting occurs.

The degree of specialization in a species affects how vulnerable it is to extinction. Species that cannot adapt their behaviors or their diets to accommodate changes in their environment are at the greatest risk. This is because a highly specialized mammal or bird may depend on a single source of food or shelter. When that source

Figure 4-15: The whooping crane is an endangered species. *Photo courtesy of Utah Agricultural Experiment Station.*

of particular food or shelter is gone, a highly specialized bird or animal will likely be unable to adjust and face extinction. Failure of a species to adapt to a changing environment is called **nonadaptive behavior**.

Most surviving species in the world today are able to adapt to modest changes in their environments. However, abrupt changes in an environment allow no time for living organisms to adjust, and in some cases adjustment is not possible.

A slow rate of reproduction contributes to extinction by reducing the recovery rate of an endangered species. Biologists refer to this problem as **low biotic potential** or fecundity. Examples of animals and birds that fall into this class include the California condor which lays only one egg every two years, and the whooping crane which takes several years to reach breeding age and then lays only two eggs per year, see Figure 4-15. More successful species of birds often lay 8 to 10 eggs or nest several times per season, see Figure 4-16.

MANAGING ENDANGERED AND THREATENED SPECIES

Humans are the only species that can make conscious decisions to destroy or preserve other forms of life. They also have a moral responsibility to preserve other organisms with which they share

Figure 4-16: A slow rate of reproduction reduces the recovery rate and can contribute to extinction. More successful species of birds lay 8-10 eggs or nest several times a year.

the environment. People in some parts of the world take this stewardship seriously, but other cultures place little value on preserving threatened and endangered species. Even in societies that accept responsibility for protecting these organisms, there are strong differences of opinion concerning how much protection should be provided.

One of the most difficult problems facing fish and wildlife biologists is managing the surviving members of a population that is found to be endangered or threatened. They must learn to identify and understand relationships between the organism and its environment. Relationships of this kind are often difficult to define because the species preferred habitat may no longer exist.

Effective management of endangered species of organisms must be based on reliable research. Those who are responsible for the recovery of endangered or threatened species should explore as many alternative management strategies as they can identify. Innovative ways of restoring acceptable shelter and providing appropriate food sources have been successfully used in critical management situations, see Figure 4-17.

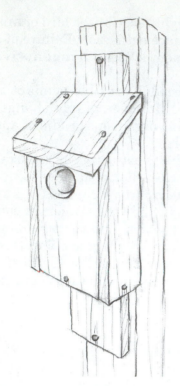

Figure 4-17: Nesting boxes are an innovative way to restore shelter in critical management situations.

Career Option

FISH AND WILDLIFE BIOLOGIST

A **biologist** who works with fish and wildlife is a person who makes a career of learning about the basic needs of animals. He/she studies the living habits of different animal species to determine the kinds of food and shelter that are needed. They also study other characteristics of organisms such as reproductive habits and territorial ranges.

Fish and wildlife biologists must have a strong background in the biological and environmental sciences. A person planning a career in this field will need a four-year degree from a good university with graduate study recommended. Often this type of biologist will need to conduct field studies to determine management alternatives for wild animals.

LOOKING BACK

Wild animals are an important natural resource to the human race. They have provided food, clothing, and shelter to many cultures around the world. Wild animal populations are seldom stable. They change as civilization and other species of animals expand into their environments. Environments often change as humans and other dominant species move into an ecosystem. Some species of organisms are favored by changes in their environments, but in other cases even minor changes make it difficult for some wild animals to survive.

Species of animals that are declining in numbers are often protected by government regulations when they become classified as endangered or threatened species. Introduction of competing or predatory species contributes to extinction due to loss of habitat, predation, starvation and disease.

Serious controversies frequently occur over the management of protected species of animals because human use of the natural resources in such areas is restricted or stopped entirely. Human management of protected species should be innovative and based on sound research.

REVIEW QUESTIONS

1. Describe how wild animals have contributed to the survival, comfort, and growth of human populations.
2. How is responsible stewardship different from abusive stewardship as it relates to the land and the environment?
3. Explain how the U.S. Endangered Species Act might be used to protect species of organisms that are in danger of becoming extinct.
4. Identify a controversial issue that has surfaced as a result of implementing the Endangered Species Act, and discuss the pros and cons of the issue.
5. List some species of wild animals and birds in your region that are protected by the Endangered Species Act, and describe the steps that have been taken to improve their chances of survival.
6. Identify some factors that are known to contribute to the extinction of wild animals.
7. Explain how a high degree of specialization in a species makes it more vulnerable to becoming extinct.
8. How does low biotic potential and nonadaptive behavior contribute to population declines of some species?
9. Explain why good scientific research is needed as a management tool for protected wildlife species.

LEARNING ACTIVITIES

1. Divide the class into two groups for the purpose of debating the issues that arise when the Endangered Species Act is invoked. Assign one group to debate in favor of restricting the use of resources that are part of the environment of an endangered species. Assign the second group to defend the rights of people who depend on those resources to earn a living.

2. Divide the United States (or the world) into geographic regions, and assign teams of students to research the species in their regions that are considered to be threatened or endangered. Describe the factors that are contributing to the problem, and offer solutions for restoring the populations that are in danger of extinction. Each team should report back to the class.

SECTION II

Ecology of Mammals

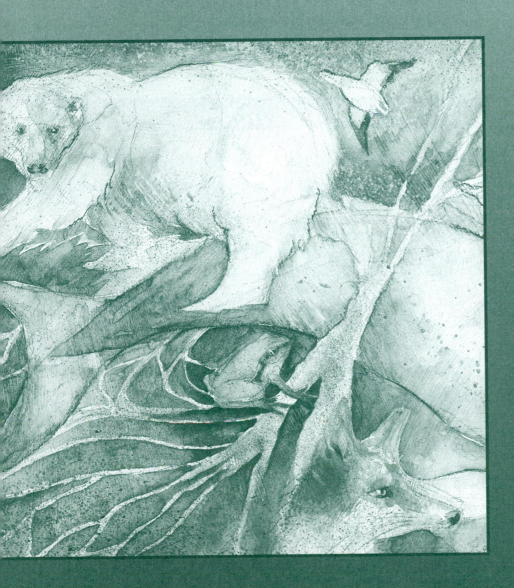

Mammals are members of a class (Mammalia) of warm-blooded animals with bony skeletons. They give birth to live young and nourish them with milk secreted from mammary glands. They usually have a protective hair coat.

5 Gnawing Mammals

THE BEST KNOWN ANIMALS in North America are the **mammals**. These animals are warm-blooded, and they usually have a protective coat of hair and a bony skeleton. Their babies are fed milk secreted by the **mammary glands**, which are the milk-producing organs of females. Mammals are also **vertebrates** along with birds, reptiles, amphibians and fish. Vertebrates have back-bones composed of many segments that enclose the spinal cord, see Figure 5-1.

OBJECTIVES

After completing this chapter, you should be able to

- identify the physical characteristics that distinguish mammals from other animals
- describe how scientists classify, organize, and define relationships among living organisms.
- distinguish between primary and secondary consumers
- define the roles of rodents and other gnawing animals in the ecosystems of North America
- predict the effects of declining or expanding populations of gnawing animals on the populations of predators that depend on them for food
- predict the effects of declining or expanding populations of predators on the populations of rodents and other gnawing animals.
- illustrate the distribution of gnawing animals on the North American continent
- profile the life cycles of specific gnawing mammals.

CLASSIFICATION OF ANIMALS

Mammals are classified into related groups based on their genetic and structural similarities to other animals. The field of science that classifies organisms and defines their relationships to one

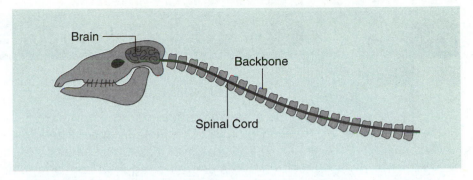

Figure 5-1: Vertebrates are identified by backbones composed of bone segments surrounding the spinal cord that extend from the tail to the skull.

another is called **taxonomy**, see Figure 5-2. The animal **kingdom** includes all of the animals, and the plant kingdom includes all of the plants. The next smaller division is the **phylum**. Below the phylum is a smaller unit called a **class**. Phylum Chordata includes the class Aves (birds), the class mammalia (mammals) and the class Reptilia (reptiles). Examples of classes include the division of birds, mammals, and others. Continuing down the taxonomic scale, classes can be divided into more than one order, an order into more than one **family**, and family into more than one **genus**. A genus is divided into **species** based on differences as simple as those between a dog and a fox.

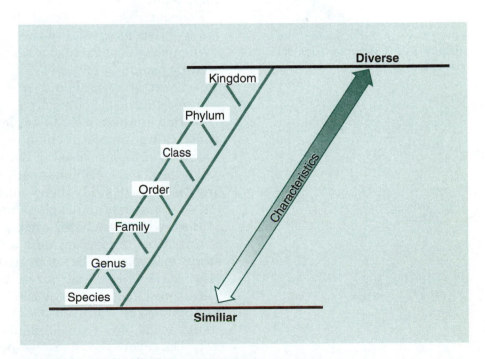

Figure 5-2: The taxonomy of living things.

Career Option

TAXONOMIST

A **taxonomist** is a scientist who classifies living organisms into related groups. This is a highly specialized field that requires a person to be able to observe and distinguish small but distinct differences among organisms. An advanced graduate degree is required from a reputable university. Much of the work of a taxonomist involves collecting specimens of organisms, and accurately observing the features that makes one organism distinctly different from other similar organisms.

Taxonomists often find careers at colleges and universities and in museums. Some are also involved in field research and collecting expeditions.

Figure 5-3: Characteristics of rodents.

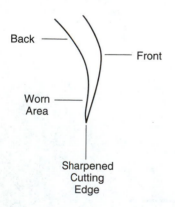

Figure 5-4: Rodent incisor tooth structure.

Some gnawing mammals are called **rodents**. This group of animals is identified by the four large incisor teeth in the front of their mouths, see Figure 5-3. These teeth never stop growing, and rodents must gnaw on wood or other materials just to keep their teeth worn down. The front edge of a rodent's teeth is composed of harder material than the back edge, causing the back edge to wear faster than the front. The result is that the incisor teeth become chisel-shaped, and they are sharpened as they wear down, see Figure 5-4.

Rodents make up the most diverse group of mammals. There are many different species of these animals, and they are capable of living in a wide variety of environments. Rodents are often considered by people to be pests. This is due to the large numbers and destructive gnawing habits of some species, but these animals are an important food source for a large number of other animals.

In this chapter we will explore the role of rodents and other gnawing mammals in the ecosystems of North America. Some species will be discussed in detail. Scientific and common names will be used to identify the species that are featured in the mammal profiles. We will study their interactions with other living things in the environment, and explore how they fit into the food web.

Most rodents are herbivores, meaning they eat plants as a source of nutrition. These animals are also known as **primary consumers.** Two examples of primary consumers are deer and squirrels. Herbivores are eaten by carnivores, or meat-eaters. Carnivores, such as the weasel and the mountain lion, are classed as **secondary consumers**. A field mouse that feeds upon the stems, seeds, and roots of plants is a primary consumer in a food web. The coyote that eats the mouse is a secondary consumer, see Figure 5-5. We learned in chapter one that the food plants that are eaten by the mouse are known in the food chain as producers.

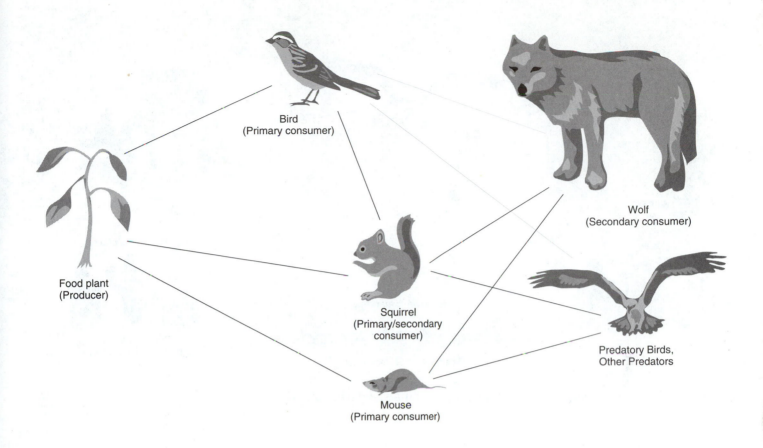

Figure 5-5: The roles of gnawing mammals in an ecosystem.

MICE, RATS, VOLES, AND LEMMINGS

There are many different species of mice, rats, voles, and lemmings in North America, but only a few of the most commons ones will be discussed in this chapter. These small rodents are very important as food animals. This group of animals is the most numerous class of rodents.

The mouse is the smallest of the rodents. Rats, voles, and lemmings are similar to mice in many ways, but they are larger in size. All of the animals in this group are efficient at converting plant materials into flesh, and they are a dietary staple for many predatory animals and birds.

Rodents are the main source of food for predatory birds such as hawks and owls, see Figure 5-6. The birds of prey have excellent eyesight, and they can see small rodents from great distances. Nearly everyone has observed a hawk riding the wind currents high above the ground while it scans the fields in search of mice. When a mouse is spotted, the hawk dives to the earth and grabs its prey in its strong talons. One blow from the bird's beak kills the mouse almost instantly, and it soon becomes a meal.

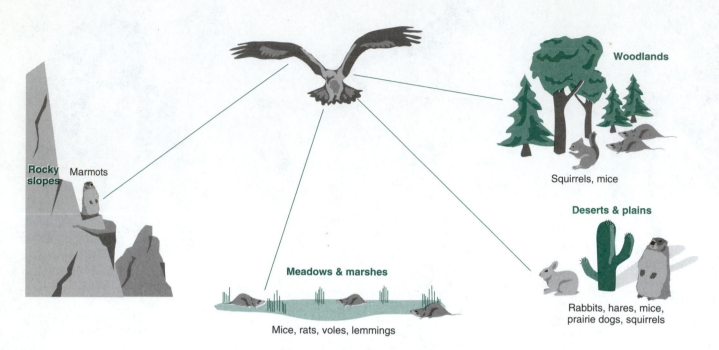

Rocky slopes — Marmots

Woodlands — Squirrels, mice

Deserts & plains — Rabbits, hares, mice, prairie dogs, squirrels

Meadows & marshes — Mice, rats, voles, lemmings

Figure 5-6: The relationship between gnawing mammals and birds of prey.

MAMMAL PROFILE

House Mouse (*Mus musculus*)

House mice are small rodents that range from 6 to 8 inches long from nose to tail, and they weigh from 0.5 ounce to 1 ounce. They are usually gray or brown in color, and they have long, hairless tails. During good weather they often live outdoors, but they prefer to live in buildings during the colder seasons. They live anywhere that food and shelter can be found in the same place. This is often inside houses and other buildings close to humans, see Figure 5-7, and its range extends through most of North America.

The house mouse begins reproducing at two months of age, and is capable of having up to eight litters each year. Litter sizes range from three or four to ten or eleven young, and they are born naked, helpless, and blind. They grow up in a nest lined with soft material that has been placed there by the female. They are an important food source for many kinds of predatory animals and birds.

Figure 5-7: A brown house mouse. *Photo courtesy of Mario Bogo.*

Skunks, foxes, coyotes, bobcats, lynx, mink, weasels, and shrews are just a few of the mammals that eat large numbers of small rodents. These animals are patient and skillful hunters. They crouch motionless near an area that is inhabited by rodents, and wait for the prey to come close enough to catch. Weasels and shrews will also go into the burrows of small rodents to pursue and kill their prey. Most predators have keen hearing and sharp eyesight, along with a highly developed sense of smell. They usually know where a rodent is because they can smell it and hear it as it moves about. Once the prey is seen, the hunter carefully chooses the moment when it will strike.

MAMMAL PROFILE

Meadow Jumping Mouse (*Zapus hudsonius*)

Figure 5-9: Distribution map of the meadow jumping mouse.

The meadow jumping mouse is light brown in color with long hind legs and tail. It measures from 7.5 inches to 9 inches long including its nose and tail, yet it usually weighs less than an ounce. It is sometimes called a kangaroo mouse because it hops about on its long hind legs like a tiny kangaroo, see Figure 5-8. When it is frightened, it can take leaps of up to 8 feet at a time. The ability to jump so far and fast makes this mouse difficult for predators to catch.

Figure 5-8: A meadow jumping mouse.

Meadow jumping mice are **nocturnal** which means that they prefer to sleep in the day and move about at night. They feed mostly on seeds and insects. Most days are spent sleeping in grass nests located in underground burrows. Three to six young are usually born in June, with a second brood born in September. Their range extends from Alaska to the east coast of the U.S., see Figure 5-9.

Meadow jumping mice **hibernate** during the winter. While hibernation is referred to as sleeping, it is different from sleep because the animals body processes slow down and they use the energy stored in body fat as their only source of nutrition. By late fall, animals that hibernate must store enough body fat to nourish them through the winter. When it gets cold, they enter deep burrows well below the frost line where they hibernate until spring.

Snakes and other kinds of reptiles use small rodents as sources of food, see Figure 5-10. Mice and rats are never safe when a snake is hunting for a meal. Snakes can go anywhere that small rodents can go, and they frequently do their hunting inside the dens of the mice and rats.

Some snakes paralyze rodents with their poisonous venom before swallowing them whole. Some nonpoisonous snakes kill or stun small rodents by biting them. Other kinds of snakes coil around

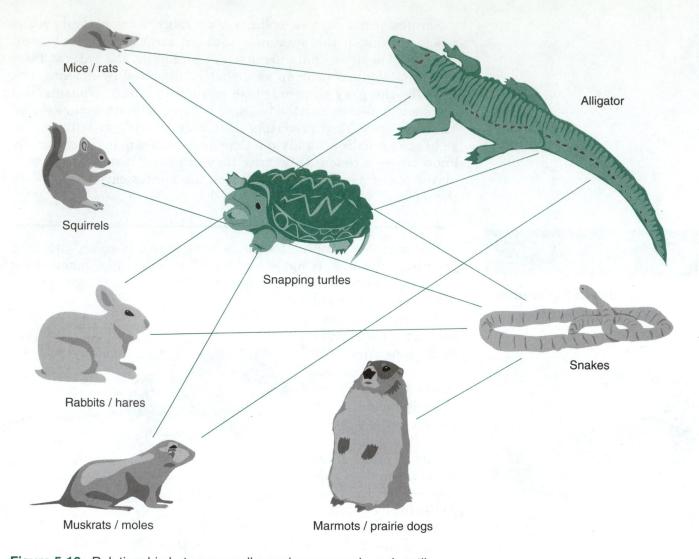

Figure 5-10: Relationship between small gnawing mammals and reptiles.

their prey and squeeze it to death before eating it. Snakes are not able to tear their prey into small pieces, so they always eat it whole. This is possible because they have hinged jaws that separate to allow large meals to pass through.

<table>
<tr><td>

MAMMAL PROFILE

Pine Vole (Pitymys pinetorum)

</td><td>

As the name suggests, the pine vole lives in the woods, see Figures 5-11 and 5-12. This vole digs shallow tunnels beneath the leaves on the forest floor, and it spends most of its time underground. It is a small brown vole with a stubby tail, and it measures from 4 to nearly 6 inches in length and weighs from 1 to 1.5 ounces. It ranges in the eastern third of the U.S.

The pine vole eats seeds, berries, roots, and tubers,

</td></tr>
</table>

Figure 5-11: A pine vole.

Figure 5-12: Distribution map of the pine vole.

and it stores food in underground larders. Several pine voles are often found living together. They emit a warning call when danger is near.

Pine voles are born in small litters of two to four young. Because they spend most of their time in underground tunnels, they have few enemies in comparison with other small rodents. They do not reproduce nearly so rapidly as house mice and field voles. Shrews and snakes sometimes venture into their tunnels, and other predators might occasionally dig up their burrows, but pine voles are usually quite safe from their enemies.

Bears are the largest of the predatory animals that prey upon mice, voles, rats and lemmings. Although these rodents don't make a very big meal for a bear, it is not uncommon for a bear to dig up a large area in an attempt to catch one of these small animals.

MAMMAL PROFILE

Norway Rat (Rattus norvegicus)

The Norway rat is not native to North America, but arrived here on ships from Europe, see Figure 5-13. It is a hardy rodent that adapts well to many different environments and its range extends to most regions of North America. This rat is brown in color with a long, hairless tail. It weighs 0.5–1.5 pounds, and measures 13–18 inches long from its nose to the tip of its tail. It is attracted to garbage dumps and sewers, and usually lives near humans.

Norway rats will eat almost anything that can be chewed. A single rat can eat up to 50 pounds of grain in a year, and they ruin even more grain than they eat by polluting it with their droppings. They are also predators, and kill small birds and animals of many kinds.

Figure 5-13: A Norway rat. *Photo courtesy of Leonard Lee Rue III.*

The gnawing habits of rats cause damage to many materials in their habitats. Much of the food supply for humans and domestic animals is damaged by rats when storage facilities are poor. Rats can also transmit diseases to humans and animals, including bubonic plague and typhus.

Norway rats live in colonies anywhere that they can find shelter and food. They reproduce at a rapid rate. Litters usually average eight or nine young, but litters as large as twenty have been observed. Rats have many enemies including cats, dogs, snakes, birds of prey, and humans, but their high reproductive capacity helps compensate for losses to predators.

Owls depend almost entirely on mice, rats, voles, squirrels, and rabbits for food. They hunt primarily at night when many small rodents are most active. The eyes of an owl are adapted to night vision, and they have excellent hearing. Once a meal has been located, the owl swoops silently down on its unsuspecting victim and carries it away.

MAMMAL PROFILE

White-Throated Woodrat (*Neotoma albigula*)

Figure 5-15: Distribution map of the white-throated woodrat.

The white-throated woodrat is a vole. It is unusual in that it builds a house of spines. Using cactus spines, it builds large clumps of debris in which it lives protected from most of its enemies except for snakes. Outside of its home, however, this woodrat is still vulnerable to other predators. It is nocturnal, and it spends the nights looking for seeds, berries, and succulent plants to eat. It gets moisture from plants and seldom drinks water.

Some woodrats build homes in trees or live in caves. All of these woodrats have the strange habit of collecting things. It may take a spoon or other interesting item, and leave a small rock or other object in its place. For this reason, woodrats are sometimes called "pack rats" or "trade rats."

Figure 5-14: A white-throated woodrat.

Woodrats live solitary lives except during the time when mating occurs. The male leaves before the female has her young. Two or three litters of young are born each year. At about two months of age, the female kicks the young out of the nest, and they begin life on their own.

White-throated woodrats are brown in color with white throats and feet, see Figure 5-14. They have hairy tails, and they range from 13 to 16 inches in length. They weigh between 6 and 8 ounces, and their range extends through the desert region in the southwestern U.S. and northern Mexico, see Figure 5-15.

A **vole** is a small rodent with a stout body and short tail; it is often confused with a mouse or a rat. It is different from mice and rats due to its blunt face, small eyes, large ears and hairy tail. The most common voles in North America are the meadow vole, field mouse, and the muskrat.

Figure 5-16: Distribution map of the meadow vole.

A meadow vole is frequently called a field mouse. It is a small brown rodent that ranges in length from 5 to 8 inches from nose to tail and weighs from 1 to 2.5 ounces. It has a stocky body and a short tail, and it is constantly on the move looking for food. It eats its own body weight in seeds, berries, and roots every twenty-four hours.

Meadow voles thrive in many locations, but they do best in meadows and grasslands. Their range extends across Alaska, Canada, and the Northern U.S., see Figure 5-16. They construct a network of burrows and trails for protection from the many predators that feed on them. Nearly every kind of predator feeds on meadow voles. Its enemies include birds of prey, weasels, coyotes, foxes, skunks, bobcats, and fish.

Five to nine offspring are born in each litter, and females are capable of producing up to seventeen litters each year. By the time a vole is three weeks of age, it is on its own, and its mother is busy raising another litter. Under favorable conditions, huge numbers of meadow voles may live in an area. Large populations can destroy large acreages of crops. They are especially destructive to fruit orchards. Meadow voles can kill trees by gnawing the bark all the way around the tree trunks.

The muskrat is a large vole that is sometimes called a marsh rabbit. It lives in marshes and streams, and spends much of its time in the water. It builds a house of reeds and grass in shallow water which it enters beneath the surface of the water. It also digs burrows in stream banks with entrances located beneath the level of the water.

The muskrat is one of the most important furbearing animals in North America. Several million pelts are harvested each year, and the meat of the muskrat is considered a delicacy by some people. It has been exported to several European countries in which it now thrives in the wild.

The muskrat is a large vole that grows as large as 24 inches from nose to tail at maturity. It is a nocturnal animal that inhabits many of the freshwater biomes of North America, see Figure 5-17. It is active throughout the year and does not hibernate. It has two scent glands located near the tail from which it gets its name and from which it obtains a musky oil to waterproof its fur. It has a flat hairless tail that it uses as a rudder while it swims. Its range extends through much of the U.S. and Canada, see Figure 5-18.

Figure 5-18: Distribution map of the muskrat.

Three or four litters are born each year in the northern regions; in the southern areas, mating continues throughout the year. Each litter consists of seven or eight offspring. Young muskrats begin to swim when they are about three weeks old. They seldom live to be more than four years of age. The most successful predators of muskrats are eagles, otters, and mink.

Figure 5-17: The muskrat. *Courtesy of U.S. Fish and Wildlife Service. Photo by R. Town.*

A lemming migration is a spectacular and unusual event. In much the same way that bees swarm from the hive when they become too numerous for the colony, entire populations of lemmings sometimes leave their tunnels and move overland. Millions of these small rodents advance across the land during a migration, like a forage-devouring tidal wave. They eat every plant they come to. Most of the forage is eaten by the migrating lemmings, and the caribou sometimes begin to eat the lemmings that have been crushed beneath their feet.

Prior to the migration, the predator population increases along with the lemming population. As the migration begins there is such an abundance of food for predators that even the trout eat small lemmings as they swim across streams.

In the end, the migrating lemmings die. Many are killed by predators. Others die from diseases that have become more common as the lemming population has grown too large. Some die from exhaustion, and many drown in rivers and lakes. In coastal areas large groups of lemmings have been observed swimming in the ocean until they drown.

When the lemmings are gone, the predators die too, but enough lemmings always survive to replenish the population. Eventually the whole cycle is repeated.

MAMMAL PROFILE

Brown Lemming
(*Lemmus trimucronatus*)

The brown lemming is a small rodent that lives in the tundra regions of Alaska and Canada, see Figures 5-19, 5-20. It has long brown fur and a stubby tail, and it lives in colonies much like prairie dogs. Brown lemmings are 5–6.5 inches long and weigh 1.5–4 ounces.

Figure 5-19: A brown lemming.

Figure 5-20: Distribution map of the brown lemming.

Lemmings live in tunnels connected by runways. They eat a variety of grasses, sedges, and mosses, and they store a supply of harvested forage underground for winter use. They prefer lowland areas where snow cover insulates the colonies during the cold winters.

Between April and September, a new litter of three to eleven young is born to mature females each month. They are born in grass nests shaped like balls and lined with soft materials like feathers, moss, or hair. Such high biotic potential results in huge numbers of lemmings in years when the environment is favorable to their survival.

Tree Squirrels

Many common species in many environments

SQUIRRELS

Squirrels are broadly divided into two groups. Tree squirrels spend most of their time in trees, and they are usually more appreciated than ground squirrels by their human neighbors, see Figure 5-21. The ground squirrels live in burrows in the ground, and are generally considered to be pests, see Figure 5-22.

Flying squirrels

Figure 5-21: There are two types of squirrels. One type is the tree squirrel which consists of many common species, such as the grey squirrel, and tree squirrels.

Ground Squirrels

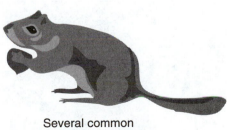

Chipmunks of several species

Several common species and habitats

Figure 5-22: The other type of squirrel is the ground squirrel. This Group consists of several common squirrel species and chipmunks of several species.

Douglas Squirrel
(Tamiasciurus douglasii)

Figure 5-23: Distribution map of the Douglas squirrel.

The Douglas squirrel is also known as the chickaree. It makes its home in pine forests, and ranges from British Columbia to California, see Figure 5-23. It spends much of its time in the tree-tops, clipping pine cones loose from the trees. It comes down to the forest floor to gather its harvest and to cache the cones in safe places for winter use. In addition to pine seeds, this squirrel eats mushrooms, nuts, and fruits.

Douglas squirrels are brown with rust-colored belly, legs, and feet, and long furry tails fringed with yellow. They chatter vigorously as they scold any creature that invades their territory. Foxes, coyotes, fishers, and martens are primary predators of the Douglas squirrel.

Females give birth to a single litter of three to seven young in the spring and summer. They are born in a hollow tree, a burrow, or in a tree nest made of twigs. Adult squirrels of this species, range from 12 to 14 inches long and weigh between 5 and 11 ounces.

Tree squirrels live in forested areas and depend on trees for both food and shelter. They are excellent climbers, and they move swiftly from branch to branch with ease. Their homes in the trees provide protection from many of the predators that kill and eat ground squirrels. Tree squirrels tend to have smaller litters of young than ground squirrels, but they sometimes have more than a single litter in a year. The late litters are able to survive because tree squirrels stay awake through the winter, and they eat food that they have hidden away during the summer.

Eastern Gray Squirrel
(Sciurus carolinensis)

Figure 5-24: Distribution map of the Eastern grey squirrel.

The eastern gray squirrel is a native of the hardwood forests located in the eastern U.S., see Figure 5-24. Its numbers declined under extreme hunting pressure and in response to the loss of habitat as the hardwood forest that once covered much of the eastern part of the continent was cleared away to make room for farms and towns.

The passage of game laws to protect this squirrel helped the population to recover. Many areas of the eastern U.S. are forested once again after being extensively cleared for agricultural use. One or two litters, each consisting of five or six young, are born to mature females in the spring and summer months each year. The squirrels are born in treetop nests of twigs where they remain until they are about six weeks of age.

Figure 5-25: The Eastern grey squirrel. *Photo courtesy of Leonard Lee Rue III.*

The main diet of the eastern gray squirrel is hickory nuts and acorns. These squirrels stay busy during much of the summer gathering and hoarding a winter supply of food. They are between 16 and 21 inches in length and weigh from 1 to 1.5 pounds; western gray squirrels are a little bigger. The Gray squirrels have long tails completely covered with hair. Despite their names, they are actually black or black and tan in color, with white tips on the fur, see Figure 5-25.

Some of the larger species of tree squirrels are classed as game animals. Fish and game departments in different states and provinces set the hunting seasons and bag limits based on population numbers of the squirrels and the **carrying capacity** of the habitat. Carrying capacity is the largest population that the resources of an environment, habitat, or ecosystem can support without causing damage. It is based on the amount of food and shelter that is available for the animals that depend on a particular habitat. Since hunters are the most effective predators that tree squirrels face, protection from hunting during critical survival periods has allowed these squirrel populations to increase. Regular seasons are established for hunting squirrels in many areas.

MAMMAL PROFILE

Fox Squirrel (*Sciurus niger*)

Figure 5-26: Distribution map of the fox squirrel.

The fox squirrel is the largest squirrel in North America. An adult measures from 19 to 28 inches long, and weighs from 1.5 to 3 pounds. These squirrels are slow compared to their smaller cousins, and they spend a lot of time on the forest floor gathering and hiding their stores of nuts for the winter.

This squirrel lives in hardwood forests and cypress swamps in the eastern United States, with the exception of New England, see Figure 5-26. In addition to nuts, it also eats fruits, corn, roots, and insects. It even taps maple trees and laps up the sweet sap that oozes from the damaged area.

Two to four young are born in the early spring, and they sometimes live as long as six years. Bobcats and foxes are their natural predators, but they are also hunted by humans as game animals. They range in color from rusty yellow to black, with white ears and nose, see Figure 5-27.

Figure 5-27: The fox squirrel. *Courtesy of U.S. Fish and Wildlife Service. Photo by Jon Nickles.*

Evolution is a natural process in which the genetic make-up of a population of organisms changes in response to changes in the environment. Physical changes become evident in an organism over a long period of time, see Figure 5-28. These physical changes

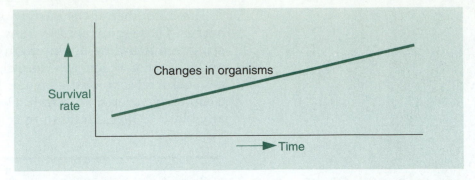

Figure 5-28: Physical changes in an organism become apparent over time. These changes can help an organism survive in a changing environment.

help the organism survive in its modified environment. An example of an evolving species is found in the region of the Grand Canyon. A single species of squirrel whose population was divided long ago by the Colorado River has developed into two similar, but distinctly different, subspecies.

MAMMAL PROFILE

Tassel-Eared Squirrels
(Sciurus aberti* and *Sciurus Kaibabensis)

Figure 5-29: Distribution map of the tassel-eared squirrel.

The tassel-eared squirrels are found on the north and south rims of the Grand Canyon, see Figure 5-29. Separated by the Colorado River and isolated by the desert, these squirrels have evolved into two distinctly different species. The Abert squirrel is found on the southern rim, but ranges into Colorado and New Mexico. It has white undermarkings and a gray-topped tail. The Kaibab squirrel lives on the northern rim of the Grand Canyon, see Figure 5-30. It has a black chest and a snowy tail.

Both squirrels are gray with chestnut-colored markings over the tops of their backs. During the winter they grow long tufts of hair from their ears. They share the same diet consisting of acorns, pine seeds, roots, mushrooms, and young birds. Unlike most other tree squirrels, they do not store winter food. During the winter, they sometimes eat the bark from pine twigs.

These squirrels prefer mountain slopes and forested areas covered with aspen, fir, and yellow pine trees. Their range extends into parts of Arizona, New Mexico, Colorado, and Utah. Three or four young are born during the summer

Figure 5-30: The Kaibab Squirrel. *Photo courtesy of U.S. Fish and Wildlife Service.*

months. As adults, they measure 19–21 inches in length and weigh 1.5–2 pounds. They are preyed upon by hawks and a variety of other predators.

Squirrels are generally considered to be primary consumers of plant materials. While this is true most of the time, they become predators when the opportunity presents itself. Some species prey heavily on bird eggs and young birds during the nesting season. Ground squirrels are known to eat the carcasses of other squirrels that have been killed on roads and highways.

MAMMAL PROFILE

Red Squirrel
(Tamiasciurus hudsonicus)

Figure 5-31: Distribution map of the red squirrel.

The red squirrel is easily recognized by its distinctive red color and light belly and throat. Its range includes parts of Alaska, Canada to the Appalachian Mountains, and the southern Rocky Mountains, see Figure 5-31. This squirrel's home is in the forest. It eats a variety of foods including acorns, spruce and pine nuts, and wild mushrooms. During the spring it also eats bird eggs and young birds.

Red squirrels are at home in hollow trees as well as in twig nests anchored in the tree branches. Sometimes they even live beneath the roots of trees in underground burrows. Like other tree squirrels, the red squirrel gathers and stores a supply of food for winter use. It defends up to one-half acre of forest territory, and becomes cross when any creature ventures into the area. It will chatter and scold all trespassers until they leave.

A litter of four or five offspring are born in the spring or early summer, and a second litter is sometimes born in September. The young squirrels learn to climb even before they are weaned, and they are soon moving about in the trees. Mature red squirrels measure from 11 to 14 inches long and weigh from 5 to 11 ounces.

Ground squirrels tend to have a single large litter in the spring. These large litters are needed because the ground squirrel is prey to many different predators, some of which enter right into the underground dens of squirrels to pursue their meals. Most ground squirrels hibernate during the winter months, and they depend on their fat reserves for nutrition during this time. If ground squirrels were born late in the season, they would have a hard time accumulating enough fat reserves to survive through the following winter.

MAMMAL PROFILE

Richardson's Ground Squirrel
(Spermophilus richardsonii)

The Richardson's ground squirrel is one of twenty-seven species of ground squirrels recognized in North America. It is brownish gray in color with short legs and a short, white-edged tail. It is between 10 and 14 inches long and usually weighs between 11 and 18 ounces. The Richardson's ground squirrel lives in long underground tunnels. It eats seeds and other plant materials and is especially fond of grain. This has led to efforts by farmers to reduce populations of this rodent. Its range extends through inland areas of northwestern U.S. and southwestern Canada. Its preferred habitat is grasslands and sagebrush areas that are near water, see Figure 5-32.

Figure 5-32: Distribution map of Richardson's ground squirrel.

Figure 5-33: Richardson's ground squirrel. *Photo courtesy of Len Rue, Jr.*

This squirrel is curious and stands erect beside its hole to keep watch on intruders. When it becomes alarmed, it whistles a warning and dives into its hole. Soon it comes back out to take another look, and a patient predator like a coyote or fox simply waits for the squirrel to come back out of its hole, see Figure 5-33. Most predatory birds, snakes, and mammals are enemies of the ground squirrel.

Litters of six to eleven young are born to adult females in late spring. They are born underground in a nest lined with grass and other soft materials. During the heat of mid to late summer, mature squirrels begin to estivate or sleep in their burrows until the temperature declines during the fall season. The young squirrels continue to eat and grow until early fall, after which they hibernate until spring.

MAMMAL PROFILE

California Ground Squirrel (*Spermophilus beecheyi*)

Figure 5-35: Distribution map of the California ground squirrel.

The California ground squirrel looks very much like a tree squirrel with its large, curved, and bushy tail, see Figure 5-34. It is mottled brown in color, and lives in long underground tunnels. These squirrels live in colonies much like marmots do. They prefer to live in grain fields, meadows, or orchards, where their feeding and tunneling activities often cause severe damage to crops and irrigation ditches. They eat green vegetation, seeds, fruit, berries, eggs, birds, and insects. The range of this species extends inland along the coastal region from Mexico to Washington State, see Figure 5-35.

This squirrel is usually 15-20 inches long and weighs 1-1.5 pounds. Between four and eleven young are born in the spring of the year.

The enemies of this squirrel include most predatory mammals, birds, and humans.

Figure 5-34: The California ground squirrel. *Courtesy of U.S. Fish and Wildlife Service. Photo by Luther Goldman.*

Chipmunks are small ground squirrels that are found in many regions of North America. There are numerous species of these little rodents living in a variety of different environments. They eat insects and seeds, and they gather supplies of food that are eaten when they awaken during the winter. They are recognized by the alternating light- and dark-colored stripes on their backs, and by their frisky behavior.

MAMMAL PROFILE

Least Chipmunk (Eutamias minimus)

Figure 5-36: Distribution map of the least chipmunk.

There are sixteen different species of western chipmunks in this genus. They are 6½–9 inches in length from head to tail and weigh 1–2 ounces. The range of these small ground squirrels extends from Alaska to Mexico, and from the West coast to the Great Lake region. It occupies a variety of habitats from sagebrush deserts to mountain coniferous forests and northern hardwood forests, see Figure 5-36.

Four or five young are born in the spring. By the fall, these chipmunks have constructed burrows of their own. They spend the summer and early fall seasons gathering supplies of seeds for winter food. Their diets consist of seeds, fruits, nuts, meat and insects.

MARMOTS AND PRAIRIE DOGS

Several species of closely related rodents fall into this group of animals. They occupy terrain from flatlands and prairies to rolling hills and high alpine meadows, see Figure 5-37. They are burrowing animals that build underground dens lined with dry grass.

Figure 5-37: A prairie dog community. *Courtesy of U.S. Fish and Wildlife Service. Photo by G. R. Zahm.*

Marmots and prairie dogs are vegetarians that eat a wide variety of plants. They gorge all summer long, and by fall they are fat and pudgy; this body fat sustains them through their long winter hibernation.

Figure 5-39: Distribution map of the woodchuck.

The woodchuck is probably the only rodent for which a holiday has been declared, see Figure 5-38. The woodchuck is also called a groundhog, and it is rumored that if a groundhog sees its shadow on the second day of February, it will go back into hibernation for six more weeks of winter. This forecast is seldom reliable and the woodchuck probably does not even wake up that early in the year, but most people know the woodchuck because of this tradition.

The woodchuck is 20 to 27 inches long from head to tail. It weighs 5-10 pounds, and it is brown in color. It eats large amounts of green plants during the spring and summer. In September it goes into hibernation and sleeps until springtime. The woodchuck prefers open woodlands and meadows,

Figure 5-38: A woodchuck. *Courtesy of U.S. Fish and Wildlife Service. Photo by Dan Hultman.*

and it can be found as far north as Alaska and Northern Canada, and as far South as Georgia and Alabama, see Figure 5-39. It causes problems for farmers by eating crops, and the mounds of dirt beside it burrows can damage farm machinery. Large animals sometimes suffer leg injuries when they step in the burrows.

Marmots and their relatives communicate effectively by calling with a shrill whistle to warn of danger and to challenge rivals during the mating season. A marmot will chirp for as long as it can sense that danger exists. This chirping call has resulted in the names "whistler" and "whistle pig," by which it is known in some communities.

The yellow-bellied marmot is also known as a rockchuck because it has a habit of building its den among the crevices of rocks. It lives in the mountains, and it lines its rocky den with dried grass. Once inside the den, this animal is safe from almost any enemy. Its diet

Figure 5-40: Distribution map of the yellow-bellied marmot.

consists of grasses and other vegetation that is located close to its den.

Much of the life of this marmot is spent sleeping. When the weather gets hot, it takes a long nap. This habit is called **estivation**. Early in the fall it begins a long hibernation period that lasts until spring. As many as eight young are born in May.

Rockchucks range from 19 to 28 inches in length, and they usually weigh between 4 and 12 pounds. They are usually yellowish brown or black in color, with a band of white across a dark face. Its range extends from Canada to California in the mountains of the western U.S., see Figure 5-40.

Most marmots have the unusual habit of biting off green vegetation and laying it out in the sun to dry. In this manner they harvest hay which is used as bedding, and as an emergency supply of food when they awaken before food is abundant in the spring.

Marmots often live in areas where water is scarce. Most of the food that they eat consists of fresh plants. They are able to obtain enough water in this manner to satisfy most of their needs for moisture.

MAMMAL PROFILE

Hoary Marmot (*Marmota caligata*)

Figure 5-41: Distribution map of the hoary marmot.

This marmot lives high in the mountains among cliffs and rock outcroppings, see Figure 5-41. It is the largest of the marmots found in North America, and can weigh as much as 15 pounds. Its range extends from the northwestern U.S. to Alaska. Its fur is dark gray with white tips on the hair, and matches the rocks on which it suns itself between meals, see Figure 5-42.

Most predators that inhabit the high mountain ranges will attempt to prey upon the hoary marmot, but the one most likely to succeed is the eagle. The marmot's home in the cliffs usually provides protection from most other predators. It feeds on a variety of plants that grow at high altitudes.

Up to five young are born in the spring, and they usually stay with their mother until the following spring.

Figure 5-42: A hoary marmot. *Courtesy of U.S. Fish and Wildlife Sevice. Photo by Steve Moore.*

Prairie dogs are closely related to marmots, and many of the behaviors of these animals are similar. Both of these animals belong to the same family (sciuridae) as the woodchucks, ground squirrels, chipmunks, and tree squirrels. Their tails are covered with hair, and most of them are active during the day. Many of them store food during the summer for later use. Prairie dogs eat grasses and other vegetation along with insects such as grasshoppers. Their diets consist of plants which they convert to flesh. In this way plant materials are converted to a suitable form of food for predatory animals.

MAMMAL PROFILE

Black-Tailed Prairie Dog (Cynomys ludovicianus)

& White-Tailed Prairie Dog (Cynomys leucurus)

■ White-Tailed Prairie Dog

■ Black-Tailed Prairie Dog

Figure 5-44: Distribution map of the black-tailed and white-tailed prairie dog.

Two kinds of prairie dogs are common to North America, see Figure 5-43. The black-tailed prairie dog is yellowish-brown in color with a black-tipped tail. It is a plains dweller who builds dikes around the entrance to its burrow to prevent flooding from heavy rains. Its range extends in a narrow band from Texas to Canada. The members of this species live in closely-knit social groups. They are 14–17 inches long from nose to tail and weigh 2–3 pounds. They greet and groom one another and generally graze peacefully together. They remain active throughout the year. Young are born in litters of 3–5 in March and April.

White-tailed prairie dogs live at high elevations in upland meadows where natural slopes eliminate flood danger, see Figure 5-44. These prairie dogs do not build dikes around the burrow entrances. They also demonstrate fewer social behaviors within their colonies than their black-tailed relatives. They are 12–14.5 inches in length from head to tail and weigh 1.5–2.5 pounds. They are yellowish-brown in color with white-tipped tails. They hibernate during the winter months. Litters of five or six young are born after the weather warms in early May. Their range extends from Montana to Arizona and New Mexico in the Rocky Mountain Range.

Figure 5-43: A black-tailed prairie dog. *Courtesy U.S. Fish and Wildlife Service. Photo by Claire Dobert.*

Prairie dogs sound alarms each time a member of the colony spots something that might be dangerous. Their worst natural enemies are rattlesnakes and ferrets that enter the burrows, and badgers that dig them out of their burrows. They are also eaten by birds of prey, coyotes, and other predators.

RABBITS, HARES, AND PIKAS

Not all of the gnawing mammals are true rodents. Rabbits and hares have many similar characteristics with rodents, and they

CHAPTER 5 GNAWING MAMMALS ❧ **95**

Figure 5-45: Characteristics of rabbits, hares, and pikas.

were once classed as rodents because of their large incisor teeth that are so well adapted to gnawing. Further study revealed that this group of animals is different from the rodents because they have a second set of upper incisor teeth behind the front set of teeth, see Figure 5-45. The **pika** has this same tooth arrangement, and it is classified in the same order (Lagomorpha) with rabbits and hares.

Distinct differences exist between rabbits, hares, and pikas. Hares have longer hind legs and longer, wider ears than rabbits. Their offspring are born with a full coat of fur and their eyes are open at birth. Young rabbits are born without any hair and their eyes are closed for several days; they are completely helpless at birth. Pikas have short legs, ears, and no visible tail. Rabbits, hares, and pikas are found in many different environments.

MAMMAL PROFILE

Eastern Cottontail Rabbit (*Sylvilagus floridanus*)

Figure 5-46: Distribution map of the Eastern and Mountain cottontail.

There are thirteen species of cottontail rabbits, see Figure 5-46. They live in many different environments in North America ranging from deserts to wooded areas. Their diet consists almost entirely of grass and other succulent vegetation, see Figure 5-47.

The Eastern cottontail has short legs and cannot run very fast. They are generally brown to grey in color, and they measure 14–17 inches in length. They hide from their enemies by sitting very still, and they prefer brushy habitats. They are a major source of food for such predators as skunks, foxes, snakes, and birds of prey. Litters consist of 4–7 young with females giving birth to 3–4 litters per year. Their diets consist of green vegetation in the summer with bark and twigs making up their winter diets. The range of this rabbit extends across the eastern two thirds of the U.S. and Mexico. Most states protect the cottontail and manage it as a small game animal that is hunted for sport.

Figure 5-47: A cottontail rabbit. *Courtesy U.S. Fish and Wildlife Service. Photo by James C. Leupold.*

Rabbits and hares are preyed upon by many predators, including humans. The reproduce rapidly, and even though many of them survive less than a year a population of rabbits or hares can expand very rapidly. Females have up to five litters of offspring each year with several young in each litter. When conditions are favorable, a few rabbits can multiply into many rabbits in a short period of time.

**Pygmy Rabbit
(Sylvilagus idahoensis)**

Figure 5-48: Distribution map of the pygmy rabbit.

This rabbit is found in the Great Basin region of the western United States, see Figure 5-48. It is a burrowing rabbit that digs a hole in which it lives and it raises 5–8 young that are born during the summer months. Its habitat is the desert, and it prefers to live near a clump of rabbitbrush or sagebrush which hides it from predators. It is a small, brownish gray rabbit that usually measures 8½ x 11 inches long, and weighs no more than a pound. Its diet consists mostly of sagebrush and it seldom moves more than 30 yards from its burrow or den. The range of the species is restricted to the northwestern U.S.

Most rabbits make grass nests for their young and line them with fur that is pulled from the female's belly. The fur-lined nest is warm and protects the young rabbits during the early days of their lives. Some kinds of rabbits build their nests in underground holes. Others build them in depressions on the ground. Some even make their nests on piles of reeds and grass in the middle of marshy areas.

**Arctic Hare
(Lepus arcticus)**

Figure 5-50: Distribution map of the arctic hare.

The arctic hare is large, weighing between 5 and 12 pounds, see Figure 5-49. In the far North it is white in color throughout the year and blends in well with the snow-covered ground. In southern areas it changes color in the summer to a brownish gray. Litters as large as eight young are born in June and July with their eyes open and full coats of fur.

This hare is equipped with long course hair on its feet that makes it possible to move about on the surface of the snow without sinking in. It prefers the treeless tundra habitat of the arctic region where its diet consists of tundra plants, see Figure 5-50. Like other hares, it can leap high into the air as it runs. This allows it to see over brush and watch its enemies during its attempt to escape.

Figure 5-49: The arctic hare.
Photo courtesy of U.S. Fish and Wildlife Service.

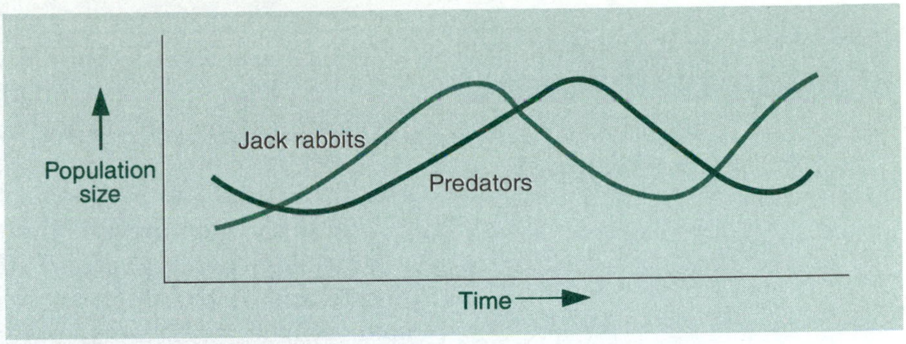

Figure 5-51: Population cycles of jack rabbits and their predators.

Hares are faster and more mobile than rabbits, and they depend upon their speed to avoid predators. Speed is important because they tend to live in wide-open spaces where cover is not always available and hiding from predators may be difficult. Hares have acute hearing, and they turn their ears toward the slightest sound as they are approached.

Populations of hares and rabbits tend to cycle up and down in a regular and predictable manner. After rabbit or hare populations become high, diseases usually kill off many of the animals. When populations drop, the predators that depend on them for food also decline in number, see Figure 5-51.

MAMMAL PROFILE

Black-tailed Jackrabbit (*Lepus californicus*)

Western settlers shortened the name of this well-known hare to jackrabbit, see Figure 5-52. It is well adapted to life in the desert and plains. Its gray coat blends in with the environment in which it lives, and its acute hearing helps it to detect the movements of its enemies long before it sees them. It reproduces at a rapid rate, and every few years the population becomes too great to be supported by the resources that are found in its environment. When this happens, the jackrabbits become susceptible to diseases that reduce the population.

Figure 5-52: A jack rabbit. *Photo courtesy of Leonard Lee Rue III.*

Rabbits and hares occupy most of the North American ecosystems. They are found in hardwood forests, plains, deserts, tundra, and marshes. In all of these regions they are important food animals for carnivores of many kinds.

Figure 5-53: Distribution map of the pika.

The pika is a small animal weighing between 4 and 6 ounces. It has short legs, ears, and tail and resembles a guinea pig more than its relatives, the rabbits and hares, see Figure 5-53. It has two sets of upper incisor teeth, however, and it is this characteristic that results in its classification with the rabbits and hares.

The pika lives high in the mountains in piles of rocks located near areas where grass and other forage are abundant. It harvests plants and stores them for winter food in much the same way as some marmots. Its range extends through the mountain region of the western U.S. and Canada. These small animals remain active throughout the winter and they survive by eating the hay that they stored during the summer. Three or four young are born in the spring, and by the time the grass is ready to harvest they are big enough to help harvest the hay that will feed them through the next winter. The pika is preyed upon by hawks, eagles, and other animals that are lucky enough to find it outside its home in the rocks.

PORCUPINES

The porcupine is a rodent that is best known for its sharps quills, which it uses to defend itself against its enemies. It cannot throw its quills, but it is capable of imbedding them into the flesh of any creature that comes close enough to be within reach of its tail. It has been reported that the mountain lion and the bobcat prey on the porcupine, and that they avoid injuries from quills by flipping the animal over and attacking its unprotected throat and belly. Other predators are sometimes successful, but they frequently gain nothing except a painful muzzle filled with quills.

Porcupines can cause considerable damage to trees and shrubs as they gnaw the bark for food. Trees that have been gnawed on by porcupines become scarred where the bark has been removed. When the porcupine eats the bark all the way around a tree, the flow of nutrients in the tree stops and the tree dies.

Like other rodents, a porcupine must gnaw constantly to keep its teeth worn down. The front teeth grow in length throughout its life. The gnawing habits of the porcupine, along with its craving for salt, have resulted in confrontation with humans. It has been known to chew on nearly anything that is salty, including boots, saddles, and tool handles.

This North American porcupine is found in the northern forests of Alaska and Canada, and in coniferous forests of the northeastern and western states, see Figure 5-54. It is active throughout the year, eating tree bark when the snow is deep and feeding upon a variety of other plants when the snow has melted.

Figure 5-54: Distribution map of the porcupine.

A mature porcupine may weigh as much as 25 pounds and measure up to 34 inches in length, see Figure 5-55. A single offspring is born in the spring. It has soft quills that soon harden. After nursing for four to six weeks, young porcupines eat a diet of plants and the inner layer of tree bark.

Porcupines spend much of their time in thickets and trees. They are nocturnal animals.

Figure 5-55: A porcupine.

GOPHERS AND BEAVERS

Pocket gophers include 272 species and subspecies that are only slightly different from one another. They spend their lives alone in underground tunnels where they eat the roots of plants and dig new tunnels. During the night they gather grasses and other plants that are eaten or carried in their cheek pouches to storage areas located in the underground tunnels.

They come out of their tunnels to seek a mate, but most of their time is spent digging in complete darkness. They remove the dirt from their burrows by pushing it to the surface where it is seen in mounds near the tunnel entrances. Their tunneling activities get them into trouble when they burrow into ditch and canal banks causing them to leak or break. Their mounds also damage the blades of farm machinery when gophers take up residence in hay fields.

MAMMAL PROFILE

Pocket Gopher (*Geomys bursarius* and related species)

Figure 5-56: Distribution map of the pocket gopher.

The pocket gophers include several species that range across the western half of the U.S. from Canada, Mexico and Panama, see Figure 5-56. Their presence is evident from a series of earth mounds that have been pushed to the surface. They prefer habitats where the earth is soft and easy to dig in, and they spend most of their time underground. Coyotes and foxes sometimes catch gophers by waiting at the entrances to open tunnels until the gopher pushes out a load of dirt from the hole. A badger is less patient and simply digs the gopher out.

As many as seven young are born in the spring. As soon as they are weaned, they dig burrows of their own and begin life alone. Mature gophers are 6–13 inches long and weigh up to a pound. They are covered with brownish gray fur except for a short stubby tail, and they are equipped with cheek pouches that are used to carry food. Their diets consist of roots, tubers, and some surface vegetation.

Figure 5-57: Distribution map of the mountain beaver.

The name mountain beaver is somewhat of a misnomer since this rodent behaves much more like a pocket gopher than a beaver. It tunnels like a gopher, and it harvests plants like a marmot or pika, carrying them to an underground storage area after they have cured in the sun. This rodent weights 2–3 pounds, and it remains active during the winter. Its preferred habitat is in forests and thickets where soil is moist and digging is easily done, see Figure 5-57.

Two or three young are born in the spring, and they set off on their own soon after they are weaned. The mountain beaver has brown fur that was used for clothing by Indians. They dig with long claws, and, unlike true North American beavers, they have short, stubby tails. This animal ranges inland from the coastal regions from California to Canada.

The true beaver (not to be confused with the mountain beaver) is the largest rodent found in North America. It lives in tunnels constructed in the banks of streams or inside lodges constructed in ponds. It feeds on the bark of willows and other trees, and constructs dams from the trees it cuts down. Using it four large incisor teeth, it fells trees. It is capable of chewing small trees into lengths that can be moved. It anchors the branches to the floor of its pond to be used as winter food, and adds some materials to its dam to strengthen it in preparation for winter.

Beavers are considered to be valuable animals due to their dam building activities and manipulation of water. They can transform a small stream into a series of ponds that reduce spring flooding and erosion. The ponds also store fresh water for livestock and wildlife, and they raise the water table in the area creating meadows. While beavers change the environment to meet their own needs, they also improve the habitat for many other wild animals and birds.

When beavers become neighbors to people, however, their instincts to cut trees and to build dams often create problems. They gnaw down trees that are valued by people, and they build dams that flood property. When this happens, the beavers are often live-trapped and moved to new locations where their dam building is appreciated.

MAMMAL PROFILE

Beaver
(Castor canadensis)

The beaver is a hard-working animal whose dam-building skills are valuable in preventing soil erosion and in storing water, see Figure 5-58. Its fur coat is waterproofed with oil secreted from two oil glands located near its tail. The beaver was trapped for its fur until it was no longer found in many of its native areas. The beaver has webbed hind feet and a long, flat tail, both of which aid in swimming. When danger approaches, beavers slap their tails on the surface of the water as a signal to other members of the colony.

Figure 5-58: A beaver. *Photo Courtesy of NORTHWESTREK, A Wildlife Park, Eatonville, Washington.*

Mature beavers measure 3–4 feet in length and weigh up to 70 pounds. A litter contains three or four young, and they are born with a full coat of fur and with their eyes open. They are not mature until they are about two years old. Their waterproof fur is reddish brown in color and it covers the entire body except for the tail.

Natural enemies of the beaver include the wolverine, bear, wolf, coyote, lynx, and mountain lion. A beaver is vulnerable when it is on land or when the pond surrounding its lodge dries up. Most of the time, however, it is quite safe from predators.

LOOKING BACK

Taxonomy is the branch of science that classifies mammals and other living organisms into related groups. Mammals are warm-blooded animals with protective coats of hair and bony skeletons, and nourish their young with milk. Gnawing mammals are identified by their large incisor teeth that grow continuously and that require constant gnawing to keep them worn down.

Rodents make up the largest group of mammals, and they are distributed in nearly every ecosystem in North America. They include all of the gnawing mammals except rabbits, hares, and pikas. Rabbits, hares and pikas have a second set of upper incisor teeth located behind the front incisors. Each of the gnawing mammals is distinctly different from other animals, and each fills an important role in the habitat that it occupies.

Gnawing mammals are primary consumers that convert plants to meat. Predators are secondary consumers. They obtain energy from plants indirectly by eating primary consumers. Populations of rodents and other gnawing animals expand or decline in response to the size of the predator population, the abundance of their food supply, and environmental factors. Populations of predators depend on the supply of food animals, and they decline rapidly when the populations of rodents and other prey animals decline.

REVIEW QUESTIONS

1. List the characteristics that distinguish mammals from other animals.
2. Describe the system that scientists use to classify and define relationships among different kinds of animals.
3. Explain how primary consumers are different from secondary consumers.
4. What roles do gnawing mammals fill in the ecosystems of North America.
5. Predict the effects on populations of gnawing animals when predator populations increase or decline.
6. Predict the effects on populations of predators when populations of gnawing animals increase or decrease.
7. Prepare maps that illustrate where populations of specific gnawing animals are distributed in North America.
8. Select at least one animal from each group of similar animals profiled in this chapter, and describe a typical life cycle.

LEARNING ACTIVITIES

1. Prepare a written report on a gnawing animal describing its habits and life cycle, and identifying a major predator. Explain how the predator is adapted to successfully prey upon the animal you have chosen.
2. Locate an area near your home or community and observe it carefully to detect the presence of gnawing animals and their predators. Prepare an oral report for the class describing the evidence you found that helped you determine which animals were living in the area.

6 Hoofed Mammals

KEY TERMS

antler
billy goat
bison
buck
bull
calf
cow
cloven-hoofed
cud
doe
ewe
fawn
javelina
kid
lamb
nanny goat
peccary
pronghorn
ram
rumen
ruminant
rut
symbiosis
ungulate
velvet

HOOFED MAMMALS include deer, pronghorns, peccaries, goats, sheep, horses, and cattle. Wild species of all these animals are found in North America. They have tough coverings on their feet made of a horn-like material. Another name for a mammal with hooves is **ungulate**.

OBJECTIVES

After completing this chapter, you should be able to

■ name the hoofed mammals that are found in North America
■ describe the process by which ruminant animals digest their food
■ recognize and describe members of the deer family
■ distinguish between wild sheep and goats
■ identify similarities and differences between pronghorns and members of the deer family
■ evaluate the roles of ungulates in the ecosystems of North America
■ speculate about why the musk ox developed the instinct to form a defensive circle with other members of the group during times of stress
■ define the role of the peccary in the ecosystem in which it lives
■ conclude whether wild horses and burros should be managed as wild animals.

All of the hoofed mammals except horses and burros are **cloven-hoofed**. This is a condition in which the hoof is divided into two parts. The hoof of a horse is different because it consists of a single hoof on each foot, see Figure 6-1.

Some of the hoofed mammals are well known and easily recognized by most people. The deer family belongs to this group along with wild horses, burros, cattle, sheep, goats, and pronghorns. A group of other hoofed animals that is less well known is the

103

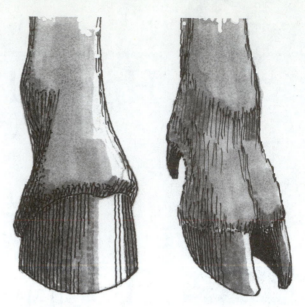

Figure 6-1: Single and Cloven hoof structures.

javelina. It is sometimes called a wild pig, but it belongs to a different family than the true pigs. Most hoofed animals are native to North America, but the wild horses that we know today were introduced by the Spaniards during their conquest of the New World as they established territories in America.

Sheep, goats, pronghorns, wild cattle, and members of the deer family all belong to a group of hoofed animals known as **ruminants**. These animals have a series of four stomach compartments that are capable of digesting food that contains a large amount of fiber, see Figure 6-2. Most grasses, brush, twigs, and other forage plants are high in fiber.

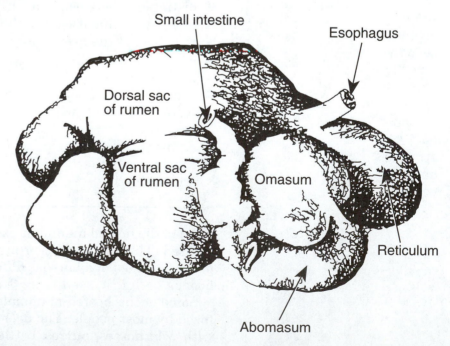

Figure 6-2: A ruminant digestive system. Note the four compartments of the ruminant stomach. *From Warren, Small Animal Care and Management, copyright 1995 by Delmar Publishers.*

Ecology Profile

RUMINANT DIGESTION

Digestion of high-fiber plant material by ruminant animals is possible because bacteria digest the food. After the food is swallowed, it enters a large stomach compartment called the **rumen** where it is warmed and soaked, creating a favorable environment for certain bacteria. Bacteria reproduce rapidly in the rumen and begin to digest the fiber in the food. Most of this fiber consists of complex carbohydrates known as cellulose.

When the ruminant animal has eaten its fill, it usually lays down to chew its **cud**. Cud is a small portion of the material from the rumen that is regurgitated and chewed thoroughly. After this material is chewed, it passes through the other stomach compartments where bacteria continue to digest the food. In the last compartment, the bacteria are broken down by the digestive juices, and they become food for the animal.

Symbiosis is defined as a relationship between two organisms in which each organism receives benefits from its association with the other. A symbiotic relationship exists between ruminant animals and the bacteria that digest the fiber in their food.

All of the ruminant animals are primary consumers. They play important roles in nature by converting nutrients stored in forage plants to meat. They are prey for large carnivores and for humans.

Populations of ruminant animals have increased considerably during the last half of the twentieth century. This is due in part to government regulation of hunting, and to management practices to improve habitat.

PRONGHORN

The **pronghorn** is the only remaining member of a family of animals that was once found in large numbers in North America. It is often called an antelope, but it is not a member of the same family as the true antelopes that are found in other parts of the world.

MAMMAL PROFILE

**Pronghorn
(Antilocapra Americana)**

Pronghorns live in deserts and plains where the terrain is relatively flat, see Figure 6-3. They depend on grasses, forbs, domestic crops, and plants such as sagebrush for food. Their best defenses against predators are their eyesight and speed. They can run faster than any other animal in this part of the world. A pronghorn can sustain a speed of up to fifty miles per hour for a short distance.

The pronghorn is a brown animal of moderate size with white undermarkings and a white rump patch. In the early part of the twentieth century, its numbers were threatened by overharvesting. It is now protected by hunting regulations, and it can be found from

Figure 6-4: Distribution map of the pronghorn.

Canada to Mexico, see Figure 6-4. It is managed as a game animal, and hunting is allowed only during short seasons. Restrictions on hunting have made it possible for this animal to increase in numbers, and an annual harvest is considered necessary to keep herds in balance with the food supply.

The horns of pronghorns are hollow except for a core of bone over which the horns grow. They are shed after the fall mating season, and new horns grow from the bony core. The horn splits into two separate prongs, and it is this characteristic that gives the pronghorn its name. Both males and females have horns; those of the **buck** or male are somewhat larger than those of the **doe** or female.

Young pronghorns are born in late spring or early summer. Twins are quite common. They are able to run well by the time they are a few days old. Coyotes and wolves are the most common predators of pronghorns. When a pronghorn gets excited, the white rump patch becomes visible and acts as a warning for other animals.

Figure 6-3: Pronghorns in their habitat. *Photo courtesy Dr. Jay McKendrick, University of Alaska, Fairbanks.*

DEER

Members of the deer family in North America include moose, elk, caribou, and several species of deer. They are found in a variety of widely distributed environments. They are also popular game animals. Members of the deer family are similar in that they have cloven hooves and ruminant digestive systems. Males have scent glands and **antlers** or bony horns that are shed annually. In some species, the females also have antlers. Populations of these animals have recovered well since they were badly depleted by excessive hunting pressure in the early part of the twentieth century.

Male elk and moose are called **bulls**, females are called **cows**, and the young of either species are called **calves**. The males of the other deer species are called bucks, females are called does, and their young are called **fawns**.

There are four different subspecies of moose in North America. They are the eastern moose, Manitoba moose, Shiras moose, and Alaskan moose. These animals are the largest of the game animals in America. Despite their size, moose are preyed upon by wolves, bears, and other large predators.

MAMMAL PROFILE

Moose
(Alces alces)

Figure 6-5: Distribution map of the moose.

The moose is the largest member of the deer family, see Figure 6-5. A mature Alaska bull moose may be nearly 8 feet tall at the shoulder and weigh 1,800 pounds. His massive set of antlers sometimes measures 6 feet across. The animal that Americans call a moose is referred to as an elk in Europe, see Figure 6-6.

Moose usually prefer to live alone in areas where there is abundant water. They feed on lush vegetation and water plants when they are available, and they eat small branches and tree bark during the winter months.

Figure 6-6: A moose. *Photo courtesy of NORTHWESTREK, A Wildlife Park, Eatonville, Washington.*

Bulls battle one another for cows during the breeding season, and calves are born in the spring. They stay with their mothers for nearly a year and are driven away just before the new calves are born. Although bulls are very aggressive during the breeding season, they are quite docile during the rest of the year. Cows with calves are very protective and they will attack nearly anything that seems to threaten the safety of their young.

The elk that are found in North America are large deer that once ranged across the continent from northern Canada to Mexico and from coast to coast. As settlers moved to the western lands, the vast herds of elk were depleted. The eastern elk were hunted to extinction, and other elk populations were greatly reduced as wintering grounds were converted to farms, and as the elk were slaughtered as food for people.

Elk management is much improved today in comparison with the recent past. Their populations have grown larger. Elk have

Ecology Profile

STARKEY RESEARCH STATION

The Starkey Research Station is operated by the U.S. government, and it is located in Oregon. It is a good example of science at work. A captive herd of deer and elk live inside a fenced area of more than 40,000 acres. A number of experiments are being conducted there to determine the best range management methods for multiple use of rangelands by elk, deer, and cattle. Data from these experiments that is published in research journals indicates that rangelands can be effectively used under controlled conditions by livestock without damaging habitat for these other wild game species. Research conducted at the Station includes birthrates of deer and elk when young males are used for breeding in comparison with mature males. Another research project is measuring metabolism of different feed materials by elk.

been introduced back into some areas that have not had any elk for many years. Preserves have also been established, and winter feeding programs have been instituted in some areas, see Figure 6-7.

Elk are migratory animals that move to the high mountain meadows during the summer and to lower elevations during the winter. They are fast runners and strong swimmers. They were called wapiti by Native Americans, and the name is still used.

Figure 6-7: A herd of elk. *Courtesy of NORTHWESTREK, A Wildlife Park, Eatonville, Washington. Photo by Robert Ferris.*

"]

MAMMAL PROFILE

Elk
(Cervus elaphus)

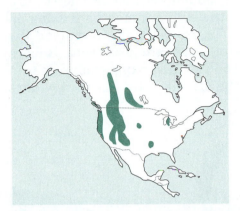

Figure 6-8: Distribution map of the elk.

Elk are popular game animals whose range extends through Canada and the western region of the United States, see Figure 6-8. They can reach 5 feet at the shoulder and mature bulls weigh more than 800 pounds, see Figure 6-9. Bulls have large, branched antlers that they shed each year. Their new antlers grow from buds and they are larger in size with each year until a bull is mature.

Bulls use their antlers to protect themselves from predators. They also battle each other for harems of cows during the **rut** or mating season in the fall. During this time, the bulls bugle a whistling call to challenge rivals and to attract cows. Hunters imitate this call, and bulls frequently forget caution as they rush to challenge what they think is another bull.

Figure 6-9: A bull elk. *Photo courtesy of U.S. Fish and Wildlife Service.*

Calves are born in May or June. They are camouflaged from predators by white spots on their backs. When a calf is lying perfectly still, the spots look like splashes of sunlight on the forest floor. Calves are strong enough to follow their mothers when they are only a few days old. Elk are most vulnerable to predators during the first few weeks after birth. Wolves, bears and mountain lions are capable of killing elk even after they reach maturity. Elk prefer mountains and forests, and they graze on grasses, forbs, twigs and bark.

Caribou are closely related to domesticated reindeer. The males and females of both species have large, branched antlers. Caribou are adapted to northern climates, and their range rarely extends south of Canada. They are large migratory deer that move along established migration routes from their southern ranges in woodlands of southern Canada to the northern tundra regions and back again. Large herds of migrating caribou, numbering in the hundreds or even thousands, make the same trip every year.

MAMMAL PROFILE

**Barren-Ground Caribou
(*Rangifer tarandus*)**

Figure 6-10: Distribution map of the caribou.

Barren-ground caribou are so named because of the tundra region which they occupy during the summer months feeding on lichens, mosses, and shrubs, see Figure 6-10. The does lead the annual migration north each spring. They are heavy with unborn fawns, and they arrive at the traditional calving ground just in time to deliver their young. The bucks follow the does a few days later.

Figure 6-11: A bull caribou. *Courtesy of Alaska Division of Tourism. Photo by Mark Wayne.*

When the fall season arrives, the migration turns south and the caribou return back to the shelter of the boreal forest for the winter. Mature caribou measure as tall as 4.5 feet at the shoulder and weigh up to 375 pounds, see Figure 6-11.

Caribou numbers declined seriously due to loss of food supplies, hunting pressure, and other factors during the early part of the twentieth century. Populations are recovering as a result of conservation practices and better management of the herds. Caribou have been moved with moderate success by wildlife biologists into some Rocky Mountain areas located south of the Canadian border.

The main predator of the caribou is the wolf. Packs of wolves follow the caribou herds along the migration routes feeding on them along the way. Young and weak animals are most vulnerable to predators. Bears and some other large predators also use caribou for food.

MAMMAL PROFILE

White-tailed Deer
(Odocoileus virginianus)

Figure 6-12: Distribution map of the white-tailed deer.

The white-tailed deer is more widely distributed in North America than any other deer. With the exception of California and the Great Basin, they are found throughout the United States, southern Canada, Mexico, and as far south as Central America, see Figure 6-12. They are approximately 4 feet tall at the shoulder and a large buck can weigh in excess of 300 pounds. Several races of white-tailed deer are evident. Among these are the Florida Keys white-tails, which are the smallest in size. Their range is restricted to several islands off the coast of southern Florida.

The white-tailed deer is named for the white color of the hair on the lower surface of its tail. When it is excited, it waves its tail like a white flag and snorts a call of alarm as it bounds away to safety. This deer is at home in woodlands, fields, deserts, and plains. It has a reputation among hunters for being extremely crafty when it doesn't want to be found.

Natural predators of deer include the coyote, mountain lion, bear, and wolf, but some deer are undoubtedly killed by lynx, bobcats, wolverines, foxes, and even eagles. Mountain lions often prey exclusively on deer. Domestic dogs are among the most important predators on deer, especially in late winter and early spring. One reason that deer populations have grown in the recent past is because predator populations have been controlled to protect domestic livestock. This has also provided protection to deer populations. Deer populations today are controlled more by hunters than by natural predators.

Career Option

GAME WARDEN

A game warden is a law officer who is responsible for patrolling an assigned area to prevent fish and game law violations, investigating complaints about crop damage by wildlife, and gathering biological research data. Game wardens also apprehend violators of fish and game laws, issue citations, and make arrests when appropriate. They present evidence in court hearings, investigate hunting accidents, offer educational programs to the public, and work with community groups to improve fish and game habitat. Education requirements for this career include training in law enforcement, and a strong curriculum in biological sciences.

As deer antlers grow and develop, they are covered with skin and hair. This covering is called **velvet**. A rich supply of blood is pumped to the growing antlers through a network of blood vessels in the velvet. When the antlers are fully developed, the blood supply stops flowing and the antlers harden. The animals then rub their antlers on objects to remove the velvet before the beginning of rut. Antlers are shed each year following the breeding season.

Figure 6-14: Distribution map of the mule deer.

Mule deer are distinguished from other deer by the branched beams of the antlers and a white rump and tail, see Figure 6-13. The tail is short with a black tip, and the mule deer tends to hold its tail down as it runs. The ears of a mule deer appear to be quite large in comparison with the ears of other deer, and bucks have dark markings shaped like horseshoes on their foreheads.

The diet of the mule deer consists of grasses, herbs and forbs during the summer, and shrubs, fungi, twigs, acorns, and lichens during the winter.

Figure 6-13: Mule deer. *Photo courtesy of U.S. Fish and Wildlife Service.*

A mature buck often weighs close to 300 pounds, and a doe weighs around 200 pounds. Mating occurs in the fall, and the young are born in early summer. Most births are twins with some single and triplet births. Mule deer prefer high mountain elevations and wooded habitat, but they also thrive in desert areas in the western regions of North America. Their range extends across the western U.S. from South Yukon to Mexico, see Figure 6-14.

Bucks get their first set of antlers when they are one year old. They consist of single spikes; an occasional antler has a single fork. These young bucks are often called spikes. Antlers grow bigger as the young buck or bull mature. A young bull elk develops some branching in its antlers when it is two years old. A young bull in this stage of antler development is sometimes called a raghorn.

The black-tailed deer is a variety of mule deer that is named because of its black tail, see Figure 6-15. The antlers of mature bucks separate on the main beam to form two forks on each side. This deer species is found in the coastal mountains from central California to British Columbia.

Figure 6-15: A black-tailed deer. *Photo courtesy of NORTHWESTREK, A Wildlife Park, Eatonville, Washington.*

Bison and Musk Ox

The North American **bison** or buffalo is closely related to domestic cattle. It is estimated that the bison population once approached sixty million animals, but it was nearly hunted to extinction between 1865 and 1885. During the peak of their slaughter, bison were killed by the millions just for their hides, and the meat was often left to rot.

In 1893 only twenty-one bison remained in the wilderness of Yellowstone, and a few hundred others were scattered in other areas. These were the conditions that existed when a bill to protect the bison was signed by President Grover Cleveland in 1893.

The decline of the bison is a classic example of human abuse of a wildlife resource. In only 20 years in the late 1800s, millions of bison were hunted until only a few of them remained in the wild. Most of the wild bison that remain are living in Yellowstone

National Park and in other game preserves, such as federal and state parks, refuges, and preserves. Some also survive in domesticated herds on farms and ranchs where they are raised as meat animals.

Ecology Profile

YELLOWSTONE BISON CONTROVERSY

The bison living within Yellowstone National Park in Montana and Wyoming do not always stay within established park boundaries. As the bison population in the park increases, they move beyond Yellowstone's borders in search of food. Much of this park is already severely overgrazed by the resident herds of bison and elk, see Figure 6-16. The lack of natural predators in the park to control the size of the ruminant population has also generated controversy. Some people favor the introduction of wolves into this area to control the size of the bison population. This is opposed by ranchers and some wildlife groups, however, who fear for the safety of livestock and game animals outside the park boundaries.

Some ranchers suspect bison that leave Yellowstone National Park carry with them a serious disease, brucellosis, that threatens the cattle industry in those states that surround the park. Brucellosis causes cattle to abort their calves and the bacteria can be passed on to humans as undulant fever.

Many of the Yellowstone bison are infected with brucellosis, and ranchers fear that it will be passed to domestic cattle by the bison. Research is being conducted to determine whether the same brucellosis organism that infects bison also infects cattle.

The state of Montana recently conducted a bison hunt to remove bison from public and private lands outside the park. Public sentiment against this control measure has caused animal control officers to consider other ways of solving this problem.

Figure 6-16: The Yellowstone National Park bison. *Photo courtesy U.S. Fish and Wildlife Service.*

The decline of the great bison herds caused serious problems for the Native American tribes that depended on the bison for many of their needs. They used the bison for food, shelter, and clothing. The destruction of this wildlife resource caused great hardships for their people and contributed to the wars that were fought during this period of time.

MAMMAL PROFILE

Bison (Bison bison)

Figure 6-17: Distribution map of the wild bison.

Bison are among the largest big game animals in the world. Mature bulls stand six feet tall at the shoulder and weigh over a ton. These large ungulates feed on grasses, forbs, sedges and shrubs. They occupy a variety of habitats including woodlands, forests and prairies. Bison traveled in herds that numbered several hundred thousand before their population declined to near extinction, see Figure 6-17.

The natural range of bison used to include the plains, grasslands, and woodlands of much of North America. It is now restricted to small areas that are scattered through its ranges. Cows give birth in the spring to single calves weighing 30-40 pounds. Mature animals have short necks, distinct humps on the shoulders, crescent-shaped horns, and shaggy, brown coats. Another race of bison is the wood bison found in parts of Canada. They are smaller than the bison described here, but they are similar in most other ways.

The musk ox is a member of the same family as domestic cattle, sheep, and goats. Like the deer and pronghorns, they were nearly exterminated by humans during the late 1800s. They are well equipped for life in the arctic with long, wooly coats that protect them from the cold, and sharp horns to fight off predators such as wolves.

MAMMAL PROFILE

Musk Ox (Ovibos moschatus)

Musk oxen are arctic animals whose range is restricted to the arctic tundra of North America, most of the arctic islands, and Greenland, see Figure 6-18. During the summer season they graze on lichens, grasses, sedges and forbs. They also browse on willows. During the winter they stay close to the wind-swept high ground where they can dig through the snow to graze.

They huddle together for

Figure 6-18: Musk oxen. *Courtesy U.S. Fish and Wildlife Service. Photo by Jerry Hout.*

warmth, with the calves in the middle, during arctic blizzards. A herd also forms a circle for defense with bulls facing out, and calves in the middle when they are attacked by predators. The bulls use their sharp horns and their feet to battle anything that attempts to attack.

Mature bulls give off a strong, musky odor from the glands on their faces that can be smelled from a distance of 100 yards or more. Bulls battle one another for cows during the breeding season, and the losers join other bulls to form their own groups. When the breeding season is over, bulls usually join a group composed of cows, calves, and other bulls.

Calves weigh around 20 pounds at birth, and they huddle under the shaggy hair of their mothers for protection from cold and predators. Mature bulls can weigh as much as 900 pounds and stand 5 feet tall at the shoulders.

WILD SHEEP

Several types of wild sheep range among the high mountain peaks of North America. They can be found from Alaska and western Canada down to Mexico. The most common wild sheep are the Bighorn sheep and the Dall's sheep. Male sheep are called **rams**, female sheep are called **ewes**, and young sheep are called **lambs**.

MAMMAL PROFILE

Bighorn Sheep
(Ovis canadensis)

■ Bighorn Sheep
■ Dall Sheep

Figure 6-19: Distribution map of bighorn and Dall's sheep.

Bighorn sheep occupy habitats that include deserts, prairies, and alpine meadows. They spend much of their lives among the cliffs and ledges of some of the most rugged mountains on the North American continent. They eat many kinds of vegetation including sedges, grasses, brush, shrubs, and cacti.

Figure 6-20: Bighorn sheep. *Photo courtesy of U.S. Fish and Wildlife Service.*

Bighorn sheep band together in small groups with ewes and lambs living in separate groups from the rams during much of the year, see Figure 6-19. During the breeding season the most dominant rams battle one another for the ewes. The winners guard the ewes jealously and take on any ram that tries to interfere.

Rams are polygynous and mate with several ewes. Lambs are born in the late spring with mostly single or twin births. Lambs soon learn to follow their mothers as they travel to water and return to the high ledges. An old experienced ewe provides leadership to the other ewes, guiding them to water and feeding areas.

Bighorn sheep are brown in color with white rump patches. Mature rams weigh up to 300 pounds and stand close to 3.5 feet at the shoulder, see Figure 6-20. Rams are equipped with massive curled horns. During the heated battles of the breeding season, they back up and charge each other at full speed. The crash of their horns can be heard for miles.

Natural predators of wild sheep include mountain lions and wolves. Mountain lions are as much at home on the high ledges as the sheep. Lynx and wolves take advantage of the need for water that brings the sheep down from the cliffs. Young lambs are also preyed upon by eagles and other predators. Sheep are endowed with large curling horns that flare out from the head. By counting the growth rings on their horns, the age of the rams can be determined.

MAMMAL PROFILE

Dall's Sheep (Ovis dalli)

Northern Dall's sheep are white in color, see Figure 6-21. A gray-colored race known as Fannin sheep live in the Yukon region, and further south a race of nearly black Stone sheep can be found.

Dall's Sheep live in rocky, mountainous terrain where their diet consists of grasses, sedges and shrubs. Their range is restricted to Alaska and western Canada.

Dall's sheep are slightly smaller than Bighorn sheep. Mature rams get up to 40 inches tall, and they weigh as much as 200 pounds.

Figure 6-21: Dall's sheep at Mt. McKinley National Park. *Courtesy of U.S. Fish and Wildlife Service. Photo by Mark Rauzon.*

Mountain Goats

Mountain goats travel together as families with an experienced billy goat as the leader. They are extremely sure-footed and they climb carefully. They seldom have trouble crossing cliffs and ledges. Thirty minutes after kids are born they are able to jump about, and it doesn't take them long to get used to their environment. Male goats are called **billy goats**, female goats are called **nanny goats**, and young goats are called **kids**.

MAMMAL PROFILE

Mountain Goat
(Oreamnos americanus)

Figure 6-22: Distribution map of mountain goats.

Mountain goats occupy habitats at high altitudes among the rocks above timberline. Their range extends from Alaska to northern Idaho and Montana, see Figure 6-22. Their diets consist of grasses, forbs, sedges and shrubs. Billy goats are equipped with sharp, spiked horns with which they defend themselves from predators and fight for females during the breeding season. Large mature billy goats are often close to 3.5 feet tall at the shoulder. A nanny goat gives birth to a single kid in May or June, and it is active soon after it is born, see Figure 6-23.

Figure 6-23: A nanny goat with her kids. *Photo courtesy of NORTH-WESTREK, A Wildlife Park, Eatonville, Washington.*

Mountain goats are white in color, and they weigh up to 300 pounds. They are extremely agile and coordinated. Only a few hunting permits outside of Alaska are issued for these animals each year.

Peccary

The **peccary** is sometimes called a musk hog or a **javelina**. Although it resembles the wild pigs found in Europe, it is not a true pig. A peccary is smaller than a wild pig. It stands about 20 inches tall and may weigh up to 65 pounds.

MAMMAL PROFILE

Collared Peccary
(Dicotyles tajacu)

Figure 6-24: Distribution map of the peccary.

Peccaries are native to the desert southwest and Mexico, see Figure 6-24. Its bristled hair coat is gray in color, with a distinct white stripe or collar around the neck. It has musk glands in its back that give off strong odors when the animal becomes excited or scared, see Figure 6-25.

Males are called boars and females are called sows. They breed throughout the year, and a normal litter consists of two piglets. Newborn peccaries are able to run quite fast within a few hours of birth. This is an important survival skill because the peccary is preyed upon by a number of animals, including the jaguar, bobcat, wolf, coyote, and ocelot.

Peccaries eat nearly anything from snakes and birds to cactus. They dig or root in the soil for grubs, worms, and roots of plants. They can cause serious damage to a farmer's crops by rooting in cultivated fields.

Figure 6-25: The collared peccary.
Photo courtesy of Len Rue, Jr.

Horses and Burros

When the Spanish conquerors invaded the region that is now Texas and Mexico, they left horses and donkeys behind. In the ensuing years, descendants of these animals have multiplied in the wild. Mustangs, or wild horses, inhabit deserts, rangelands, and plains, see Figure 6-26. Their diet consists of grasses, forbs, sagebrush, saltbrush, and shrubs. Their range is restricted mostly to

Figure 6-26: A wild mustang.

Figure 6-27: Distribution map of the wild mustangs.

MAMMAL PROFILE

Wild Burro

federally owned forests and rangelands in the western U.S.

They are not native to North America, but they are protected by the Wild Horses and Burros Act of 1971. Congress requires that they be managed by government agencies as though they were wild, native species, see Figure 6-27. It also allows them to be destroyed humanely if their population exceeds the available habitat.

Since they gained protection under the law some herds have multiplied until their ranges have become damaged by overgrazing. In recent years, the Bureau of Land Management has started a program that allows people to adopt wild horses for a modest fee. This has tended to ease the conflict between ranchers and environmentalists over grazing rights in areas where free-roaming mustangs and burros live.

The wild horses and burros that inhabit the western ranges are primary consumers. They convert plants to meat, and they are sometimes preyed upon by large predators such as mountain lions.

The Grand Canyon is the home to a species of wild donkey that has adapted very well to the rugged terrain of the region, see Figure 6-28. Donkeys and burros are generally considered to be domesticated offspring of the wild asses found in North Africa. The wild burro found in the Grand Canyon is an example of a domesticated animal that has returned to the wild state.

Figure 6-28: A wild burro. *Courtesy of Kent A. Vliet.*

LOOKING BACK

Hoofed mammals, also called ungulates, include the wild horse, burro, sheep, goat, bison, musk ox, deer, and peccary. Some of these animals are also ruminants that benefit from symbiotic relationships with the bacteria that digest the fiber in their diets. All of the hoofed mammals are primary consumers or herbivores with the exception of the peccary. All of the hoofed mammals fill roles in the environment as food animals for meat-eating carnivores.

REVIEW QUESTIONS

1. Name the hoofed mammals of North America that are described in this chapter.
2. Describe how ruminant animals digest their food.
3. List the characteristics that distinguish species within the deer family from one another.
4. Describe the differences between wild sheep and goats.
5. How are pronghorns and members of the deer family similar?
6. List the differences between pronghorns and members of the deer family.
7. What roles do hoofed mammals fill in the ecosystems of North America?
8. Describe the defensive behavior of the musk ox, and speculate on ways that this behavior may have evolved.
9. Describe the habitats and diets of wild peccaries.
10. Describe the kind of protection that wild horses and burros gained from the Wild Horses and Burros Act of 1971.

LEARNING ACTIVITIES

1. Choose a species of hoofed mammals that is native to your region. Learn from experts such as fish and game management officials who have been invited to the class to discuss how these animals have been managed in the past. Take field trips (or locate and view videos) to observe the animal in its natural habitat. Identify the problems that are associated with management of the species and develop a management proposal to resolve the issues you identified.
2. Study the wild horse issue as a class, and divide the class into two groups according to their stand on whether wild horses and burros should enjoy protection as a wild species or whether they should be treated as an intruding species. Set up a mock courtroom complete with judge, jury, defense attorneys, and prosecuting attorneys. Charge one or more of your students with criminal behavior for capturing and selling wild horses to a horse meat processing company. Allow students to use their imaginations as they testify on both sides of the issue, and have the jury render a verdict.

Note: This activity is a reenactment of an actual court case.

7 Predatory Mammals

boar
carrion
delayed gestation
dog
gestation
jaguar
jaguarundi
margay
musk
ocelot
omnivore
photoperiod
predator
puma
sow
vixen

WILD ANIMALS are part of a cycle that moves from an abundance of food to famine and back to abundance. Populations of wild animals rise and fall in response to food supplies, diseases, predation and natural disasters. Predatory animals bring stability to ecosystems by preventing populations of food animals from expanding beyond the capacity of their habitats to provide food and shelter.

OBJECTIVES

After completing this chapter, you should be able to

- explain the rise or decline of specific animal populations due to food supplies, diseases, and natural disasters
- describe how predators help to stabilize populations of primary consumers
- evaluate the roles of the wild cats in the ecosystems of North America
- evaluate the roles of the wild dogs in the ecosystems of North America
- speculate on the effects that total destruction of all predators would have on other animal populations
- assess the roles of bears in North American ecosystems
- discuss the characteristics of members of the weasel family that contribute to their success as predators
- speculate on reasons that raccoons and coyotes have been successful in adapting to human civilization.

Predators are animals that kill and eat other animals. As cruel as this may appear to be, it is really a more humane method of controlling animal populations than is starvation. Animals that become too numerous for their food supplies are in agony for weeks as they slowly starve to death.

122

Predators help to keep animal populations stable by preventing them from increasing so rapidly that food supplies are threatened. Predators do not eliminate starvation among the animals known as primary consumers, but they do reduce the frequency of mass starvation by slowing the growth rates of animal populations.

Predators are secondary consumers in the food chain. They are the meat eaters, and they are totally dependent upon abundant populations of other animals that can be killed and eaten for food. Predators eat large numbers of primary consumers, but they also eat other predators that they are strong enough to kill.

WILD CATS

The most elusive predators are the wild cats. They prefer to hunt at night, and they like to live alone. They are patient hunters that wait motionless for hours beside a game trail until an unsuspecting creature comes along. Wild cats are found in many of the wildlife habitats of North America.

MAMMAL PROFILE

Canada Lynx
(Lynx canadensis)

Figure 7-1: Distribution map of the Canada lynx.

The Canada lynx is a native to the northern coniferous forests of Alaska, Canada, and the Rocky Mountains of the U.S., see Figure 7-1. It is easily identified by its grayish brown fur, short tail with a black tip, and tufts of hair on its ears and cheeks, see Figure 7-2. A mature adult male weighs up to 40 pounds and stands nearly 2 feet tall at the shoulder.

The lynx preys heavily on hares. When there is an abundance of hares, the lynx population also increases. When a disease reduces the hare population, the lynx population soon decreases too. The lynx also preys on birds and other small animals, including foxes and small deer.

Figure 7-2: The Canada lynx. *Photo courtesy of NORTHWESTREK, A Wildlife Park, Eatonville, Washington.*

During the mating season, adult lynx yowl like domestic cats. Kittens are born in the spring with their eyes open, and in a few hours they are able to move about the den. They are spotted at birth, but the spots fade as the kittens mature. Kittens hunt with their mothers until they are almost one year old. She teaches them how to find and catch their own food before she abandons them.

Most members of the cat family are blind at birth and completely helpless. They require long periods of care by their mothers before they are prepared to survive in the world by themselves. She nurses them during the early weeks of their lives and brings food home to the den when they are old enough to eat meat. Teaching the young cats to hunt, catch, and kill their food requires a long period of training. All of this responsibility is taken on by the female.

MAMMAL PROFILE

Bobcat
(Lynx rufus)

Figure 7-4: Distribution map of the bobcat.

The bobcat is also known as a wildcat. It is similar to the Canada lynx in appearance ranging from 22-47 inches in length, see Figure 7-3. It has a mottled reddish brown fur coat and a short white-tipped tail.

Bobcats are well adapted to forests and to brushy desert environments as long as there is an abundance of cover and an adequate food supply, see Figure 7-4. The range of this species extends from southern Canada to Mexico, except for the Midwest. They feed on rabbits, rodents, and birds, but they will also kill fawns and lambs when the opportunity arises. They are good tree climbers, but they do most of their hunting on the ground at night.

Two to four young are born in the spring in a nest of leaves or grass. They are completely blind and helpless when they are born, and it is several weeks before they are strong enough to join their mother as she hunts.

Figure 7-3: The bobcat. *Photo courtesy of NORTHWESTREK, A Wildlife Park, Eatonville, Washington.*

Cats of the same species that live in the desert are usually lighter in color than those that live in the forest. The lighter color helps them to blend in better with the desert floor and with the brushy vegetation that is found in desert environments.

MAMMAL PROFILE

Mountain Lion (*Felis concolor*)

Figure 7-6: Distribution map of the mountain lion.

The mountain lion is often called a **puma,** see Figure 7-5. Other names for the mountain lion are cougar or panther. It is a large, tan-colored cat weighing up to 175 pounds and measuring 7.5 feet in length.

Pumas are able to adapt to many different kinds of environments. Their range is mostly confined to the Western U.S., but they are also found in isolated areas in the Appalachian Mountains and in Florida. They are found on high mountain tops, in the deserts, and in humid jungle regions expanding into Central and South America, see Figure 7-6. Deer are the favorite prey of the mountain lion, but it will kill and eat many other animals including elk, porcupines, sheep, cattle, and horses. Instances have also been reported where pumas attacked and killed humans.

Figure 7-5: A mountain lion. *Photo courtesy of U.S. Fish and Wildlife Service.*

Kittens are born in late winter or early spring. They are blind and helpless at birth, and they are covered with spots. Up to five young are born, and they feed only on their mother's milk for several months. Young pumas sometimes stay with their mothers until they are two years old. Adult males usually stay by themselves.

Several other wild cats are found in North America, see Figure 7-7. They include the ocelot, see Figure 7-8, the margay, and the jaguarundi or otter cat. The **ocelot** is a spotted cat that ranges from Texas and Arizona to Central America. It preys on small animals, birds and reptiles from the deserts of the southwest to the

Figure 7-7: Distribution map of the margay, ocelot, and jaquarundi.

Figure 7-8: An ocelot. *Courtesy of U.S. Fish and Wildlife Service. Photo by William Hutchinson.*

tropical forests of Central America. Ocelots breed at any time of the year, and two kittens are born several weeks later.

The **margay** and **jaguarundi** are relatives of the ocelot. They range in the extreme southern areas of the United States and south to South America. The ocelot, margay and jaguarundi are all listed as endangered species.

MAMMAL PROFILE

Jaguar (Felis onca)

Figure 7-9: Distribution map of the jaquar.

The **jaguar** is the largest cat living on the North American continent, (Figure 7-9). Its spotted coat resembles that of a leopard. Its range extends across forests, plains, deserts and mountains, and it eats deer, peccaries and a number of other large and small animals. Jaguars also prey heavily on domestic livestock such as sheep, cattle, and horses. The jaguar stands 27-30 inches tall and weighs up to 250 pounds at maturity. It ranges from Mexico to Argentina.

WILD DOGS

Wild members of the dog family include foxes, coyotes, and wolves. All of the wild dogs are meat eaters. They live in a variety of habitats, and they prey on many different kinds of mammals, birds, and fish.

MAMMAL PROFILE

Red Fox
(Vulpes fulva)

The red fox is easily recognized by its rust-colored coat and full, white-tipped tail, see Figure 7-10. Its ears, legs, and nose may be black, and some members of this species are completely black or silver in color. This fox thrives in farming areas. Four to ten cubs are born in the spring, and both parents help feed and care for them.

Figure 7-10: A red fox. *Courtesy of U.S. Fish and Wildlife Service. Photo by Robin West.*

Red foxes get as tall as 16 inches and weigh up to 15 pounds at maturity. They prey heavily on rodents and birds but will sometimes kill and eat domestic poultry and other small farm animals. Foxes also eat fruit and insects. The range for this fox includes all of North America except the Pacific coast and central plains.

A male fox is called a **dog** and a female is called a **vixen.** Mating takes place in late winter or early spring, and the young are born in dens a few weeks later. Dens may be located in rock piles, in hollow logs, or in underground burrows. Cubs remain with their parents for a few months until they have been taught to hunt for their own food.

MAMMAL PROFILE

Gray Fox
(Urocyon cinereoargenteus)

The gray fox is recognized by its grey coat with reddish, black, and white markings, see Figure 7-11. An adult of this species weighs 7.25–15.5 pounds. It is sometimes called a tree fox because of its habit of climbing into trees to escape its enemies. A single litter of 1–7 kits are born in the spring in a protected den.

Figure 7-11: A gray fox. *Photo courtesy of U.S. Fish and Wildlife Service.*

This fox ranges from southern Canada to northern South America, and it usually hunts at night. It is adept at catching rodents, lizards, insects, and birds. It is also fond of eggs, fruit, nuts, berries, and fish.

Most predatory animals will kill and eat other smaller or weaker predators. Foxes prey upon small animals, but they are preyed upon by bobcats, wolves, coyotes, and other predators that are able to catch and kill them. The most important threats to foxes are humans who hunt them for sport, and who kill them in attempts to protect their pets and livestock.

MAMMAL PROFILE

Arctic Fox
(Alopex lagopus)

As its name implies, the arctic fox is a native of the northern tundra region, see Figures 7-12 and 7-13. This fox has a brown coat in the summer that is replaced by a white or grayish blue coat for the winter. It has short ears which helps to reduce the loss of body heat in the cold climate in which it lives. It also has hair on the bottom of its feet which gives it better traction on ice.

This fox feeds on lemmings, other small rodents, and birds. When lemmings are scarce, it extends its hunting range to the south in search of other food. It also follows larger predators such as polar bears and wolves, and eats from their kills when they have finished.

Figure 7-13: Distribution map of the arctic fox.

Figure 7-12: An arctic fox with her litter. *Courtesy of the Alaska Division of Tourism. Photo by Robert Angell.*

Cubs are born in late spring in hillside burrows. Four to eight cubs are the usual number born, but when lemmings are abundant, litter size increases. This is probably due to better nutrition before and during **gestation**, the period of pregnancy.

Foxes have been long trapped for their furs. When furs were in the greatest demand, foxes were raised on special farms. Today the demand for fox furs has diminished and most commercial fox farms have gone out of business. Trapping of wild foxes is also less common now than it used to be. The foxes that are native to North America are widely distributed across the continent.

The kit fox is a small fox that is about one foot tall and weighs between four and six pounds, see Figure 7-14. This fox is reddish gray in color, with a black-tipped tail and dark markings on either side of its nose. It is a nocturnal hunter that preys mostly on insects, lizards, rodents, and rabbits. Its range extends from the southwestern U.S. to southern Oregon, and its preferred habitat is desert rangelands. The desert swift fox is closely related to the kit fox, and except for its larger ears and lighter color, it is very similar in appearance. A single litter of 3–6 kits is born in the spring.

MAMMAL PROFILE

Kit Fox
(Vulpes macrotis)

Figure 7-14: Distribution map of kit foxes.

Coyotes and wolves play a role in the environment by preying on rodents, rabbits, hares, peccaries, and large hoofed mammals. Both species are sometimes seen as enemies to humans because they will also kill pets and domestic livestock.

Bounty hunting has eliminated much of the wolf population in the United States. The coyote has adapted to living near humans, and it has even expanded its range in spite of bounty hunting and trapping in areas where sheep and cattle are grazed.

MAMMAL PROFILE

Coyote (Canis latrans)

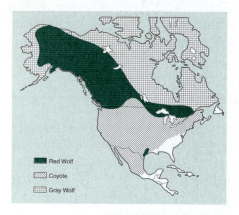

Figure 7-15: Distribution map of coyotes and wolves.

The coyote is sometimes called a prairie wolf or a brush wolf, see Figure 7-15. It ranges from Alaska to Central America, and it has not only adapted to humans, but it has learned to thrive near population centers, see Figure 7-16. It generally feeds on rodents, rabbits, hares, and birds, but it has been known to catch fish, domestic cats, and small dogs. They also damage the sheep and cattle industries by killing lambs and calves. The coyote is an **omnivore**, meaning that it eats both plants and animals. Coyotes living near population centers sometimes eat melons and other fruits and vegetables.

Figure 7-16: A coyote. *Photo courtesy of U.S. Fish and Wildlife Service.*

Coyotes are tawny colored with black-tipped, bushy tails. They weigh up to 50 pounds and large males sometimes measure as tall as 26 inches at the shoulder. Coyotes mate for life, and they generally have large litters of pups. A large hole in the ground serves as a den until the pups are old enough to follow their parents.

Career Option

PREDATORY ANIMAL CONTROL OFFICER

The killing instincts of predatory animals sometimes brings them into conflict with humans. Animals that develop appetites for domestic livestock are eliminated or removed to areas where access to livestock is reduced. Government agencies that manage public lands hire officers whose duties include controlling predation on livestock. One of their methods includes live trapping for relocation. When this fails, or when large numbers of predators are concentrated in a region, they may be poisoned or shot. A strong background and understanding of the habits and behaviors of wild animals is required.

Wolves and coyotes are related closely enough to domestic dogs that they are capable of interbreeding. This is possible only when there is a close match in the chromosomes of both species. Interbreeding is a serious concern to environmentalists and wildlife biologists who have begun to establish wild populations of wolves in Yellowstone National Park and in other remote wilderness areas. They do not want the wild populations to lose their genetic purity through interbreeding, because it could reduce their chances of survival in their natural habitat.

ENDANGERED SPECIES PROFILE

Gray Wolf (*Canis lupus*)

The gray wolf is also known as the timber wolf (Figure 7-17). Its color varies from black to white, including many intermediate shades of color. In North America the range of this wolf has been greatly reduced as areas have become populated by people. It is now restricted to Alaska, Canada, and the northern forests of Minnesota, Wisconsin, and Michigan. Some wolf sightings have also been reported in wilderness areas of Idaho and Montana where experimental populations have been introduced.

Figure 7-17: The gray wolf. *Photo courtesy of Utah Agricultural Experiment Station.*

Wolves mate for life. Litters of six to twelve pups are born in dens each spring. The male hunts for food while the female cares for the pups. After the pups are weaned, they are fed partially digested meat that the parents disgorge for them.

Mature wolves are large. They sometimes weigh as much as 175 pounds, and they stand as tall as 38 inches at the shoulder. They prey on large animals such as deer, elk, caribou, and moose. When bison were plentiful, wolves followed the herds just as they follow migrating caribou today. Their raids on domestic sheep, cattle, and horses led to the elimination of gray wolf populations in many areas where they were once abundant. The gray wolf is listed as an endangered species in the United States, except in Minnesota where it is listed as threatened and Alaska where it is not listed.

Wolves hunt together in packs made up mostly of family members. A pack of wolves working together can bring down large animals such as moose, elk, and bison. They also prey on rabbits, rodents, and birds. A single dominant wolf leads the pack, and only the dominant male and his mate are known to produce offspring.

ENDANGERED SPECIES PROFILE

Red Wolf
(Canis rufus)

The red wolf is smaller than the gray wolf, but larger than the coyote (Figure 7-18). Both red and black wolves are found in the population. Red wolves once ranged from Florida to Texas and north to Illinois and Indiana. Today the Florida race of the red wolf is extinct, and the Texas red wolf is listed as an endangered species. Its range is now restricted to a few counties in Texas.

Figure 7-18: A red wolf. *Photo courtesy of U.S. Fish and Wildlife Service. Photo by Steve Maslowski.*

BEARS

Bears are the best known and largest predators in North America. There has been an increase in bear populations in the northeastern U.S. where forests have grown up in areas of abandoned agriculture. Before the land was settled, black bears were found in most wooded areas of North America.

MAMMAL PROFILE

American Black Bear (Ursus americanus)

The black bear is the most widely distributed bear in North America, see Figure 7-19. It ranges from the arctic region to central Mexico. Adults usually weigh 200–400 pounds, and large black bears stand as tall as 40 inches at the shoulder. Its colors include black, cinnamon brown, and chocolate brown, see Figure 7-20.

Bears mate in the summer, and the cubs are born during the winter. Twins are common among black bears, and cubs stay with

Figure 7-19: Distribution map of the American black bear.

their mothers for several months while they learn to find food and to take care of themselves. At birth they are blind and completely helpless. By the time the female emerges from her winter den, the cubs are large enough to follow her about. When the cubs are in danger, the mother sends them up a tree until it passes. Unlike some bears, adult black bears are able to climb trees.

Figure 7-20: A black bear. *Courtesy of NORTHWESTREK, A Wildlife Park, Eatonville, Washington. Photo by Michael Mauch.*

Bears require large undisturbed areas covered with forest where they can find plenty of food and protective cover. Most bears are nocturnal, but in undisturbed areas they spend part of the day looking for food. Bears are omnivores, and they will eat almost any food that they find. Their diet includes grass, berries, insects, rodents, and larger animals when they can catch them. They have ravenous appetites and they often eat **carrion,** which is the rotting flesh of dead animals. Bears must get fat enough during the summer to keep them from starving to death while they hibernate during the cold winter months.

MAMMAL PROFILE

Grizzly Bear (Ursus horribilis)

The grizzly bear gets its name from the light-colored tips of its hair coat, which gives it a silvery or grizzled look. This bear is large and powerful, measuring 6–7 feet in length, measuring 3–3.5 feet

Figure 7-21: A grizzly bear. *Courtesy of NORTHWESTREK, A Wildlife Park, Eatonville, Washington. Photo by Liz Tunnell.*

tall and weighing 350–850 pounds. It is afraid of nothing, see Figure 7-21. It has poor eyesight, but its hearing and sense of smell are excellent. It can outrun a horse for a short distance, which enables it to catch some of the larger animals upon which it preys.

The grizzly mates in the summer and gives birth to its cubs during its winter sleep. One to three cubs are normal, with twins being the most common. Bears reach maturity when they are around ten years old, and they sometimes live for over thirty years.

Grizzly bears require large territories that are isolated from human activity. Conflicts between grizzly bears and humans usually occur when the bears are startled or when they are protecting their cubs. Major conflicts also arise when bears prey on livestock.

Bears live alone most of the time. They come together in groups when there is an abundance of food in a small area. This is common when large numbers of fish are moving up a stream to spawn or when a patch of berries ripens.

A mature female bear is called a **sow,** and a mature male bear is called a **boar.** Sows with cubs tend to avoid large boars because they sometimes kill and eat cubs.

The Alaska brown bear is the largest land-based carnivore in the world, see Figure 7-22. It weighs up to 1,500 pounds and stands between 4 and 5 feet tall at the shoulder when it reaches maturity. It spends the summer feeding on salmon or any other food that it can find.

Mating occurs in the summer, and up to four tiny cubs are born seven months later. A sow with cubs makes a den in late fall and they spend the winter together. She remains with her yearling cubs and teaches them to fish for salmon and to find other food.

Alaska brown bears include the huge Kodiac and Peninsula bears. Eight different races of Alaska brown bear are known, including the Shiras bear that is nearly black in color.

Figure 7-22: The Alaska brown bear. *Photo courtesy of the Alaska Division of Tourism.*

Most bears eat fish when they are available, but none is more at home in the water than a polar bear. This bear spends its entire life in or near the water, and it sometimes swims far out from the shore in search of seals or other prey.

Polar bears are found in the cold arctic regions of North America. They are large bears weighing 330–1,100 pounds, measuring 7-8 feet in length and 3–4 feet tall at the shoulder. They are completely white except for their dark eyes and nose, see Figure 7-23. Polar bears are carnivores and except for seaweed they eat meat for almost every meal. They prey mostly on seals, but they also eat lemmings, birds, fish, walrus, and whales that have died or become stranded. A hungry polar bear will attack just about any animal it can find, including humans.

Figure 7-23: The polar bear. *Courtesy of U.S. Fish and Wildlife Service. Photo by Dave Olson.*

Pregnant females go into dens to give birth to their cubs. They are the only polar bears that go into dens during the cold winters. Male polar bears and females with half-grown cubs spend the winter season on the ice looking for seals or other food. Polar bears have poor hearing, but they have excellent eyesight and a keen sense of smell.

WEASELS

The weasel family is made up of an unusual group of predators. It includes the weasel, mink, ferret, fisher, marten, otter, wolverine, badger, and skunk. They are some of the most vicious predators in the animal kingdom. Predators kill to eat, but members of the weasel family often kill many more animals than they need for food. They are small, highly efficient predatory animals that are adapted to survival in the wild.

MAMMAL
PROFILE

Short-Tailed Weasel
(Mustela erminea)

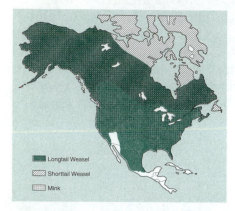

Figure 7-25: Distribution map of the long-tailed weasel, short-tailed weasel and mink.

The short-tailed weasel and its close relatives are long, slender animals with short legs and black-tipped tails, see Figure 7-24. Their coats change color with the seasons: white in the winter and brown in the summer. Such color changes help the weasel to hide more easily from its enemies and to stalk its prey without being seen.

Weasels are found in nearly every kind of habitat, where they prey on rabbits, birds, and rodents, see Figure 7-25. Large litters are born in the spring consisting of up to a dozen or more young. Their mothers begin to teach them to hunt as soon as they are old enough to follow her about. Mature weasels of this species are 7–14 inches long, and they weigh 2–5 ounces.

Figure 7-24: A long-tailed weasel. *Photo courtesy of Max H. Shroeder.*

Most members of the weasel family experience **delayed gestation** following mating. This occurs because the embryo does not become attached to the inner surface of the female's reproductive tract until chemicals in her blood prepare her body for gestation.

It has been shown that gestation is delayed until the length of the **photoperiod,** the number of daylight hours in a day, increases to signal the longer days of spring. Gestation does not begin in many of these animals until the photoperiod is long enough to assure that the spring season is near and the conditions are favorable for the survival of the young.

MAMMAL
PROFILE

Mink
(Mustela vison)

The mink is an animal that resembles the weasel in the shape of its body. It is bigger than a weasel, however, measuring 18–28 inches long with weights from 1.5–3.5 pounds. It ranges in color from light to dark brown, with a white spot on the throat, see Figure 7-26. It is an important furbearing animal that is raised commercially on farms. Several fur colors have been bred into domestic mink that are not observed in the wild.

Figure 7-26: A mink. *Photo courtesy of Leonard Lee Rue, III.*

Mink live alone in dens that are often taken from their prey. Adult mink come together only to mate. Four to eight kits are born in the den in the spring. They are completely helpless and blind at birth, and they remain in the den for several weeks.

Mink hunt birds and small mammals on land, and crayfish, frogs, snakes, and fish in water habitats. They are active day and night, and they require large amounts of food to maintain their energy.

The existence of predatory animals depends on adequate populations of animals that are acceptable as food. When a predator becomes so specialized that it depends on a single food source and it is unable to adapt to a different diet, it faces possible extinction when the population of its prey declines.

ENDANGERED SPECIES PROFILE

Black-Footed Ferret (*Mustela nigripes*)

The Black-footed ferret is a medium-sized member of the weasel family that measures 18–22 inches in length and weighs 2–3 pounds. It is yellowish-brown in color with black feet, forehead and tail tip. The population of black-footed ferrets has been small for as long as the animal has been known, see Figure 7-27. Its range is the plains region bordering the Rocky Mountains on the east. This animal eats prairie dogs as its principal source of food. As prairie dog populations have declined, a serious decline has been observed in the population of black-footed ferrets. It has been a protected species since 1967. Several times it has been thought to be extinct, but its numbers have been increased by breeding it in captivity. It will be reintroduced to the wild as its population increases sufficiently to do so.

Figure 7-27: The black-footed ferret. *Courtesy of U.S. Fish and Wildlife Service. Photo by Rich Krueger.*

The long slender bodies of the weasel, mink, ferret, and marten make it possible for these predators to enter the dens of many of the animals they prey upon. This gives them a distinct advantage against their prey. Most rodents and other small animals have no place to go for safety once predators enter their dens.

The marten is a large weasel-shaped animal with brown fur and light orange-colored undermarkings, see Figure 7-28. It is 20–25 inches long, and it weighs 1.5–2.75 pounds. It is an agile climber that pursues tree squirrels and other prey such as rodents, rabbits, birds, and insects. It also eats the fruits of some plants.

Figure 7-28: A marten. *Courtesy of NORTHWESTREK, A Wildlife Park, Eatonville, Washington. Photo by Chris Schmitz.*

Martens lives in forested areas. Their dens are often hollow trees lined with dry leaves. Martens mate in the summer, but delayed gestation results in the birth of three or four offspring in the spring. Birth occurs about nine months after mating. The young are blind and helpless when they are born, but by fall they have learned to hunt alone.

Members of the weasel family have glands at the base of the tail from which a foul-smelling fluid known as **musk** is secreted when the animal is disturbed or frightened. It is also used during the mating season to attract a mate. Nearly everyone is familiar with the odor of a skunk, but few people realize that other members of the weasel family have scent glands too.

The fisher is an efficient predator whether it is on the ground or climbing trees, see Figure 7-29. This dark-colored predator ranges across Canada with populations occurring in the Rocky Mountains of the western and northeastern U.S. where it preys on all kinds of birds and mammals. The only fish it is likely to eat is one that it has stolen from another predator. A large fisher seldom weighs more than 12 pounds, but it is capable of killing a deer. It is one of the few animals that consistently preys on porcupines. It does this by flipping the porcupine over onto its back and biting the throat or belly where there are no quills.

The female fisher usually gives birth to three or four young in the hollow trunk of a standing tree. Soon after they are born, she mates again. Delayed gestation accounts for her extended pregnancy. Mother and young usually split up when winter comes, and they establish separate hunting territories.

Figure 7-29: The fisher. *Photo courtesy of NORTHWESTREK, A Wildlife Park, Eatonville, Washington.*

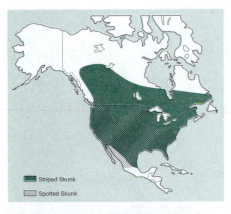

Figure 7-30: Distribution map of skunks.

Several kinds of skunks are found in North America, see Figure 7-30. Among them are the hog-nosed skunks and hooded skunks that range from the southwestern United States to Central America. Spotted skunks and striped skunks are more widely distributed.

The striped skunk is recognized by its distinct white stripes on a black body, see Figure 7-31. It is 20–30 inches long and weighs 4–9.75 pounds. The skunk odor comes from scent glands located near the base of the tail. An enemy that gets too close to the raised tail of a skunk is likely to be the target of well-directed stream of musk. A few predators such as the great horned owl, cougar, mink, and coyote will kill and eat skunks despite the pungent smell.

Skunks are omnivores that feed on berries, nuts, fruits, insects, rodents, frogs, and birds. They also kill chickens and eat eggs when the opportunity arises. They live in bur-

Figure 7-31: The stripped skunk. *Courtesy of U.S. Fish and Wildlife Service. Photo by Roger Collins.*

rows located underground or under buildings or piles of debris. Their range extends from Canada to Mexico, and includes desert, plains, and woodland habitats. Four or five young are born in the spring, and they may remain with their mother when she enters the den to sleep through the cold months of winter.

The badger and wolverine have a distinctly different body form from other members of the weasel family. They have short, powerful legs and heavy bodies combined with strong teeth and claws. They are among the strongest animals known for their size.

MAMMAL PROFILE

Badger (Taxidea taxus)

The badger is a predator that is equipped with long digging claws, see Figure 7-32. It uses them to dig for mice, gophers, and other small mammals that live in underground burrows. A badger can dig a hole fast enough to escape many of its enemies, filling the hole behind it as it digs.

Badgers live in underground burrows and make nests of dried grass at the ends of their tunnels. Up to five young are born in the nest in late winter or early spring. They are helpless and blind at birth, and they stay in the den for several weeks. When the spring season arrives, they join their mother on hunting trips.

The coat of the badger is grizzled with a silvery appearance, and its face and head are marked with white stripes. Mature males often weigh over 20 pounds, and they are some-

Figure 7-32: A badger. *Photo courtesy of NORTHWESTREK, A Wildlife Park, Eatonville, Washington.*

times as long as 30 inches. Their range includes much of the western U.S. and southwestern Canada, extending as far south as Mexico. Preferred habitat includes open areas and farmlands.

Members of the weasel family prey on nearly every animal in North America. They consider every living creature to be a potential meal. All of these animals are fearless predators that do not hesitate to attack much larger creatures.

MAMMAL PROFILE

Wolverine (Gulo luscus)

The wolverine is one of the most aggressive predators in North America (Figure 7-33). It is known to prey on animals as large as the caribou, and to drive bears and mountain lions from their kills. It has no fear of man, and it has been known to break into cabins and caches of food supplies. It usually destroys anything it finds.

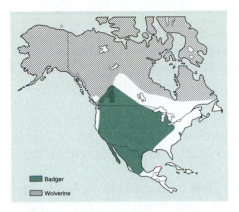

Figure 7-34: Distribution map of the badger and wolverine.

Figure 7-33: The wolverine.*Photo courtesy of NORTHWESTREK, A Wildlife Park, Eatonville, Washington.*

Anything that is left after it has eaten its fill is sprayed with musk to establish its ownership. Wolverines prefer mountain, forest and tundra habitats. They sometimes climb trees to lie in wait for prey. At other times they pursue an animal until it is exhausted and unable to get away.

Adult wolverines live alone except during the mating season. Two or three young are born in late spring. During the summer their mother teaches them to hunt, and by winter they are on their own.

Mature wolverines weigh up to 60 pounds, and they are the largest members of the weasel family. They look like small brown bears with light-colored markings around their rumps and along their sides. The range of the wolverine extends from Alaska to the forests of the northwestern and northern California, see Figure 7-34.

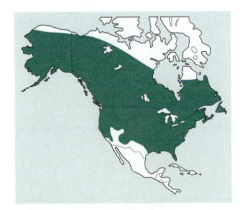

Figure 7-35: Distribution map of the otter.

Some of the most valuable pelts in the world once came from otters. Sea otter pelts were responsible for much of the exploration of Alaska and the northwest coast of North America, see Figure 7-35. Otter pelts drew trade from many parts of the world during the eighteenth and nineteenth centuries. This brisk trade nearly caused the extinction of the sea otter as they were slaughtered for their pelts with no thought of preserving the population. It is now illegal in most places to take otter pelts because populations of both river and sea otters have declined, although river otters are still abundant in some areas.

The body of a river otter is shaped much like that of a mink, and the quality of the fur is similar. The otter is much larger than a mink, weighing as much as 30 pounds and measuring as long as 55 inches. It is an expert at catching fish, snakes, and other aquatic animals.

River otters prefer wetland habitats that have cover, such as woodlands. Their range used to extend to many parts of North America, but they are now found in scattered pockets within their former range.

Pups are born in the spring in a den under a stream bank, and the female protects them carefully from every possible danger. Otters are playful animals. They build mud slides down stream banks where they play together.

Sea otters are larger than land otters, and they spend their time in the coastal waters of Alaska, Washington, Oregon, California, and Mexico, see Figure 7-36. Females give birth to a single pup in the spring. Pups are developed enough at birth to survive in the water. They have fur to keep them warm, and their eyes are open. They do not have blubber to insulate their bodies like most mammals.

Female sea otters cuddle their young and carry them about as they swim. They rest and hide from predators in kelp beds. The favorite foods of sea otter are abalone, mussels, crabs, and fish. Pups remain with their mothers for a year or more, and continue to depend

Figure 7-36: The sea otter. *Courtesy of the Alaska Division of Tourism. Photo by Robert Angell.*

upon them even after they are grown. The U.S. Fish and Wildlife Service placed this otter on the threatened species list in California.

RACCOONS

Raccoons are among the most adaptable animals in North America. They are found in nearly every habitat from deserts to woodlands. They are adept at swimming and at climbing trees, and they can find food almost anywhere. They live in housing subdivisions almost as easily as they live in the woods. Raccoons seem to find advantages in living near humans. They find shelter in human structures and they eat garbage and other food that is abundant near humans.

MAMMAL PROFILE

Raccoon
(Procyon lotor)

Figure 7-37: Distribution map of the Raccoon.

The raccoon is a nocturnal animal that ranges from southern Canada to South America, see Figure 7-37. They eat almost anything that is edible, but they are especially fond of shellfish. The habitats of raccoons include nearly anywhere that water is available, especially along wooded streams.

Raccoons are recognized by their black-ringed tails and their masked faces, see Figure 7-38. The white hair in their brown coats helps them to blend in with their surroundings. A large raccoon may weigh as much as 30 pounds, but some species are as small as 5 pounds. Three to six young are born in the spring, and they stay with their mother until the following year.

Figure 7-38: The raccoon.
Photo courtesy of NORTHWESTREK, A Wildlife Park, Eatonville, Washington.

LOOKING BACK

Predators fill important roles in ecosystems by keeping populations of primary consumers from expanding too rapidly. When populations of food animals rise, predators gain an abundant food supply. This soon prompts a rise in the predator population. When the predator population gets too large or when diseases infect the primary consumers, populations of food animals decline. As the food supply for predators is reduced, the predator population also declines due to starvation, diseases, and reduced birth rates.

The predators that were discussed in this chapter are the wild cats, wild dogs, bears, weasels, and raccoons. All of these predators feed on rodents and birds. Some of them also feed on hoofed mammals, and some predators eat other predators as well as carrion. Predatory mammals control the rates at which most animal populations expand. This helps to prevent extreme fluctuation in animal numbers. It also tends to reduce instances of mass starvation of animal populations that have outgrown their food supplies.

REVIEW QUESTIONS

1. Explain how food supplies, diseases, and natural disasters affect the rise or decline of the animal populations near your home.
2. Describe how predatory animals play a role in stabilizing populations of primary consumers in the food web.
3. What roles do wild cats fill in the ecosystems of North America?
4. Select a predatory animal and summarize the events in its life cycle and their significance, such as the timing of mating, gestation, birth, and weaning of the young, and other important events.
5. What are the roles of foxes, coyotes, and wolves in the ecosystems where they live?
6. What do bears eat and what are their roles in North America?
7. List some characteristics of members of the weasel family that contribute to their success as predators.
8. What characteristics of raccoons and coyotes have allowed them to thrive in and near human population centers?

LEARNING ACTIVITIES

1. Assign groups of students a predatory animal to research and have them prepare written and oral reports on their findings. Have each group make an oral presentation on their species to the rest of the class.
2. Identify a person in the area who has a collection of pelts and/or skulls of predatory animals. Fish and game officers are usually aware of people who have these interests. Invite them to bring their materials to class for a presentation.

8 Marine Mammals

MAMMALS are found in all types of environments including the oceans. The mammals that are found in the oceans are called **marine mammals.** They have special adaptations to their bodies that make it possible for them to survive in their water environments. They have flippers instead of feet, see Figure 8-1, and they store layers of fat on their bodies, improving their buoyancy in the water. Fat layers also insulate them from the cold.

OBJECTIVES

After completing this chapter, you should be able to

- describe how the bodies of marine mammals are adapted for living in water
- name and describe the different kinds of animals that are classified as seals
- identify differences that distinguish eared seals, true seals, and walruses from one another
- list major species of pinnipeds that make up the three seal groups
- evaluate the roles of the seals in marine ecosystems
- analyze the roles of manatees in the North American marine ecosystems
- consider the characteristics of whales that qualify them as mammals
- distinguish between toothed whales and baleen whales
- describe the characteristics of major species of whales found in the oceans of North America.

In contrast with most other life forms that are found in the sea, the marine mammals are warm-blooded animals. They also give birth to live young, and they nourish their offspring with milk.

145

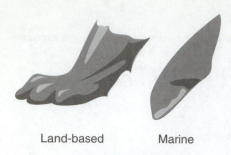

Land-based   Marine

Figure 8-1: Marine mammals have special adaptations to their bodies that make it possible for them to survive in their water environment. Land-based mammals have feet, whereas the marine mammal has flippers.

Marine mammals are very much at home in the ocean, and those that can leave the water are usually quite awkward on land. Unlike many sea dwelling animals, they obtain oxygen and eliminate waste gases through their lungs. Most other sea animals exchange respiratory gases across their gills, see Figure 8-2.

SEALS

The seals are aquatic animals that have flippers instead of feet. For this reason they are sometimes called **finfeet.** Scientists call them **pinnipeds.** Mature pinnipeds are equipped with strong molars that are used to break the shells of mollusks and crustaceans. The different species of seals, sea lions, and walruses are all classified as seals. Fourteen species of these animals live along the coasts and offshore islands of North America. They are found in the Atlantic, Pacific, and Arctic Oceans.

Seals are predators. They eat **crustaceans** such as lobsters and crabs, which have hard outer shells on their bodies, see Figure 8-3. **Mollusks** also make up a part of their diets. These organisms include clams, oysters, mussels, and snails. Mollusks and crustaceans are types of **shellfish** because they have shells for protection.

Pinnipeds are divided into three major groups, the eared seals, the true seals, and the walruses, see Figure 8-4. Eared seals have small external ears, and this group includes sea lions and fur seals. They are able to travel more easily on land than the true seals because they are capable of rotating their rear flippers forward

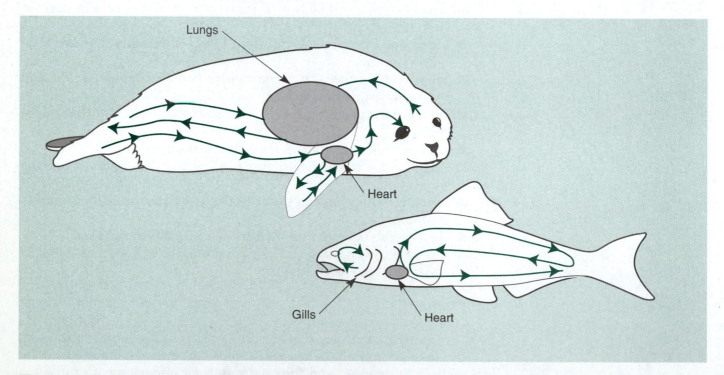

Figure 8-2: Exchange of respiratory gases in marine mammals.

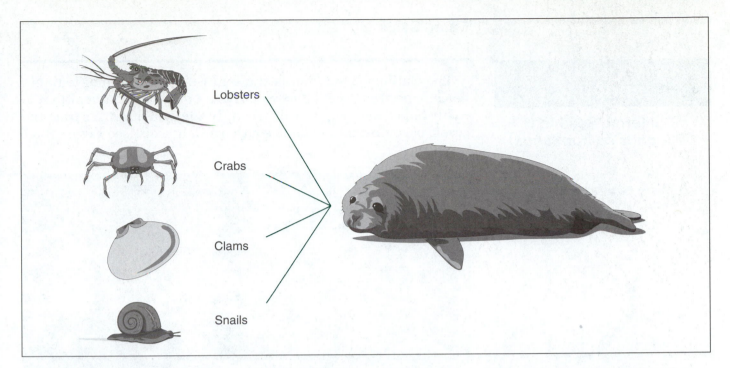

Figure 8-3: Pinnepeds are predators, and their diets include shellfish such as lobsters, crabs, clams, and snails.

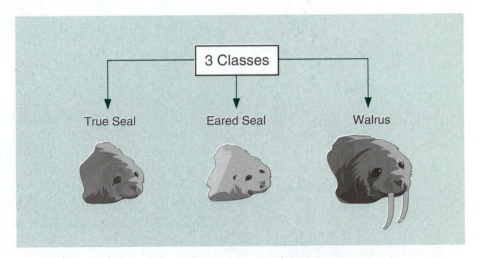

Figure 8-4: Types of pinnepeds.

enough to walk and even run with them. This group of seals includes fur seals and sea lions. The true seal has no external ears, and its hind flippers cannot be moved forward for travel on land. These seals must drag their bodies on land, but they are very mobile in water. Members of this group include harbor seals, harp seals, ribbon seals, and elephant seals.

The walrus is different from either the eared or the true seals. It is the largest seal and the only seal species in this group. Its flippers rotate forward allowing it to travel on land, and it has no external ears. Its skin is hairless, and it has large tusks.

Eared Seals

California Sea Lion
(Zalophus californianus)

Figure 8-6: Distribution map of the California sea lion.

The California sea lion is the seal that most animal trainers teach to perform, see Figure 8-5. It is an intelligent animal that is easily tamed and taught to do tricks. In the wild, it ranges from the coasts of southern Canada to southern California, see Figure 8-6.

Figure 8-5: California sea lions. *Courtesy U.S. Fish and Wildlife Service. Photo by Byrd.*

Cows are mature at about three years of age, and they give birth to a pup every year. Bulls establish breeding territories on beaches that they vigorously defend. Cows mate again soon after their pups are born, and the young stay near their mothers for nearly a year. The diet of adult sea lions includes octopus, squid, crabs, and fish.

Sea lions are eaten by sharks and killer whales. Adults grow to as much as 8 feet long and weigh up to 600 pounds.

Seals feed on sea life much as a bear feeds on rodents and other animals on land.

Steller's Sea Lion
(Eumetopias jubatus)

The Steller's sea lion occupies much of the same territory as the California sea lion, but it is bigger, see Figure 8-7. Its diet consists mainly of fish, and it competes with commercial fishermen for salmon. It is capable of diving to depths of several hundred feet in search of food.

Mature males measure as long as 12 feet, and they weigh as much as a ton. Females are much smaller than the males. A single pup is born to a female each year on the rocky shores of the Barren, Pribilof, and Aleutian Islands and along the West Coast of North America from the Gulf of Alaska to the California Channel Islands.

Figure 8-7: Steller sea lions. *Photo courtesy of Leonard Lee Rue, III.*

The pinnipeds are graceful and move easily through the water. They are somewhat awkward on land, however, and this makes them easy prey for land-based predators such as polar bears, and for hunters. The young animals of all have been hunted for their meat, oil and furs for centuries, and some species have been hunted to near extinction. The Guadalupe fur seal has twice been incorrectly thought to be extinct, because the known breeding herds were destroyed by hunters.

MAMMAL PROFILE

Northern Fur Seal (*Callorhinus ursinus*)

Figure 8-9: Distribution map of the Northern fur seal.

The northern fur seal is a migratory animal, see Figure 8-8. It ranges from the Bering Sea islands to California and Japan. These seals migrate to southern regions during the winter, and return primarily to the Pribilof Islands in the Bering Sea, see Figure 8-9, each spring where the bulls fight to establish harems of cows.

The female has a single pup, and she mates sometime during the summer months to conceive next year's pup. Dominant bulls guard their harems through

Figure 8-8: Northern fur seal bull. *Photo courtesy of Leonard Lee Rue, III.*

the entire summer. They don't dare leave even to eat or their cows will be taken by another bull. By fall they are thin and exhausted from the difficult summer.

True Seals

True seals have sleek, smooth heads because they have no external ears. Some of them also have little or no hair on their bodies. They must drag themselves with their front flippers when they come ashore because their rear flippers cannot rotate forward for walking.

MAMMAL PROFILE

**Harp Seal
(Phoca groenlandica)**

The harp seal is considered to be valuable for its oil and hide, see Figure 8-10. Newborn harp seals are sometimes called **white coats** because of their white fur. The white fur is harvested when the pups are just a few days old. A lot of controversy has been generated by the practice of killing the pups with clubs for their pelts. Mother harp seals nurse their pups for only two or three weeks before abandoning them on the ice. During the nursing period, these young seals often gain 50 pounds or more.

The abandoned pups live for a week or more on their fat reserves before going into the water in search of food. Soon they are eating fish, squid, mollusks, and crustaceans, and they join the adults again. Harp seals are migratory animals that drift south with pack ice to the Atlantic Coastal region around the Hudson Bay from their northern range in the Arctic. Mature harp seals weigh as much as 500 pounds, and they measure 5–7 feet long.

Figure 8-10: The harp seal.

The polar bear depends on seal meat for survival. Polar bears and seals are at home on the sea ice of the far northern regions of the continent. The polar bear is profiled in Chapter 7 in the discussion of predatory land mammals. Seals live in the water, and they seldom leave the water except to give birth and to mate. Polar bears are capable of swimming long distances in the cold water. They are able to venture out on the ice to hunt seals and still get back to the shore when the ice pack breaks up in the spring. Seals must come to the shore when the ice pack breaks up in the spring.

Seals must come to the surface to breathe, and polar bears wait at their blowholes in the ice or at the edge of open water to catch them.

<table><tr><td>

MAMMAL PROFILE

**Harbor Seal
(Phoca vitulina)**

Figure 8-12: Distribution map of the harbor seal.

</td><td>

The harbor seal is the most widely distributed seal in North America, see Figure 8-11. It is found near the shore on both coasts in the Bering Sea, Arctic Ocean, and along major coastal rivers, see Figure 8-12. Its wide range results from its adaptability and wide tolerance of temperature and water salinity. These seals live in small groups of 30–80 animals with groups being larger in areas where food is plentiful. Pups are born in the spring, and within a few days they shed their newborn white coats for spotted ones.

Seals prefer diets consisting of fish, crustaceans, and mollusks that are abundant on the seabed. Pups grow quickly due to the richness of their mother's milk. In a few weeks they are weaned, and they learn to feed with the adults.

Figure 8-11: The Harbor seal. *Photo courtesy of the Alaska Division of Tourism.*

</td></tr></table>

Female seals usually have one or sometimes two pups each year. Mature harbor seals average 220 pounds and measure 5.5 feet in length.

Walrus

Walrus are found in the Atlantic, Pacific, and Arctic regions of North America. They are large sea mammals that are often grouped collectively with the seals. One way that they are different from other pinnipeds is that both sexes have tusks.

<table><tr><td>

MAMMAL PROFILE

**Walrus
(Odobenus rosmarus)**

</td><td>

The walrus is the largest pinniped in the Arctic region, see Figure 8-13. The walrus ranges from the Arctic Ocean to areas of the Bering Sea and Hudson Bay. The arctic ice breaks up as the weather warms, and floats south, carrying the walrus to the Bering Sea during the summer. In the spring, they migrate back to the Arctic Ocean where the cycle begins once more.

</td></tr></table>

Figure 8-13: Distribution map of the walrus.

Figure 8-14: A group of walrus bulls and cows. *Photo courtesy of U.S. Fish and Wildlife Service.*

Bulls measure up to 13 feet long and weigh up to 3,700 pounds. Cows are much smaller. Adults have tusks up to 3 feet long that are used to dig for food, see Figure 8-14. Large bulls also use them to fight for cows during the mating season and to pull themselves from the water onto the ice pack.

Cows have young every two years. Walrus calves are born on the ice in the spring, and they live on their mother's milk for about two years. When the calf gets tusks, it is able to dig mollusks from the bed of the sea to feed itself. The enemies of walrus are humans, killer whales, and polar bears.

MANATEES

The **manatee** is a large aquatic mammal that resembles a small whale in shape. It is a primary consumer in the food chain, grazing on vegetation that grows in and near the water. It performs a valuable service to humans by keeping coastal waterways from becom-

Career Option

OCEANOLOGIST

A career as an oceanologist includes **oceanology,** which is the study of the environment in oceans. Areas of interest include the water, plants, animals, reefs, and other features in the ocean environment. It also includes exploration of the underwater environment.

A person working as an **oceanologist** will need a university degree in the sciences with a strong emphasis in biological science. A graduate degree is usually required.

ing restricted by growing plants. The manatee was once hunted for oil and meat, and its numbers have been reduced.

MAMMAL PROFILE

**Manatee
(*Trichechus mantus*)**

Figure 8-15: Distribution map of the manatee.

The manatee ranges from the warm coastal waters of Florida to South America, see Figure 8-15. Its whale-like body is equipped with paddle-shaped forelimbs, a large flat tail instead of rear limbs, and an upper lip covered with bristles, see Figure 8-16. The manatee can get as large as 1,200 pounds, and as long as 15 feet. It is a close relative of the Steller's sea cow that was hunted to extinction by Russian hunters in the late 1700s. It is now classed as an endangered species and protected by law. Only one species now exists in North America.

Figure 8-16: The manatee. *Photo courtesy U.S. Fish and Wildlife Service.*

The manatee gives birth to a single offspring every 2–3 years. These mammals cannot survive temperatures below 46 degrees. Their greatest threat of death comes from motorboat props that injure them before they can dive to safety.

WHALES

Whales are the largest animals on earth. Like other mammals, they are warm-blooded, give birth to live offspring, and nourish their young with milk. They have lungs instead of gills, and they must come to the surface of the water to breathe. They have **blowholes** or nostrils on top of their heads. Whales are called **cetaceans** by scientists due to the location of their blowholes, their horizontally flattened tails, lack of external hind limbs, and the presence of paddlelike forelimbs.

Whale hunting has long been important to humans because huge amounts of oil can be rendered from the thick layer of **blubber** or

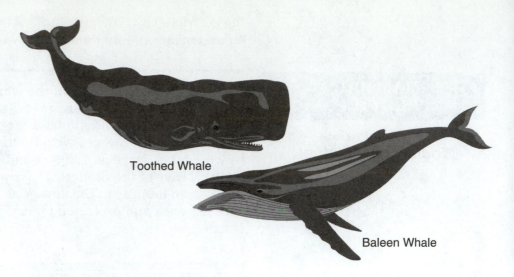

Toothed Whale

Baleen Whale

Figure 8-17: Types of whales.

fat that insulates the bodies of whales against the cold waters in which they live. Whale oil has long been used to provide heat and light for the dwellings of native Eskimos. Blubber provided the high energy food that was needed to provide body heat in cold climates.

Many of the whale populations of the world have been reduced to dangerously low levels due to harvesting practices that kill whales faster than they are able to reproduce. Declining whale populations have caused enough concern among conservation-minded people that a large movement has started to protect whales. "Save the whales!" a common slogan among these groups, is widely promoted, and even young school children have been exposed to the plight of the whales. Many of the whales are protected in North American coastal waters by the U.S. Fish and Wildlife Service.

Whales are of two types, see Figure 8-17. The **baleen whales** have toothless jaws, and their mouths are equipped with **whalebone.** This is a comblike sieve composed of horny, flexible material with which whales strain small food organisms out of the water. The throats of these whales are small, and they are only able to swallow small organisms. Baleen whales also have two blowholes through which they breathe.

Toothed whales have large throats and sharp teeth with which they are able to attack large prey and bite it into pieces that are small enough to swallow. The toothed whale has a single blowhole for breathing.

Baleen Whales

The baleen whales include a number of species that strain their food from sea water. These whales swim with their open mouths through masses of plankton or small, shrimplike crustaceans known as **krill.** When a sufficient amount of food has been gathered, they expel the sea water through the whalebone sieves that hang from their upper jaws and swallow the trapped organisms.

MAMMAL PROFILE

Blue Whale
(Balaenoptera musculus)

The blue whale is the largest mammal ever to live on earth, see Figure 8-18. Large members of this species reach 100 feet long, and they weigh as much as 150 tons. This whale has blue coloring mottled with gray. It spends the summers feeding in arctic waters and goes south in the winter. Its only enemy other than humans is the killer whale.

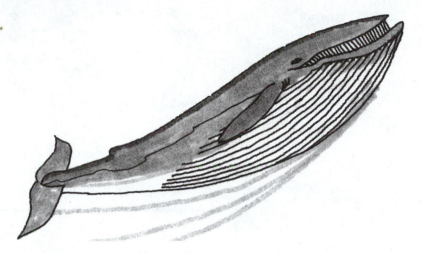

Figure 8-18: The blue whale.

A blue whale calf weighs 2 tons at birth. Whale milk is high in fat and protein compared to the milk of land mammals, and a whale calf consumes up to 200 pounds of milk every day until it is weaned at about seven months of age. By then the calf will weigh around 23 tons and measure 52 feet long. The blue whale matures sexually in four to six years, and has a life expectancy of thirty to forty years. The blue whale is protected as an endangered species.

The baleen whales have been important to the whaling industry for many years. During the nineteenth century, the whalebone or baleen was as valuable to whale hunters as the oil or meat. As much as 1.5 tons of whalebone can be obtained from some species. Only two of the baleen whales are discussed in this chapter. Other important baleen whales are the bowhead whale, the gray whale, the fin whale, and the humpback whale.

MAMMAL PROFILE

Right Whale
(Eubalaena glacialis)

The right whale was once the most common species of whale, see Figure 8-19. It is shorter and more stocky than many of the other whales. It measures up to 55 feet in length at maturity. The right whale is black in color, and it is found in the North Atlantic and Pacific Oceans, see Figure 8-20.

Cows give birth to single calves, to which they are extremely devoted. A cow will stay with her calf even if it is dead. Whale

Figure 8-20: Distribution map of the right whale.

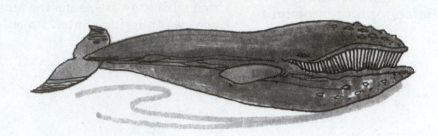

Figure 8-19: The right whale.

hunters used this devotion against female whales by killing the calf first. The cow was then an easy target because she refused to leave. The right whale is protected today as an endangered species.

Toothed Whales

The toothed whales include the sperm whales, the killer whales, narwhals, porpoises, and dolphins. They are of many sizes and shapes. Small toothed whales are sometimes called **dolphins** and the large ones are called whales. These names are based more on size than upon genetic relationships.

The largest of the toothed whales is the sperm whale, see Figure 8-21. This whale measures up to 50 feet in length and has a large head and mouth. It feeds on squid, octopus and **cuttlefish.** The cuttlefish is a squid-like mollusk with ten arms and a hard internal shell. These creatures are found in deep water, and sperm whales

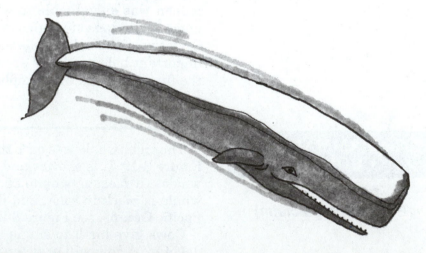

Figure 8-21: The sperm whale.

consume as much as a ton of food every day. Sometimes sperm whales also eat fish such as herring.

Sperm whales live together in **pods** or herds of females and calves. Each group is led by one or two older cows. Bulls win dominant positions by defeating their challengers, and dominant bulls enter the herds during the breeding seasons. Calves are born about sixteen months after conception. They are nursed by their mothers for six months during which time they double in size. Cows are very protective of their calves, and they remain with them even when they die. Sperm whales are protected by law as an endangered species. They range in all oceans.

Most species of whales are highly intelligent animals. They learn simple tasks quickly, and animal trainers find some of them particularly the dolphins to be willing performers. They band together for protection from enemies, and some whales hunt together.

MAMMAL PROFILE

Killer Whale (Orcinus orca)

Figure 8-23: Distribution map of sperm and killer whales.

Killer whales are black and white in color, and they are the largest of the dolphins, see Figure 8-22. They grow as long as 31 feet and weigh as much as 8 tons. This whale has a reputation as the most feared predator in the oceans of the world (Figure 8-23). Killer whales frequently hunt together and they are known to prey upon some of the largest of the marine animals. Their prey includes seals, penguins, walrus, squid, fish, and other whales.

Figure 8-22: Killer whales. *Courtesy of U.S. Fish and Wildlife Service. Photo by Bob Jones.*

These giant dolphins live together in groups, and they are known to coordinate their movements in attacking large prey. They range in all oceans, and scientists believe that they are able to communicate effectively with one another. A mature female gives birth to a single calf at 2–3 year intervals.

Whales communicate with each other using musical sounds that carry through the water for long distances. The brains of whales appear to be quite highly developed in comparison with most other animals.

The narwhal is an unusual whale with a long, hollow spiraled tusk in the left upper jaw of the male, see Figure 8-24. Adult narwhals are 13–16 feet long, and the single tusk of males can reach 9 feet. This whale ranges in the far northern latitudes of the arctic seas. It eats small fish.

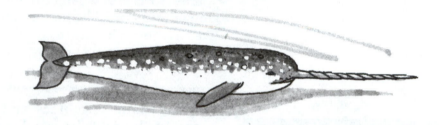

Figure 8-24: A narwhal.

Narwhals have been hunted for food by Eskimos who also prize the ivory tusk. They use the oil obtained from this whale to provide heat and light for their dwellings.

Dolphins make up another group of the toothed whales. North American species included in this group are porpoises and dolphins. A characteristic that is used to distinguish dolphins from porpoises is the shape of their mouths. Most dolphins have beak-like mouths lined with teeth. The snout and mouth of a porpoise is quite blunt in appearance. These animals are mostly fish eaters that fill roles as secondary consumers in marine environments. The dolphin appetite for fish includes the tuna. These fish are caught using large nets. Dolphins get caught in the nets along with the tuna, and they drown. Modified nets allow many of the dolphins to escape, but they also yield reduced catches of fish. Controversy has developed over the deaths of dolphins in the nets of fishermen.

Related species of this dolphin are found all over the world in warm and temperate seas, see Figure 8-25. This dolphin is common on both the Pacific and Atlantic coasts of North America, see Figure 8-26. These dolphins mature at lengths up to 8 feet and weigh up to 180 pounds. Their long, thin jaws that look like pointed beaks

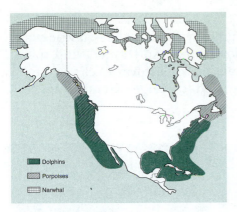

Figure 8-25: Distribution map of dolphins, porpoises, and narwhals.

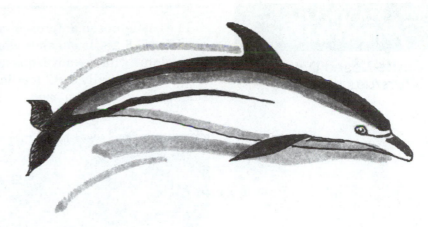

Figure 8-26: A common dolphin.

lined with sharp teeth distinguish them from porpoises. They are adept at catching the fish that make up their diets.

These dolphins are intelligent animals that are often seen swimming with boats and ships at sea. They often live together in large herds. They are very streamlined in their appearance, and the colors that distinguish this species are black backs with brown or yellow sides. A single offspring is born to mature females every 2–3 years.

Dolphins and other whales are born tail first to keep them from filling their lungs with water and drowning, see Figure 8-27. They are fully developed at birth, and they live on a milk diet during their first several months. Mother whales are sometimes seen lying on their sides in the water, making it possible for their young to breathe while they nurse.

Figure 8-27: Whales and dolphins are born tail first to keep them from filling their lungs with water and drowning. They are fully developed at birth.

Bottle-Nosed Dolphin
(Tursiops truncatus)

The bottle-nosed dolphin is commonly seen on the Atlantic coast, see Figure 8-28. It is easily domesticated and trained, and this is the dolphin that is most often taught to perform tricks in public shows. It is usually 9–12 feet long, and its range extends from New England to South America.

Figure 8-28: A bottle-nosed dolphin. *Photo courtesy of Kent A. Vliet.*

These dolphins mate in the spring or summer, and they give birth the following spring. The young dolphins nurse for about sixteen months before they are weaned to diets of fish. They are known to rescue other dolphins that are sick by lifting them to the surface to breath.

Harbor Porpoise
(Phocaena phocaena)

The harbor porpoise lives along the Atlantic, Pacific, Bering and Arctic coasts of North America, see Figure 8-29. It has a medium-sized dorsal fin, a black body, and a light-colored belly. Its mouth is

Figure 8-29: A common porpoise.

shaped differently from that of a dolphin in that it does not have a beak-like appearance. Porpoises give birth in late spring or early summer. By fifteen months of age the porpoise is sexually mature.

This porpoise eats fish of several types, and is preyed upon by killer whales and sharks.

The Sea Otter is a mammal that lives in a marine environment, but it is included as part of Chapter 7 along with the land otters. It is not included as part of this chapter.

LOOKING BACK

Marine mammals are adapted to living in a water environment. They have flippers instead of feet, and they have fat layers in their bodies that insulate against the cold and improve their buoyancy. Fourteen species of seals (pinnipeds) live in the coastal waters of North America. They are classed as eared seals, true seals, or walruses, and they eat fish, and shellfish such as mollusks and crustaceans.

Other marine mammals are the whales and manatees. Whales are the largest mammals on earth. They are of two types known as baleen whales and toothed whales. Baleen whales strain small organisms out of ocean water using the whalebone that lines their jaws. Toothed whales have larger prey, including fish and other marine animals. Dolphins and porpoises are toothed whales. Manatees are primary consumers in the food chain, but all of the other marine mammals discussed in this chapter are secondary consumers, see Figure 8-30.

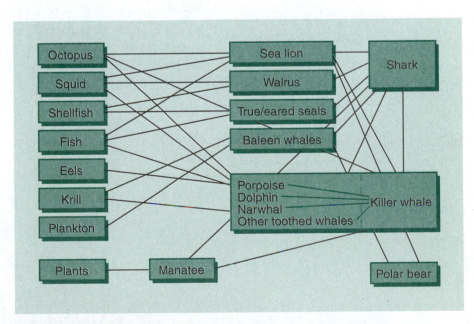

Figure 8-30: The place of marine mammals in the food chain.

REVIEW QUESTIONS

1. Describe how the bodies of marine mammals that are adapted to living in the water are different from the bodies of land-based mammals.
2. Make a chart listing the three different kinds of seals and the distinguishing characteristics of each.
3. Identify the major differences between eared seals, true seals, and walruses.
4. List the species of pinnipeds that belong to each of the three seal groups.
5. Discuss the roles of seals in the marine ecosystem in which they live.
6. Analyze the role that the manatee plays in the North American marine ecosystem.
7. Describe the characteristics of whales that identify them as mammals.
8. List the characteristics that distinguish toothed whales from baleen whales.
9. List the species of whales discussed in the chapter that are found in North America, and describe the characteristics of each that distinguish them from other whales.

LEARNING ACTIVITIES

1. Obtain a video from an educational supplier that describes the habitats and behaviors of one or more species of marine mammals. Watch it in class and identify the critical survival needs, such as habitat, diet, and reproduction, of each animal that is featured.
2. Obtain photographs of the different marine mammals found in North America, and have the class prepare a marine mammal photo display. Assign different class members to prepare a short summary of the behaviors and needs of each animal. Invite elementary school classes to hear the presentation and to view the display.

9

Unusual Mammals

KEY TERMS

carapace
insectivore
marsupial
marsupium
metabolism
ovum
placenta
placental mammals
prehensile
prolific
uterus

SOME OF THE MAMMALS that are found in North America are different in unique ways from all of the other native mammals. Some are highly specialized for living under unusual circumstances. The animals discussed in this chapter are unrelated to any of the other mammals that we have studied in earlier chapters.

OBJECTIVES

After completing this chapter, you should be able to

■ describe the reproductive characteristics that make marsupials different from other mammals
■ explain how placental mammals nourish their developing babies during gestation
■ analyze the roles of opossums in the ecosystems of North America
■ speculate on the possible relationships between the rates of metabolism in moles and shrews and their short life spans
■ identify the roles of moles and shrews in the ecosystems of North America
■ predict the life span of a mammal based on its rate of reproduction
■ explain how bats are able to navigate safely as they fly in darkness
■ identify how bats are able to match the birthdates of their offspring to favorable survival conditions
■ evaluate the roles of bats in North American ecosystems
■ appraise the roles of armadillos in the ecosystems in which they live.

OPOSSUMS

The opossums found in North America belong to an unusual class of pouched animals known as **marsupials,** see Figure 9-1. Most of

163

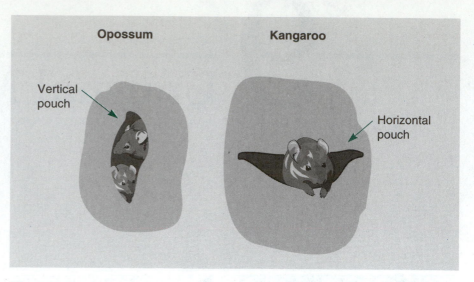

Figure 9-1: Both opossum and kangaroos are marsupials. Their young are born not fully developed and must continue to develop in the safety and warmth of their mothers marsupium.

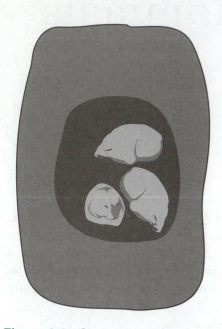

Figure 9-2: Since a marsupial female lacks a placenta to nourish her young, they must obtain nourishment by nursing on teats located in the mothers pouch.

the surviving animals in this group are located in Australia. They include the kangaroo, wombat and bandicoot.

Marsupials give birth to their young before they are fully developed. The tiny newborn must crawl into the external pouch or **marsupium** of its mother to continue its development. The teats of the mother are located in the marsupium, and the partially-developed young nurses to obtain nourishment, see Figure 9-2. It rides along with its mother in her pouch, but it also uses the pouch for warmth and protection. As a marsupial matures, it is able to leave the pouch and obtain nourishment from other sources.

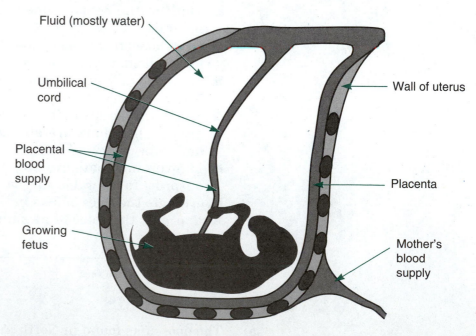

Figure 9-3: The placenta nourishes the unborn offspring of placental mammals.

A marsupial female lacks a **placenta** with which to nourish its unborn offspring. This organ consists of blood vessels and protective tissues that attach to the inner surface of the female organ called the **uterus,** see Figure 9-3. Nutrients that are needed by an unborn animal flow across the membranes that separate the blood supply of the mother from the blood supply of her offspring. Once the nutrients reach the blood of the fetus, they are circulated through its body and used as food. Waste materials flow across the membranes to the blood supply of the mother. Her body eliminates these waste materials along with those produced by her own metabolism. Mammals that nourish their unborn young in this manner are called **placental mammals.**

MAMMAL PROFILE

**Opossum
(Didelphis marsupialis)**

Figure 9-5: Distribution map of the opossum.

Opossums (commonly called possums) are the only marsupials in North America, see Figure 9-4. Females give birth after a gestation period of only eleven to thirteen days; the young are barely the size of beans when they are born. The number of young in a litter sometimes exceeds twenty. They compete with one another for a teat to nurse on, and since females only have 11–13 nipples, the weakest ones die. The mother's warm pouch serves as an incubator until the young are mature enough to survive without its protection.

The mother carries her offspring about on her back from the time they leave the pouch until they are strong enough to follow her. Usually, only one litter per year is produced, although more than one litter may be born in the same year in warm climates. Opossums range south to Mexico from Southern Canada, the eastern half of the United States, and the Pacific coastal states, see Figure 9-5.

Mature opossums look somewhat like rats but they are larger, measuring between 12 and 20 inches in length, and weighing from 4 to 12 pounds. They have long, bare, **prehensile** tails, adapted

Figure 9-4: The opossum. *Photo courtesy W.L. French.*

for grasping, that they use like an extra hand to hang upside down from branches.

They are nocturnal animals that move about at night. They are omnivorous, and they eat insects, frogs, snakes, snails, small birds, voles, seeds, berries, grass and other green vegetation. They prefer wooded habitats near water, but they can survive in a wide range of habitats. They are preyed upon by bobcats, foxes, coyotes, birds of prey and other predators. They are also hunted and eaten by humans. They are known to play dead when they are in danger.

SHREWS AND MOLES

Shrews and moles are small predatory mammals that prey on worms, insects, grubs, crustaceans, and small rodents. They are classed as **insectivores** by scientists because they are animals that feed heavily on insects. Shrews and moles have large appetites, and they spend nearly all their time eating or looking for food. They seldom sleep.

Metabolism is the process by which food is digested and used by the cells of the body to release energy. Moles and shrews expend a tremendous amount of energy. Their heart rates are very high compared with those of other mammals, and they are extremely active. They require large amounts of food to sustain their energy levels. A mole will eat its weight in food every day, while a shrew sometimes consumes its weight in food in three hours.

The rapid rates of metabolism in moles and shrews are thought to explain why these animals have very short life spans in comparison with other mammals. Moles that are not eaten by predators will probably die of old age before they reach three years. Shrews live half that long.

The Townsend's mole has a long nose and a strong, muscular body with hand-like feet and claws for digging, see Figure 9-6. It builds tunnels constantly as it searches for worms, insects, and grubs. Female moles give birth to three or four offspring in March, and by June they have matured enough to live on their own.

This mole is the largest North American mole, measuring up to 9 inches long. Most of its life is spent underground, but it does come out on the surface occasionally. Its range is generally limited to the

Figure 9-6: A Townsend mole.

coastal areas of the Pacific Northeast. Its preferred habitat is moist areas in meadows, flood plains, fields and coniferous forests. Other relatives include the eastern mole and the star-nosed mole.

Shrews are not as well equipped for digging as are moles, but they still spend a considerable amount of time underground. Shrews have small beady eyes that are not covered with skin as the eyes of some moles are. Shrews have small external ears that are concealed by fur, and moles have no external ears. The front feet of shrews are equipped with much smaller digging claws than those of moles. They are ferocious predators that can kill mice and other animals that are larger than themselves in size. They will also kill and eat other shrews. Some shrews, such as the short-tailed shrew, are equipped with poison glands in their lower jaws with which they paralyze mice and other prey.

Shrews and moles live alone most of the time, and they are widely distributed in North American ecosystems. They can be found from Alaska to Mexico.

MAMMAL PROFILE

Short-Tailed Shrew (Blarina brevicauda)

Shorttail Shrew
Townsend Mole

Figure 9-8: Distribution map of the short-tailed shrew and Townsend mole.

The short-tailed shrew is gray with a pointed nose and a short tail, see Figure 9-7. It may be the most deadly predator of its size in the world. It is a vigorous hunter that is capable of poisoning its victims with its bite. It eats several times its weight in food every day. Its diet includes insects, snails, worms and small mice. This shrew is found in the eastern United States and southeastern Canada. Its habitat includes grasslands, forests, brushy areas and marshes. Females give birth to 2–3 litters each year and each litter consists of 5–8 offspring, see Figure 9-8.

Figure 9-7: A short-tailed shrew. *Photo courtesy of Leonard Lee Rue, III.*

MAMMAL PROFILE

Masked Shrew (Sorex cinereus)

The common shrew is a small, gray brown mammal with a pointed nose that is found in underground tunnels or hunting for food under the cover of vegetation, see Figure 9-9. Its most critical need is food. A shrew may starve to death in a single day if it doesn't have adequate food, see Figure 9-10.

Female shrews give birth to litters of up to ten young, and a female often becomes pregnant again while she is still nursing a

Figure 9-9: A masked shrew.

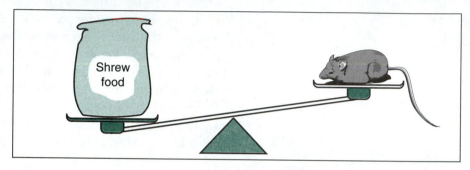

Figure 9-10: A shrew must eat several times its weight in food per day in order to survive.

litter. Because they have such large numbers of offspring, these shrews are described as being very **prolific.**

This shrew has adapted to a wide range of climates and conditions, and it can be found from Alaska to the southern United States. If prefers damp areas in brushlands, forests, open grasslands and fields.

Shrews and moles are active night and day, and this makes them vulnerable to predatory birds and animals when they venture above the ground. In spite of their short life spans they are still plentiful in many regions.

Career Option

WILDLIFE TECHNICIAN

A career as a wildlife technician will involve in-depth study and education to develop specialized skills that deal with wildlife problems. Most careers in this field require a bachelor of science degree. Technicians are usually skilled in gathering appropriate testing materials such as blood or tissue samples, and in conducting laboratory tests. They may specialize in the causes and treatments of diseases, or finding ways to overcome reproductive problems in certain animals. Whatever the problem might be, technicians try to find ways to identify the causes and to discover solutions.

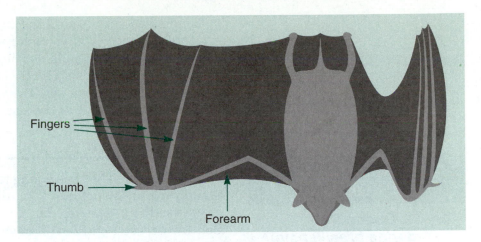

Figure 9-11: The wings of bats are made up of skin stretched over the arms and fingers of the world's only flying mammal.

BATS

Bats are the only known mammals that have the ability to fly under their own power. The wings of a bat consist of skin stretched over the bones of their arms and fingers like the webbed foot of a duck, see Figure 9-11. They are furry mammals that bear live offspring and nurse them when they are young.

Bats are warm-blooded animals, but during winter hibernation in a cold cave their body temperature may fall nearly as low as the surrounding air. Their body processes all slow down during hibernation helping them to conserve energy. If the cave temperature is too high, the bat will starve to death during hibernation because its body uses too much energy at the higher temperature. If the temperature drops below freezing, hibernating bats must find a warmer area or die.

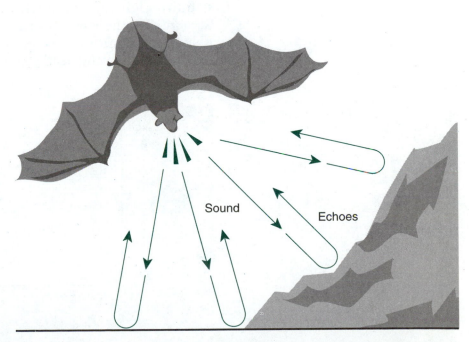

Figure 9-12: Bats are able to navigate by listening to the echoes from their high-pitched cries as they fly.

A high frequency sound is emitted by a bat during flight to help it locate objects and navigate around them. They are guided by the echoes from their signals as they fly, see Figure 9-12, and in this way are able to precisely locate obstacles and the insects that they usually feed on. The effect of the echoes from their sound waves is similar in many ways to the use of radar to navigate aircraft.

MAMMAL PROFILE

Guano Bat
(Tadarida brasiliensis)

Figure 9-13: Distribution map of the big brown bat.

The free-tailed bat has a long tail, which extends beyond the skin membrane that encloses the tails of most species of bats. It is a cave-dwelling species that is abundant in the South and Southwest, see Figure 9-13. Several million of these bats leave and return to the Carlsbad Caverns in New Mexico after dusk every day. They hang upside down during the day from the rock surfaces of the cave roof. They emerge from the cave to search for the insects that make up their primary source of food. A single offspring is born in June or July, and the mother leaves it with other young bats in the roosting area while she gathers food.

Based on tag and recapture studies bats are known to live twenty years or longer. Only a few predators such as owls, hawks, snakes, and cats are successful in catching bats. Their survival rate is quite high compared with most other small mammals. This accounts for the low birth rates in bats, see Figure 9-14. They are able to maintain their population numbers with the birth of only one or two offspring each year.

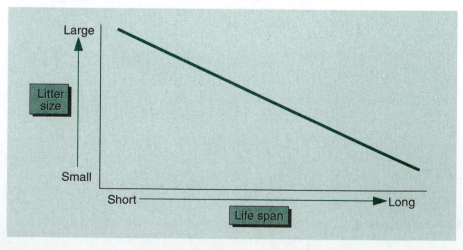

Figure 9-14: The relationship between the life span and litter size of bats.

Although many bats mate during the fall before they hibernate, the male sperm is stored in the body of the female until spring when the **ovum** or egg is fertilized. Young bats sometimes cling to the nipples of their mothers as they fly to new roosting areas.

The little brown bat is widely distributed in North America, see Figure 9-15. It ranges from Alaska and Labrador south to California and Mexico, except for the southeast and the southern

Figure 9-15: A little brown bat. *Courtesy of U.S. Fish and Wildlife. Photo by W. D. Fitzwater.*

states. It hibernates in the winter in nearly any cave that has the right temperature.

This bat eats primarily soft insects, such as flies and moths. Female bats leave their caves to give birth in buildings and near water. Mature bats of this species have wing spans of approximately 10 inches, and weigh up to a third of an ounce.

Even though several species of bats are widely distributed in North America, they generally do not live in the polar regions or in high mountain areas. In addition to the bats that are discussed in this chapter, the red bat and the big brown bat are quite widespread and well known. Nearly all of the bats in this part of the world eat insects. Some bats located in other parts of the world prefer fruit, pollen, nectar, fish, or even blood.

ARMADILLO

The armadillo is found only in North and South America. It is related to the anteaters and sloths that live in Central and South America. Except for its belly, it is covered by a **carapace** or coat of bony armor.

Armadillos are nocturnal animals. They are also omnivorous, eating grubs, ants, roaches, tarantulas, grasshoppers, and other insects along with some plant materials. They live in burrows in the ground. Coyotes and domestic dogs prey upon them by turning them over and attacking their soft underparts. They are also killed by hunters and cars.

MAMMAL PROFILE

Nine-Banded Armadillo (*Dasypus novemcinctus*)

The nine-banded armadillo is the only animal of its kind found in North America, see Figure 9-16. Its range is restricted to the South central and Southeastern regions of the United States extending south to Mexico. Its preferred habitat is open fields where it can dig. Its nine bands are telescoping joints that help the animal move about quite freely. Like its southern hemisphere relatives, it is well named. The Spanish meaning of armadillo is "little armored thing."

Figure 9-16: The nine-banded armadillo. *Photo courtesy of Leonard Lee Rue, III.*

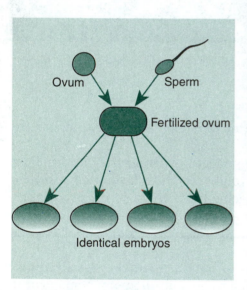

Figure 9-17: Armadillo quadruplets are quite common. The ovum splits into four embryos after being fertilized.

Quadruplets are common in armadillo litters. This is because the female usually produces a single egg or ovum that splits into four embryos after it is fertilized, see Figure 9-17. Each of the offspring in a litter is identical to each of its littermates because all of them were produced from a single fertilized egg cell.

Young armadillos live with their mothers until after they are weaned at about two months of age. She teaches them to root in the ground to find grubs and worms. Mature armadillos of this species may weigh a little over a pound.

LOOKING BACK

Some North American mammals have traits that are different from those of all the other native species. Opossums are marsupial animals. They give birth prematurely, and the offspring continue their development in the mother's marsupium or pouch where they nurse to obtain nourishment. Most other mammals nourish their unborn offspring during this period of development through an internal organ called the placenta.

Moles and shrews are small predatory mammals whose metabolic rates are extremely high. They require large amounts of food to maintain their active lifestyles. They consume their own body weight in food one or more times every day.

Bats are the only mammals that are able to fly under their own power. They have long life spans and low birth rates. Females store the sperm from fall mating until spring, when they conceive and give birth to live offspring.

The armadillo is the only mammal of its kind that is found in North America. It is covered by a carapace or coat of bony armor that surrounds its entire body except for the belly. It usually gives birth to identical quadruplets that originate from the same ovum.

REVIEW QUESTIONS

1. Describe how reproduction is different in marsupial mammals than it is in placental mammals.
2. Explain how the placenta exchanges nutrients and waste products between the circulatory systems of the mother and her offspring.
3. Identify ways that opossums are unique among the mammals of North America.
4. Explain how the metabolic rates of moles and shrews might be related to the short life spans of these mammals.
5. Identify the roles of moles and shrews in the ecosystems of North America.
6. Describe how an animal's life span might be predicted based on the reproductive rate of the species.
7. Explain the method that bats use to navigate safely as they fly in darkness.
8. Identify how the birthdates of young bats are timed to assure that they are born when conditions are favorable to their survival.
9. Discuss the roles of bats in the environments where they live.
10. Describe the characteristics of armadillos that make them different from other mammals.

1. Assign students to prepare group reports on the mammals discussed in this chapter. Have them prepare visuals and charts for oral presentation in a panel discussion format. Allow time for the groups to answer questions from the class.
2. Obtain commercially prepared specimens of one or more of the animals that are discussed in this chapter, and examine its unique physical characteristics in the laboratory. Invite a fish and game expert to talk to the class about the specimens that have been obtained.

SECTION  III

Ecology of Birds

Birds are members of a class (Aves) of warm-blooded feathered vertebrates, most of which are capable of flight. They reproduce by laying and incubating eggs. They care for their young by providing food and warmth. They are widely distributed in North American ecosystems.

10 Waterfowl

SOME BIRDS depend upon a water habitat for protection and for food. These include both swimming and wading birds. This chapter focuses on the swimming birds that are classified by scientists as **waterfowl,** including ducks, geese, and swans.

OBJECTIVES

After completing this chapter, you should be able to

- identify birds classified as waterfowl
- define the term oviparous
- clarify how hunting regulations contribute to the protection of waterfowl during critical periods in their life cycles
- describe how the restoration of wetlands contributes to maintaining and increasing populations of migratory waterfowl
- distinguish differences between ducks classified as divers and as surface feeders
- explain how ducks are different from swans and geese in their devotion to their mates
- evaluate the roles of ducks in North American ecosystems
- identify major species of North American ducks, geese and swans
- suggest ways that the migratory instincts of waterfowl contribute to their survival
- evaluate the roles of geese and swans in North American ecosystems
- identify characteristics of swans that make them more susceptible than geese to declines in their populations.

Birds play important roles in all of the biomes of North America. They belong to a zoological class called **Aves,** and for this reason birds are sometimes referred to as **avians.** Some birds are predators while others eat fleshy fruits, seeds, or other plant parts. All birds are **oviparous,** meaning that they produce eggs that hatch

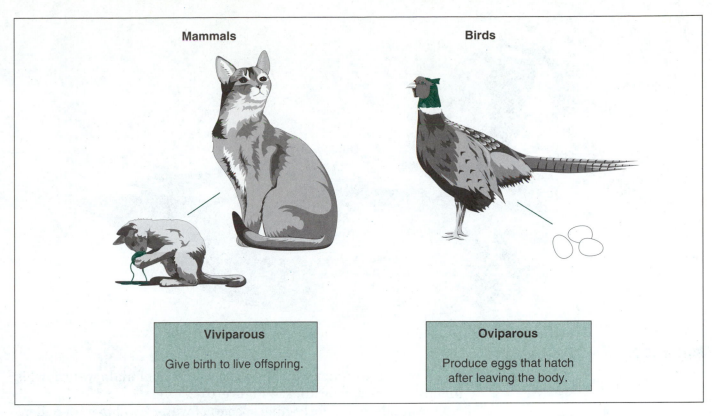

Figure 10-1: Reproduction in animals and birds.

after they leave the body of the female. The eggs hatch only after a period of **incubation** during which time they are warmed by the body heat of a parent (usually the female). In contrast animals that give birth to live young are **viviparous,** see Figure 10-1.

Waterfowl are adapted to living in water. They produce oil that is used to waterproof their outer feathers. They also have an insulating layer of **down** or soft, fluffy feathers beneath this outer coat. The sum total of all of the feathers of a bird make up its **plumage.**

Career Option

ORNITHOLOGIST

An **ornithologist** is a scientist who specializes in the branch of zoology that deals with birds, called **ornithology.** A career in this field requires a graduate degree in zoology, biology, or a related science, and a specialty in ornithology.

The work of an ornithologist will require gathering research data through fieldwork, and using the data to learn more about the relationships of birds to the environments in which they live. Ornithologists also use their knowledge of bird migration and behavior to assist in the evaluation of hunting regulations and the restoration of endangered species.

Figure 10-2: Harlequin duck.

Of all the North American birds, the plumage of male waterfowl is among the most colorful, see Figure 10-2.

One hundred and forty-nine species of waterfowl have been identified in the world. Ten of these species are restricted to North America. Fifty more species found in North America are also found on two or more continents.

Figure 10-3: Nearly all waterfowl are migratory and migrate biannually.

Hunting regulations protect waterfowl during the critical stages of their life cycles when they are nesting or raising their young. Since nearly all waterfowl are migratory, another crucial period is during their biannual migrations every spring and fall, see Figure 10-3. Organized groups such as The Nature Conservancy, Izzak Walton League, U.S. Fish and Wildlife Service and State Fish and Game Departments contribute to the improvement of water habitats for waterfowl by raising funds that are used to restore wetlands along active migration routes, see Figure 10-4. The prairie pothole region includes some of the most important wetland habitat in North America. This is because it supports migrating waterfowl and provides ideal conditions for duck reproduction.

Figure 10-4: Wetlands are being improved and maintained along active migration routes.

DUCKS

The best-known waterfowl are probably the ducks. These birds have thick, waterproof plumage. The males of most species are brightly colored and the females have plain or dull coloring. The beak of a duck is flat with comblike **lamellae** with which it can strain small food particles from the water. They have webbed feet and short legs that are placed far back on the body for swimming. On land a duck walks with a waddle due to the placement of its legs.

Feed sources for ducks

Divers:	Surface Feeders:
Mollusks	Insects
Shellfish	Grass
Aquatic plants	Seeds
Worms	Aquatic plants

Figure 10-5: Feed sources for ducks.

AVIAN PROFILE

Common Merganser
(Mergus merganser)

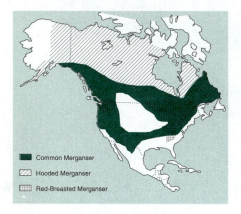

- ■ Common Merganser
- ▨ Hooded Merganser
- ▦ Red-Breasted Merganser

Figure 10-7: Distribution map of the common, hooded, and red-breasted merganser.

Many species of ducks migrate south from their breeding grounds to warmer climates for the winter season. This is necessary because these birds must be able to find open water. Since ducks obtain nearly all their food from the water, they would starve if they remained in the northern regions of Canada and Alaska after the water freezes over.

Ducks are divided into two types known as divers and surface feeders. Divers feed on mollusks, shellfish, and other aquatic life found on the seabed and the bottoms of lakes and ponds, see Figure 10-5. Surface feeders eat insects, grass, seeds, weeds and aquatic plants in addition to small animals.

The common merganser is a diving duck with a spikelike bill that feeds on small fish and aquatic insects in freshwater streams, rivers, and lakes, see Figure 10-6. It uses its wings as well as its feet to swim underwater when it dives for food. This duck is found in northern regions of North America during the breeding season, but it migrates southward to open waters for the winter, see Figure 10-7.

Figure 10-6: Common merganser.

Female mergansers lay 6–12 eggs in hollow trees or in protected areas on the ground. Its preferred habitat is cold mountain rivers and lakes, but it may rear its young in coastal inlets and bays when nesting near the ocean. Other relatives include the hooded and red-breasted mergansers. These ducks are also known as sawbills.

Unlike geese and swans that usually mate for life, most ducks remain in pairs only during the breeding season. The pair bond is usually broken when the female begins to incubate her eggs.

Nesting ducks usually seek protected areas for their nests. This might be under a bush or in an area protected by thick vegetation located near water. Dry vegetation is used to form the nest, and the female lines it with downy feathers plucked from her own breast.

AVIAN PROFILE

Mallard
(Anas platyrhynchos)

The mallard is probably the most important duck to humans. It is a colorful duck distinguished by a head and upper neck area that is green in males, and by blue speculum (wing stripe) in both sexes, see Figure 10-8. Wild mallards are found in most areas north of the

Figure 10-8: The mallard.

Figure 10-9: Distribution map of the mallard.

equator, see Figure 10-9, and domestic mallards have been raised for eggs, meat, and feathers for thousands of years. Nesting females lay 6–10 eggs early in the season in nests built of grass, leaves, and other vegetation and then lined with down.

Figure 10-10: A mother duck keeps watch while her ducklings rest. *Photo courtesy of Wendy Troeger.*

Young ducks are called **ducklings.** They are self-feeders, meaning that they gather their own food. They are active and able to follow their mothers within a few hours after they hatch, see Figure 10-10. They are also natural swimmers and have no fear of water.

Some ducks care for their young for long periods of time; others abandon them almost as soon as they hatch. Most female ducks remain with their young only until they can fly. The mother duck keeps them warm beneath her wings during cold weather. Ducklings are sometimes observed riding on the back of their mother as she swims.

Some female ducks will attempt to lure enemies away from their young by pretending to be injured. They will drag a wing on the ground and stay just out of reach of a predator as they lead it away from their young. Once they think their ducklings are safe, they fly away and circle back to gather up their families.

AVIAN PROFILE

**Northern Shoveler
(Spatula clypeata)**

Figure 10-12: Distribution map of the Northern shoveller.

The northern shoveler is a large marsh duck with a green head like a mallard and a white lower neck and breast, see Figure 10-11. It has a large spoon-shaped bill with which it strains food from the water and mud in shallow marshes. Its main diet consists of worms, leeches, snails, insects, tadpoles, and aquatic plants.

Figure 10-11: The Northern shoveller.

The preferred habitat for this duck is shallow freshwater marshes and ponds. Females lay 8–12 eggs in ground nests beneath bushes or weeds. Nests are located near water. These ducks generally range across the western half of the continent from Alaska to Mexico, see Figure 10-12. During the winter season they inhabit tidal bays in coastal regions.

Sea ducks of several species inhabit both coasts of North America. They are larger than most other kinds of ducks, and they live on diets of mollusks, shellfish, and aquatic plants that they obtain by diving to great depths. These ducks usually feed during the day and rest offshore at night.

AVIAN PROFILE

**Common Eider
(Somateria mollissima)**

The common eider is a large sea duck with a short neck and white back, neck, and breast. The lower body is black, and the head is topped with a black cap and light green nape, see Figure 10-13. This duck ranges along the arctic and Canadian coasts, see Figure 10-14.

Females usually nest in colonies on small rocky islands. Females lay 4–7 eggs in down-filled nests that are located on the ground in small depressions lined with down. In some parts of the world farmers entice eiders to nest by furnishing nesting sites from which

Figure 10-14: Distribution map of the common eider.

Figure 10-13: A common eider hen. *Photo courtesy U.S. Fish and Wildlife Service.*

they harvest the down for use in insulated clothing and sleeping bags.

Ducks that feed on the surface or just beneath the surface by up-ending with their heads submerged are called **dabbling ducks.** These ducks are usually small in size, which contributes to their ability to maneuver in flight. They migrate long distances each year. Examples of these ducks include mallards, shovelers, pintails, widgeons, and teal.

The blue-winged teal is a small marsh duck with a bluish gray head, a brown upper body, and a buff-colored lower body spotted with black, see Figure 10-15. It ranges throughout much of southern Canada and into the northern two-thirds of the United States. During the winter season it migrates to southern coastal regions and to northern Mexico, see Figure 10-16.

Figure 10-16: Distribution map of the blue-wing teal.

Figure 10-15: A blue-wing teal.

This duck prefers a diet of seeds, grasses, and aquatic weeds, but it will eat insects, snails, and tadpoles. Its habitat of choice is a small marshy pond, meadow, or boggy area. Females lay 10–12 eggs in a ground nest near water.

Ducks are most likely to become prey for predatory animals during the nesting and brooding period and again during the **molt** when they lose their old feathers and replace them with new ones. Sometimes a duck loses its ability to fly during this period, and it is caught more easily by predators such as foxes, coyotes, mink, skunks, snapping turtles, and birds of prey. Ducklings are also eaten by large fish. Thus, ducks occupy a niche in the food chain in which they feed on aquatic plant and animal life before they eventually become prey for other animals.

AVIAN PROFILE

Wood Duck (Aix sponsa)

The male wood duck is a colorful woodland bird with a green and purple head with white stripes, a white throat, and a chestnut-colored breast, see Figure 10-17. It prefers habitat close to woodland lakes, rivers, and streams. This duck is widely distributed east of the Rocky Mountains from the Gulf Coast to southern Canada, and along the Pacific coast, see Figure 10-18.

Figure 10-18: Distribution map of the wood duck.

Figure 10-17: A female (left) and male (right) wood duck keep a watchful eye for food. *Courtesy U.S. Fish and Wildlife Service. Photo by Dave Menhe.*

This duck is unusual in that it sometimes perches in trees, and it builds its nest in tree cavities. Females lay 10–15 eggs in down-lined nests from which the ducklings jump soon after they are hatched. The diet of wood ducks includes aquatic plants, grasses, aquatic animals, insects, seeds and nuts.

Only a few of the many kinds of ducks have been profiled in this chapter. Although some species of ducks share habitats and eat similar foods during much of the year, each species is unique.

GEESE

Geese are close relatives of ducks and swans, but they are usually larger than ducks and have shorter necks than swans, see Figure 10-19. Territories of ducks and geese often overlap and they exhibit many similar behaviors. Among the greatest differences is the tendency of most geese to mate for life. Several species of geese are found in the northern hemisphere, and they are migratory birds that sometimes travel great distances. Most geese have northern breeding areas where they hatch and raise their young before migrating to wintering areas farther south.

Swan   Goose   Duck

Figure 10-19: Relative size and shape of waterfowl.

AVIAN PROFILE

**Snow Goose
(Chen caerulescens)**

The snow goose is pure white except for its black wing tips, see Figure 10-20. It breeds in the far northern arctic region of North America, and winters in harvested grain fields and marshes in California, the Gulf Coast, New England coast and isolated interior regions of the United States, Figure 10-21.

These geese prefer habitat in tundra and marshland areas during the breeding season. Females lay 4–8 eggs in a down-lined

Figure 10-21: Distribution map of the snow goose.

Figure 10-20: Snow geese. *Courtesy of the Alaska Division of Tourism. Photo by Robert Angell.*

nest in the tundra. After the **goslings** (young geese) hatch, the rest of the summer is spent grazing on plant shoots, other vegetation, and seeds.

Geese are vegetarians, and they gather food both on land and in water. They often gather kernels of grain from stubble fields during the fall and winter months, see Figure 10-22. They are fond of grain sprouts, and can cause serious damage to fields in which grain is sprouting and beginning to grow by plucking up entire plants as they graze.

Figure 10-22: Canada geese resting and feeding in a stubble field. *Photo courtesy Michael Dzamen.*

AVIAN PROFILE

Canada Goose
(*Branta canadensis*)

Figure 10-24: Distribution map of the Canada goose.

Figure 10-25: Nesting platforms are provided for geese in some areas to protect the nests from flooding and predators.

The Canada goose is a large bird with a gray body, black rump, tail, neck, and head, and white cheek and throat markings, see Figure 10-23. This species is probably the most wide-ranging and best-known goose in North America. It ranges during the breeding season across the northern tier of states and southern Canada as well as into some parts of northern Canada and Alaska. Its winter range includes parts of Mexico and much of the United States except the most northern states, see Figure 10-24. Some of these geese no longer migrate north, but breed locally along the Atlantic seacoast and other waterways. They are often considered to be a nuisance or health hazard when they become accustomed to living near people. The Canada goose prefers habitat where bogs, sloughs, marshes, and lakes are found. Several subspecies of Canada geese have been identified; the largest of these are often called honkers.

Figure 10-23: Canada geese land at the Aleutian Islands to feed. *Courtesy of the Alaska Division of Tourism. Photo by Robert Angell.*

Females often nest on piles of vegetation where 6–7 eggs are deposited in down-filled nests. Nesting platforms mounted on posts are provided for geese in some areas to protect the nests from predators and flooding, see Figure 10-25. Both parents take responsibility for raising the family after the goslings are hatched.

Geese go through mating ritual in the spring before they begin nesting. They become quite noisy and males become aggressive as they attempt to drive off rivals. Nesting often begins while the spring season is still wet and cold. By the time the weather warms up, the goslings have hatched and are feeding heavily. They grow rapidly and attain most of their growth in a few weeks.

Both parents assist in rearing their offspring, and they usually molt while they are raising their broods. During this period they

Ecology Profile

LEAD POISONING OF WATERFOWL

The lead shot used by hunters is sometimes found and eaten by waterfowl. All birds eat small pebbles that are used to grind seeds during the process of digestion. These pieces of rocks and gravel are called **grit**. They are swallowed whole and remain in the **gizzard** of the bird that consumed them. The gizzard is a muscular organ that grinds seeds and other food into small particles by crushing them between the hard pebbles.

Lead shot breaks down in the gizzards of birds and lead compounds in the body fluids eventually poison the bird. Death is slow and birds are unable to feed. In many instances, hunters have replaced lead shot with steel shot, which is less toxic. However, in areas where much hunting has taken place over the years, the lead shot will remain a hazard for sometime to come.

are unable to fly, and they spend their time grazing with their young in meadows and marshy areas near their nesting sites.

In addition to the geese that are discussed in this chapter, several other species are native to North America.

SWANS

Swans are large aquatic birds with graceful necks that are longer than their bodies. They are white with some black markings, and they are the largest waterfowl. Swans mate for life, and they usually nest at high altitudes or in the northern arctic regions. Their diet consists mostly of seeds and other plant materials.

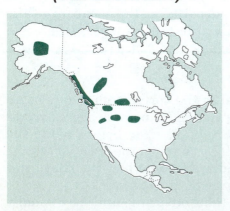

Figure 10-27: Distribution map of the Trumpeter Swan.

The trumpeter swan is a large migratory bird with a wingspan that may exceed 6.5 feet. It is white with a black bill, see Figure 10-26. The breeding range of this swan includes isolated areas in

Figure 10-26: Trumpeter swans with their young. *Photo courtesy of U.S. Fish and Wildlife Service.*

Canada and the states of Alaska, Idaho, Wyoming, Montana, and Oregon, see Figure 10-27. It spends the winter in many of these same areas where open water can be found.

Members of this species do not breed until they are four years of age or older, and they produce only 4–6 eggs. With such a low reproductive rate, their numbers do not increase very quickly even when conditions are favorable. The preferred habitat of this swan is marshlands, rivers, and lakes.

Swans build huge nests of reeds, twigs, and other vegetation, and they return to the same nest each year. Males become very aggressive in defending their breeding territory from intruders. Young swans are called **cygnets.** After the eggs have hatched, both parents help rear their young. Cygnets remain with their parents for four to five months.

AVIAN PROFILE

Whistling Swan
(Olor columbianus)

Figure 10-29: Distribution map of the whistling swan.

The whistling swan is a large white bird with a black bill. It closely resembles other swans except for a small yellow marking near each eye, see Figure 10-28. It calls with a high-pitched whistling sound as it migrates.

Figure 10-28: The whistling swan. *Photo courtesy of U.S. Fish and Wildlife Service.*

This swan nests on small islands in lakes and marshes of the northern tundra region, see Figure 10-29. Females lay 4–5 large eggs, and the young cygnets remain with their parents during the fall migration. Whistling swans winter along both the Atlantic and Pacific coasts in freshwater marshes and marine estuaries.

Swans and geese play similar roles to ducks in the habitats that they occupy. Because of their longer necks, however, they can reach into deeper water then the dabbling ducks. They are primary consumers and they are preyed upon by a number of animals that occupy both their summer and winter ranges.

LOOKING BACK

Aves is a scientific term for a class of animals that is commonly known as birds. The branch of science that studies birds is ornithology. Waterfowl are migratory birds such as ducks, geese and, swans. All birds are oviparous, meaning they produce eggs that hatch after they leave the female. Hunting regulations protect waterfowl during critical periods in their life cycles, such as when they are nesting, rearing young, and migrating.

Some species of ducks are predators while others eat only plants and seeds. Most geese and swans tend to be vegetarians. Some waterfowl obtain food by diving for aquatic animals and plants living on the bottom of streams and lakes. Others strain floating food particles out of the water with bills equipped with lamellae.

Most ducks, geese, and swans nest on the ground, but some ducks, such as the wood duck, nest in trees. Waterfowl prefer wet habitats where protective ground cover is available. They like bogs, ponds, marshes, streams, and lakes because these areas usually offer an abundance of food and protection.

Ducks, geese, and swans fill intermediate roles between aquatic plants and predators in the ecosystems and food chains of North America.

REVIEW QUESTIONS

1. Identify what waterfowl are, and name several species of birds that are classified as waterfowl.
2. Define what is meant when a bird is described as an oviparous member of the animal kingdom.
3. Clarify how hunting regulations are used as a tool to protect waterfowl during critical periods in their life cycles.
4. Describe how the restoration of wetlands contributes to maintenance or expansion of migratory waterfowl populations.
5. Distinguish some differences between ducks classified as divers and ducks classed as surface feeders.
6. Contrast swans and geese in comparison with most ducks in their devotion to their mates.
7. Describe the roles of ducks in North American ecosystems.
8. Name the major species of ducks, geese, and swans that are found in North America, and list traits that are helpful in identifying them.
9. Suggest some ways that the migratory instincts of waterfowl contribute to their survival.
10. Describe the roles of geese and swans in the ecosystems of North America.
11. Identify some characteristics of swans that make it difficult for them to maintain their populations.

LEARNING ACTIVITIES

1. Visit a marsh, bog, or lake, and do a complete assessment of the food supply that is available for waterfowl. Identify the plants that are available. Skim the surface of the water, and study the particles that you find to determine whether they might be used as a food source by ducks that are equipped with well-developed lamellae. Use screens to separate small plants and animals from the mud, sand, or gravel on the bottom and edges of the body of water. After the food supply is inventoried, identify some waterfowl that use the kinds of foods that you found. Consider whether any conditions exist that might interfere with the survival of the waterfowl species that you identified.

2. Have each member of the class choose a species of duck, goose, or swan, and prepare a short written and oral report to be shared with the rest of the students. The report should address the following: description, range, preferred habitat, diet, nesting habits, rearing of young, migration routes, natural enemies, and unusual features or habits.

11 Game Birds

KEY TERMS

beard
clutch
coveys
crop
dove
fledge
gallinaceous
gizzard
monogamous
pigeon milk
plume
polygamous
polygynous
scrape
squab
wattles

THE GAME BIRDS discussed in this chapter include birds whose habitats include the deserts, fields, mountains, and forests. These birds are often hunted for human food, and hunting regulations have been implemented to protect them during critical periods in their life cycles. Their diets consist mostly of seeds, insects, and plant materials.

OBJECTIVES

After completing this chapter, you should be able to

- list the kinds of game birds that are found in North America
- describe the nesting behaviors of gallinaceous birds
- distinguish between polygamous and monogamous birds
- define the relationship between doves and pigeons
- evaluate the roles of game birds and pigeons in the ecosystems of North America
- assess the similarities and differences among quails, partridges, and pheasants
- identify major species of North American game birds and pigeons
- speculate on the effects that a reduced population of game birds might have on populations of birds of prey that live in the same region
- consider the importance of camouflage coloration in most female game birds
- explain why most game birds must produce large clutches of eggs
- describe the importance of mating rituals among game birds.

North American game birds include quails, partridges, pheasants, grouse, and turkeys. These birds are all similar in body shape, with plump bodies and short powerful wings, see Figure 11-1. They have similar diets consisting mostly of seeds and plant shoots that are sometimes supplemented with insects.

Quail

Partridge

Pheasant

Grouse

Turkey

Figure 11-1 Relative sizes and shapes of game birds.

Most game birds are called **gallinaceous** birds meaning that they are heavily bodied and they nest on the ground. The number of eggs and young is known as a **clutch.** The clutches of gallinaceous game birds are larger than the clutches of most other birds. This is necessary to maintain game bird populations because large numbers of them are lost due to predation and weather. Some game birds are **polygamous** in their breeding habits in that both males and females may have more than one mate, and they do not bond as pairs. Others are **polygynous** meaning that the males attract and mate with several females.

The pigeon family includes eleven species found in North America. Both pigeons and doves belong to the same family, but a

dove is really just a small pigeon. Pigeons are **monogamous,** meaning that they have only one mate at a time. They usually mate for life. They are not gallinaceous birds, but prefer to nest on ledges and in trees.

QUAILS

Quails, partridge, and pheasants are all members of the same family. The terms quail and partridge are also common names for related groups of birds, and there is a tendency for birds that are called quail in one region to be referred to as partridge in another. This chapter identifies quail as the smallest of the gallinaceous game birds.

AVIAN PROFILE

**California Quail
(Lophortyx californicus)**

Figure 11-3 Distribution map of the California quail.

The California quail is a small, plump bird with a black **plume** or feather that curves forward on its head, see Figure 11-2. It has a brown back, a grayish-blue breast, and a cream-colored lower body with brown markings. It prefers habitat in brushy open areas located in foothills, canyons, deserts, and suburbs. Its range extends along the Pacific coast from Canada to Mexico, and inland to northern Nevada, see Figure 11-3.

Figure 11-2 A male California quail. *Photo courtesy Leonard Lee Rue III.*

Males defend territories during the breeding season, where they strut about and call to attract females. Twelve to fifteen eggs are incubated by the females in grass-lined depressions on the ground. Chicks are covered with a coat of down when they hatch, and by ten days of age their wing feathers have developed enough to fly. These quail roost in trees for safety, and they feed on seeds and insects. They gather in large groups called **coveys** in the late summer and fall. The California quail is the state bird of California.

Many of the gallinaceous game birds are fast runners, and they prefer running to flying. They are capable of rapid flight for short distances when they are in immediate danger. Some of these birds sometimes fly to water in the early morning and evening hours, but they usually prefer to run. Some species even travel between winter ranges in the lowlands and summer ranges at higher elevations by walking or running.

AVIAN PROFILE

Bobwhite (*Colinus virginianus*)

Figure 11-4 Distribution map of the Bobwhite quail.

The bobwhite is usually considered to be an eastern quail because it is native to the eastern half of the United States and to the Gulf Coast of Mexico, see Figure 11-4. It has also been successfully introduced into the northwestern United States. It is a small brown bird with a white throat and a white stripe across its eye, see Figure 11-5. The male has a distinctive whistle that sounds like "bob-white."

The bobwhite is attracted to farms where it eats weed seeds and insects during the growing season and gleans grain when it is available. Stubble fields make good places for bobwhites in the fall and winter because kernels of grain are scattered throughout the fields by the grain combines that are used to harvest the crop.

The male bobwhite is a family-oriented bird. Unlike most other game birds, once it attracts a mate, it helps incubate the eggs and raise the young. The clutch of this quail consists of 10–15 eggs or young birds. Family groups form **coveys** that stay together until the next breeding season. They seek cover in bushes, and during cold weather they huddle together for warmth.

Figure 11-5 A male Bobwhite quail. *Photo courtesy of Leonard Lee Rue III.*

One of the greatest threats to quail and other ground-dwelling game birds is heavy snow followed by crusting on the surface. Game birds tend to crowd together in the shelter of weeds and brush until a storm passes. When a hard crust forms on the surface of the snow, the birds may become trapped beneath it and starve to death, see Figure 11-6.

PARTRIDGES

The partridges are medium-sized game birds that have been successfully introduced to North America from Europe. They band together in coveys that are usually made up of family units. Unattached birds join the coveys, and males often leave to join other coveys as the mating season approaches. This behavior helps to reduce inbreeding.

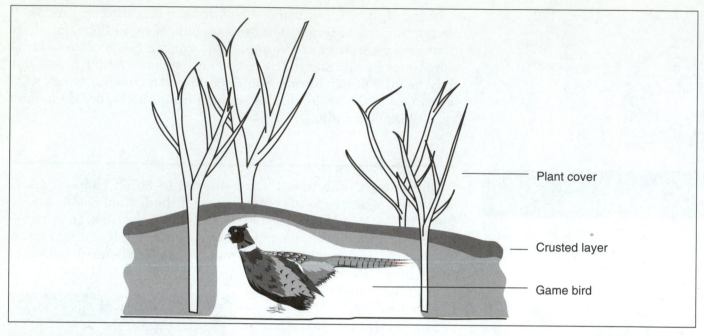

Figure 11-6 A sudden snow and crusting of the surface is a serious threat to ground-dwelling game birds who can be trapped beneath it.

Gray or Hungarian Partridge
(Perdix perdix)

■ Gray Partridge
▨ Chukar

Figure 11-8 Distribution map of the gray partridge and the chukar.

The gray (Hungarian) partridge is a brown bird with a gray neck and breast and a bar-shaped pattern on the flank region and tail feathers, see Figure 11-7. On the lower breast is a chestnut-colored marking shaped like a horseshoe. This partridge prefers a habitat of open farmland and grainfields. Their diet consists of grains, seeds, buds, and insects. Its range extends across the prairie region of southern Canada and the northern border states of the United States, see Figure 11-8.

Females lay 10–16 eggs in ground nests sheltered by brush, weeds, or crops. Both parents help to rear the chicks. They do not

Figure 11-7 A chukar partridge. *Photo courtesy of Leonard Lee Rue III.*

roost, but seek sheltered areas on the ground to rest during the night.

Partridges are considered to be a worthy challenge for hunters because they are so difficult to approach. They can usually be heard calling to one another just out of range. They are very fast in their short flights, and hunters usually find it difficult to harvest them.

Partridges are most vulnerable to predators during the nesting period and until the chicks are capable of flight. Once the young chicks learn to fly, they are able to escape most ground predators, but they are never safe from birds of prey. Hawks, falcons, and eagles are natural predators of these birds.

PHEASANTS

Pheasants are distinguished from other game birds by their long tails and bright, colorful plumage. They are not native to North America, but were widely introduced to this country from Asia. Pheasants are fast runners, and they prefer to escape from danger by running or hiding. If they are in immediate danger, however, they will burst from cover and quickly fly to safety.

Career Option

GAME BIRD FARM MANAGER

In some ways fish and game agencies and private hunting preserves supplement the natural populations of game birds with birds that have been raised on game farms. Managers of these farms are required to have administrative and managerial skills to direct the farm employees.

In most cases, a college degree related to wildlife management is necessary. A strong background in the biological sciences is the foundation of all of the careers in fish and game management. A good understanding of avian nutrition and diseases is essential to success in this career.

Ring-neck pheasant chicks hatching in an incubator. *Photo courtesy of Leonard Lee Rue III.*

**Ring-Necked Pheasant
*(Phasianus colchicus)***

Figure 11-10 Distribution map of the ring-necked pheasant.

The male ring-necked pheasant is brightly colored, distinguished by an iridescent greenish-black head, bright red facial coloring, and a white ring around the neck. The body is a deep reddish-brown with light-colored mottling on the flank area, and the tail is long and pointed, see Figure 11-9. Females in contrast, are a plain mottled brown and have a shorter tail. Pheasants range throughout much of the Pacific Northwest, the grain-producing regions of the Midwest and southern Canada, and the northeastern United States, see Figure 11-10.

Figure 11-9 A ring-necked pheasant. *Photo courtesy of U.S. Fish and Wildlife Service.*

Pheasants prefer farmland habitat with plenty of woodlands or brushy cover nearby. Males crow to attract a harem of females which they protect against other males until the eggs are laid. The clutch numbers 6–14 offspring that are cared for by the female. They feed on insects, berries, seeds, and grain.

Pheasants are polygynous birds; males form harems of several females during the breeding season. Females may nest two or more times in a single season if their eggs are damaged by predators or bad weather. Unlike quails and partridges, pheasants usually scatter after they have raised their broods and they tend to live solitary lives.

GROUSE

Grouse are among the largest of the game birds, second only to the turkey in size. In most species the legs and feet are protected by feathers, allowing them to live in cold, northern climates. Feathers also protect the nostrils of grouse.

AVIAN PROFILE

Ruffed Grouse
(Bonasa umbellus)

The ruffed grouse is one of the better-known grouse species due to its ability to adapt to a wide range of habitats including Alaska, Canada and the northern U.S. It prefers to occupy the edge zones in coniferous forests or deciduous woodlands where clearings and meadows interrupt the cover of trees and brushy plants. This grouse is mottled brown in color on its upper body with alternate gray-brown and buff crossbar markings on the lower body, see Figure 11-11. The tail of the ruffed grouse is shaped like a fan, and it has a small crest of feathers on its head.

The male of this species drums its wings during the mating season to attract females, but it does not help rear the chicks. Females usually lay 6–12 eggs in a sheltered ground nest lined with leaves and other vegetation. Ruffed grouse feed on small fruits and berries in the summer, and they gather buds and **catkins,** the flowering parts of trees and shrubs, during the winter. Several races of this grouse are recognized in the different regions that it occupies.

Figure 11-11 A ruffed grouse.
Courtesy of U.S Fish and Wildlife Service. Photo by Luther C. Goldman.

Figure 11-12 Distribution map of the grouse.

Grouse are adapted to a variety of habitats including forested areas, tundra, and plains. Several different species range over the North American continent from arctic regions to Mexico, see Figure 11-12. Their range is somewhat restricted on its northern and southern borders by the distribution of coniferous forests.

AVIAN PROFILE

Spruce Grouse
(Canachites canadensis)

The spruce grouse is a northern wilderness bird that occupies most of the wooded areas of Canada and Alaska. Its habitat is coniferous forest areas where it eats the foliage and buds from coniferous trees, along with insects, berries, and other fruits.

This grouse is gray on the upper body and black on the lower body, with a white line around the throat and white spots on the flanks, see Figure 11-13. Males display by fanning their tails and strutting about on the breeding grounds. After mating, the female lays 8–12 eggs in a shallow **scrape** or depression that she has pre-

Figure 11-13 A spruce grouse hen. *Courtesy of U.S. Fish and Wildlife Service. Photo by Mike Boylan.*

pared by scratching away some of the dirt and lining it with dry vegetation. Females raise their broods of chicks alone.

The spruce grouse is called the spruce partridge in some areas and several races of this species have been identified including a southern one that has long since been destroyed. This bird is very tame and does not fly away when it is approached. It survives in remote areas where there are few people.

With the exception of the spruce grouse, most other grouse spend the majority of their time on the ground. They find most of their food there and they always nest on the ground. They sometimes seek the protection of trees by landing in them when they are forced to fly away from danger.

AVIAN PROFILE

Sage Grouse (*Centrocercus urophasianus*)

The sage grouse is the largest gallinaceous bird except for the wild turkey. It has a gray-streaked upper body and a black lower body. The male has a long, pointed tail, a black throat, and a white collar and breast, see Figure 11-14. During the courtship display, the male is able to inflate his neck and breast region by expanding special air sacs. They make chuckling and popping sounds as they are alternately inflated and deflated.

Females wander about the breeding grounds inspecting the males before choosing their

Figure 11-14 A male sage grouse. *Photo courtesy of Leonard Lee Rue III.*

mates. They lay 7–8 eggs in shallow nests sheltered by sagebrush or other cover. Chicks are able to move about soon after they hatch.

This grouse prefers open areas in dry foothills and plains where sagebrush is plentiful as a source of food and cover. The buds and leaves of sagebrush form the main part of its diet.

Young grouse tend to eat large amounts of insects in comparison with adult birds. Insects are high in protein and are an important source of nutrition to chicks during this period of rapid growth. As they grow older, grouse increase the proportion of fruits, seeds, acorns, and other vegetable matter in their diets.

AVIAN PROFILE

Greater Prairie Chicken (*Tympanuchus cupido*)

The greater prairie chicken is a medium-sized grouse that inhabits tallgrass prairies, rangelands, and brushy fields. Its diet consists of seeds, berries, buds, and insects. The range of this grouse extends from Canada to Oklahoma. The population of the greater prairie chicken has declined as prairie habitat has been converted to farming and other uses. The Atlantic Coast race known as the Health Hen is now extinct. It is brown with buff-colored bar markings and yellow on the comb above the eyes, see Figure 11-15. Males have long, black feathers on their necks that are raised during mating displays.

Figure 11-15 The greater prairie chicken. *Courtesy of Fish and Wildlife Service. Photo by Lynn Nymeyer.*

During courtship the polygynous males make loud, booming sounds that can be heard for long distances. Females choose their own mates, and they lay 8–12 eggs in scrapes located in tall grass. After the chicks hatch, the females raise them alone.

Grouse are adapted to nearly every region of North America. Some grouse, such as ptarmigan, change their color from brown summer plumage to white plumage in the winter months. This protective coloring is important when cover is less available because of snow accumulations.

The rock ptarmigan is a grouse that is adapted to cold, northern climates. It ranges across the tundra regions of Alaska and northern Canada. This grouse changes color with the seasons. In the summer it is mottled gray and brown, with white wings and belly. In the winter months it molts to a completely white plumage, see Figure 11-16. The feathers on their feet work like snowshoes to keep them from sinking in the snow. Males have red combs over their eyes.

Figure 11-16 The rock ptarmigan. *Courtesy of U.S. Fish and Wildlife Service. Photo by Karen Bollinger.*

Females produce 6–9 eggs, and they are incubated in shallow nests on the ground. The young are cared for by the female. The digestion of woody material such as buds and twigs is aided by bacteria that digest the fiber for their own use. The grouse then digests the bacteria.

TURKEYS

The wild turkey is a native American species. It is the largest of the game birds. The Aztec civilization tamed turkeys long before the Spanish conquerors took them to Europe. Our modern domestic turkeys are descendants of the wild turkey.

AVIAN PROFILE

Turkey
(*Meleagris gallopavo*)

Figure 11-18 Distribution map of the wild turkey.

The turkey is a large bird. Males measure 4 feet in length from head to tail and females are as long as 3 feet. They are brown in color with a bronze glow to the feathers. Their heads, in contrast, are blue in color and lack feathers. Males have prominent red facial tissue called **wattles** that swell with blood taking on a bright red color when the birds are displaying in the mating ritual, see Figure 11-17. A tuft of feathers called a **beard** dangles from the breast; females are less likely to have a beard.

Figure 11-17 Gobblers at Bosque del Apache. *Courtesy of U.S. Fish and Wildlife Service. Photo by Gary Zahm.*

The turkey is a wary bird whose habitat is open woodlands, but its range has declined steadily as wooded areas have given way to farms and houses. Because of agricultural development of farmland such practices as clearing land contributed to declines in turkey populations. Other contributing factors included loss of trees due to disease, and overhunting of the turkey populations. Since the 1960s, populations have recovered in many areas due to restricted hunting, increased woodlands, and introductions of new populations to suitable areas. The range of the wild turkey extends from the eastern states to the southwestern states and Mexico. It has also been introduced successfully to the Northeast and in several western states, see Figure 11-18. The diet of the wild turkey consists of nuts, acorns, berries, seeds and insects.

Turkeys are polygynous, and dominant males gather harems of females by calling and displaying in a mating ritual. Hens usually lay 8–15 eggs in a nest on the ground. Sometimes nests are observed with large numbers of eggs, because more than one female is using it. The eggs are incubated by the female, also known as a hen. She also cares for the young.

The turkey is active only during daylight. It spends much of its time gathering food such as nuts and acorns from the forest floor. At night it roosts in trees where it is safe from its enemies.

PIGEONS AND DOVES

Distinguishing features of pigeons include rounded tails and larger sizes than the birds of this family known as doves. Doves have pointed tails, and they are the smaller members of the pigeon family.

The band-tailed pigeon is a forest bird that is larger than most domestic pigeons. It has a purple-colored head, neck, and underparts and a dark gray upper body. A small white collar extending partway around the neck is also evident, see Figure 11-19. Its range extends along the Pacific coast to Alaska and into southwestern United States and Mexico, see Figure 11-20.

Figure 11-19 Band-tailed pigeon.

These pigeons live in coniferous forest areas or oak woodlands, where acorns become an important part of their diets. It also feeds on berries and seeds. Clutches of 1–2 eggs are laid on a platform-shaped nest of twigs that is usually located in a tree or bush.

Pigeons and doves drink by putting their bills in water and sucking. Most birds drink by getting water into their bills and quickly raising their heads, causing the water to flow down their throats by gravity. Water and food are mixed and food particles are soaked in an organ called the **crop.** Once the food has become soft, it moves into the **gizzard,** which is an organ that grinds the food particles between small rocks that have been ingested for just this purpose.

AVIAN PROFILE

**Band-Tailed Pigeon
(Columba fasciata)**

Figure 11-20 Distribution map of the band-tailed pigeon.

**AVIAN
PROFILE**

**Rock Dove
(Columba livia)**

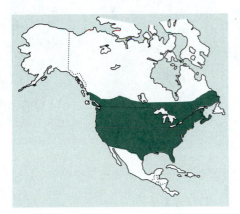

Figure 11-22 Distribution map of the rock dove.

The rock dove is not native to North America, but it is probably the best-known species of pigeon. It was introduced from Europe, and is now found throughout southern Canada and the continental United States, see Figure 11-21. It is a city bird that varies in color from pure white to brown, black, blue, or a combination of these colors, see Figure 11-22. It is a cliff dwelling bird that has adapted to

Figure 11-21 The rock dove.

city buildings and the urban environments. The diet consists of fruits, insects, grains and seeds.

Two eggs are laid in a flat nest on a sheltered building ledge, and it is common to find young birds that are nearly old enough to fly in a nearby nest. Several broods are raised each year in most locations. Both parents help to care for their young.

Young pigeons are called **squabs.** Their parents feed them a food called **pigeon milk** that is secreted in the crops of both parents. It is composed of cells that are sloughed off the inner wall of the crop, and its food value and appearance is similar to cottage cheese. The squabs place their beaks in the mouths of their parents to receive this regurgitated food.

**AVIAN
PROFILE**

**Mourning Dove
(Zenaida macroura)**

The mourning dove is a moderate-sized member of the pigeon family. It is brownish gray in color with lighter underparts and a black spot on the ear patch, see Figure 11-23. The mourning dove gets its name from the cooing call of the male. In northern regions it is a migratory bird whose range extends from southern Canada to northern Mexico, see Figure 11-24.

A pair of mourning doves normally nests two to four times each year, producing two eggs each time. Both parents care for the

Figure 11-24 Distribution map of the mourning dove.

Figure 11-23 A mourning dove with her young. *Courtesy U.S. Fish and Wildlife. Photo by J. Leupold.*

young. They usually nest in trees, but they are known to nest on the ground. Squabs grow rapidly and often **fledge** (learn to fly) by the time they are a month old. The preferred habitat of mourning doves is dry, upland areas, grainfields, deserts, and suburban areas. The diet includes mostly insects, seeds, grain, and fruits.

Pigeons and doves are primary consumers that convert plant materials to meat. They are preyed upon by a wide variety of animals during the nesting period, but adults are most often captured by falcons, hawks, and other birds of prey.

The passenger pigeon was cited earlier in this text as a bird that was hunted until it became extinct. It was probably once the most abundant bird on earth and certainly the most numerous of the North American pigeons. Nonetheless, it was wiped out in just a few years due to overhunting. The loss of this species is a powerful example of the importance we should place on proper management of our wildlife heritage.

LOOKING BACK

The game birds include quails, partridges, pheasants, grouse, pigeons and turkeys. They are managed as game birds by many state and federal fish and game agencies.

All of the game birds except pigeons are gallinaceous, meaning that they have heavy, plump bodies, and they nest on the ground. Some of them are polygynous or polygamous in their mating behavior, others are monogamous. All of the birds discussed in this chapter are generally considered to be primary consumers, although some of them eat insects in addition to vegetable matter.

Game birds are preyed upon in large numbers by a variety of predators including several of the birds of prey. They are strong runners, and they usually choose to run instead of fly to escape their enemies. They are generally able to sustain their numbers because they produce large numbers of offspring each year.

REVIEW QUESTIONS

1. Name the six kinds of game birds that are considered in this chapter.
2. Describe a nesting behavior of gallinaceous birds that distinguishes them from other birds.
3. Distinguish between monogamous, polygynous and polygamous birds, and list some examples of each.
4. Define the relationship between doves and pigeons.
5. Evaluate the roles of game birds in the environments that they inhabit.
6. List the common species of quails, partridges, and pheasants that are found in the region where you live, and record them on a chart.
7. Identify some major species of North American game birds and pigeons, and list additional species that are found where you live.
8. List the probable effects that a reduced population of game birds might have on populations of birds of prey that inhabit the same area.
9. Describe how the coloring of many game birds differs between the sexes. List some ways that color differences between males and females contribute to mating and to survival during the nesting period.
10. Explain the relationship between the survival rates of game birds and the number of eggs that the females produce.
11. Describe how mating rituals among game birds are used to attract mates.

LEARNING ACTIVITIES

1. Do a survey of the game birds in your region and map the distribution of the species that are found in your area. Obtain photographs, prints, and other visuals of these species from local fish and game offices. Research the ecology and management practices for each species. Prepare and present a conservation program to elementary school classes using the materials and information that you have gathered.

2. Obtain an incubator, and purchase fertile eggs of quail or other game birds from a game bird farm. Incubate the eggs to ensure that they will be hatching at the time you are studying the game bird unit.

12 Birds of Prey

THE BIRDS OF PREY are the predators of the sky. North American birds of prey include four families of hawk-like birds and two families of owls. They are also called **raptors.** They are important as secondary consumers in all of the ecosystems of North America. Predatory birds help to maintain a natural balance between living organisms and their food supplies. They do this by helping to control populations of birds, reptiles, rodents, and other small animals.

OBJECTIVES

After completing this chapter, you should be able to

- name the major differences between birds of prey and other kinds of birds
- list the six families of raptors that are found in North America
- describe species of raptors that are representative members of each of the raptor families
- identify similarities and differences among raptors
- evaluate the roles of raptors in the ecosystems of North America
- discriminate between hawks and owls
- name the two families of North American owls, and distinguish between them
- describe species of owls that are representatives of the two families of North American owls
- evaluate the roles of owls in North American ecosystems
- discuss the positive and negative impacts on raptors of chemicals that are manufactured for garden, yard, industrial, and agricultural uses.

All the birds of prey are meat eaters. Their toes are equipped with claw-like **talons** with which they grasp their prey. They have three toes pointed forward and a fourth toe pointing backward,

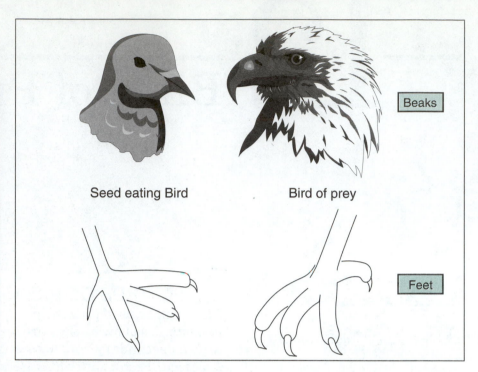

Figure 12-1 Seed eating birds use their claws to scratch at the ground in a search for food and their beaks to break open seeds. Birds of prey, on the other hand, use their claws to grasp and kill their prey and their beaks to tear it into bite-size pieces.

Raptor Families

Hawks, Eagles and Kites

- Black hawk
- Cooper's hawk
- Ferruginous hawk
- Gray hawk
- Harris hawk
- Marsh hawk
- Red shouldered hawk
- Red tailed hawk
- Rough-legged hawk
- Sharp-shinned hawk
- Swainson's hawk
- Zone-tailed hawk
- Bald eagle
- Golden eagle
- White-tailed kite
- Goshawk

Vultures

- California condor
- Black vulture
- Turkey vulture

Osprey

- Osprey

Falcons

- Aplomado falcon
- Peregrine falcon
- Caracara

Owl Families

Typical Owls

- Elf owl
- Burrowing owl
- Great gray owl
- Great horned owl
- Pygmy owl
- Long-eared owl
- Saw-whet owl
- Screech owl
- Short-eared owl
- Snowy owl
- Spotted owl

Barn Owls

- Barn owl

Figure 12-2 Raptors are important as secondary consumers in all the ecosystems of North America. Hawks, eagles, kites, vultures, osprey, falcons, and owls are all considered birds of prey.

making it possible for them to grip things. The beaks of these birds are sharp, and they are shaped to tear raw flesh, see Figure 12-1. The jaws and feet of these birds are strong and powerful, and they have exceptional eyesight. They generally kill prey with their talons and tear it into bite-size pieces with their bills. They are well fitted for their roles as predators.

HAWK-LIKE BIRDS

One family of hawk-like birds found in North America consists of the eagles, hawks, and kites. Others are the New World vultures, ospreys, and falcons, see Figure 12-2. These birds of prey belong to the scientific order known as Falconiformes. Other birds of prey are the owls. They belong to the order strigiformes that includes two families, the typical owls and the barn owls.

Eagles, Hawks, and Kites

The eagles, hawks, and kites are **diurnal** birds of prey meaning that they are active only during daylight hours. The nocturnal birds of prey such as owls are most active during hours of darkness. The eagles, hawks, and kites are all similar in shape, but there are major differences among them in size, diet, and preferred habitat.

AVIAN PROFILE

Bald Eagle
(Haliaeetus leucocephalus)

Figure 12-4 Distribution map of the bald eagle.

A mature bald eagle has a yellow bill, a dark brownish-black body, and a white neck, head, and tail, see Figure 12-3. Immature birds are brown with light-colored markings on their underparts, and they take up to five years to acquire adult plumage. The wingspan of this bird ranges from 6.5 feet to 7.75 feet. They prefer coastal habitats or rivers and lakes, where fish are available for food. For this reason, the bald eagle is classified as a fish eagle. It is also a scavenger that eats carrion. Birds (especially water birds), rodents, rabbits, and snakes are also included in the diets of bald eagles.

The bald eagle is the national bird of the United States of America. It ranges through much of the United States and Canada except for some areas in the southwest, and the grain belt along the border between the two countries, see Figure 12-4. It is most abundant in Alaska, but avoids the far northern tundra biome. Although the distribution of eagles is widespread, they are only abundant in localized areas.

Figure 12-3 The majestic bald eagle. *Photo courtesy of the Alaska Division of Tourism.*

Two or three eggs are laid each year, and a single brood of offspring is raised. Young eagles are called **eaglets**. They are covered with a whitish down when they hatch. Mated pairs of eagles first nest at 3–5 years of age, and they tend to return to the same nesting sites year after year. Nests are constructed of sticks, and new material is added during each nesting season.

A number of the members of this family, particularly the eagles, have long been thought to mate for life. Some raptor experts now say that there is limited evidence to support this claim. They prefer to say that many raptors mate with the same partners for one or more seasons. Some species of raptors are known to be polygynous, meaning that a male mates with more than one female. In a few instances, females have been known to mate with more than one male. This latter mating behavior is known as **polyandry.**

Some raptors are known to return to the same nesting sites each year. The nest of a raptor, called an **aerie,** is usually located in a high place. Both parents are typically involved in caring for their young, and some species will attempt to drive intruders away from their nests when they get too close. There are few recorded instances, however, of these birds actually attacking a human. An unusual trait among raptors is the tendency for males to help incubate the eggs.

AVIAN PROFILE

Golden Eagle
(Aquila chrysaetos)

Figure 12-6 Distribution map of the golden eagle.

Figure 12-5 The golden eagle.
Courtesy of U.S. Fish and Wildlife Service. Photo by Tom Smylie.

The golden eagle is a large bird with a wingspan of 7 to 7½ feet. It has dark brown plumage, and unlike the bald eagle, it has feathers on its legs, see Figure 12-5. The range of this eagle extends across much of Alaska and Canada, throughout most of the western United States, and south to Mexico, see Figure 12-6.

The golden eagle is a hunter that eats fresh meat, but seldom carrion. It hunts in forests and rangelands for rabbits, birds, and rodents. It is also known to occasionally kill domestic animals such as lambs, small goats, young pigs, and chickens.

Golden eagles may alternate between nesting sites, but they usually return to established aeries. They nest starting at 3–5 years of age, and females lay only two eggs per season. Breeding pairs build a platform nest of sticks lined with grass or other vegetation. Nests are located on cliff ledges or tree tops. Males assist with raising the young by hunting for food that they bring back to the nest.

Most raptors establish territories that they defend against others of their kind. Territory boundaries surround the nesting area, and the size of the territory varies depending on the raptor species and the abundance of the food supply. In a preferred raptor habitat, nests tend to be spaced fairly evenly throughout the area. Marsh hawks are exceptions because they live and nest in colonies.

AVIAN PROFILE

Red-Tailed Hawk
(Buteo jamaicensis)

The red-tailed hawk is a large bird that is dark brown on the upper body and light-colored underneath. A dark band extends across its belly. The tail is light colored in juveniles, and reddish in adults, see Figure 12-7. This hawk is widely distributed in North

Figure 12-8 Distribution map of the red-tailed hawk.

America, see Figure 12-8. Its breeding range extends northward from the border between the United States and Canada, and it ranges permanently throughout most of the United States and parts of Mexico.

This hawk is adapted to a wide range of habitats from desert to tundra. It preys mostly on small rodents and some insects. Without the red-tailed hawk and other birds of prey, rodent populations would become far too abundant for the environment to support.

Females of this species lay from 1–4 eggs in large nests made of sticks and lined with twigs and roots of plants. Their aeries are constructed in trees or on cliffs.

Figure 12-7 The red-tailed hawk. *Photo courtesy of U.S. Fish and Wildlife Service.*

A **kite** is a hawk with a long square or forked tail, and long wings that come to a point at the tips. Kites are found only in the warmer areas of North America, mostly in the southeastern and Gulf Coast states and a few in California. Kites are generally few in number. They include the swallow-tailed kite, black-shouldered kite, Mississippi kite, snail kite, and white-tailed kite.

ENDANGERED SPECIES PROFILE

Everglade Snail Kite (*Rostrhamus sociabilis*)

Figure 12-9 Distribution map of the Everglade snail kite.

The Everglade snail kite is an unusual bird in that it eats only one species of freshwater snail that is found in the Everglades of Florida, see Figure 12-9. It is this limited diet and the destruction of natural wetlands in south Florida that has resulted in this bird being included on the endangered species list by the U.S. Fish and Wildlife Service. It is bluish-gray in color, with black bill and talons, orange feet, and reddish eyes, see Figure 12-10.

This bird produces only 2–3 eggs in a season, and recovery of this population is likely to be very slow.

Figure 12-10 The Everglade snail kite.

The genus of hawks known as **harriers** are excellent hunters in tall vegetation such as marsh reeds and field grasses. They are unusual in their nesting habits because several pairs of these birds nest in close proximity to one another instead of establishing separate territories.

Figure 12-11 The Northern Harrier.

AVIAN PROFILE

Northern Harrier
(Circus cyaneus)

Figure 12-12 Distribution map of the Northern Harrier.

The male northern harrier (marsh hawk) is light gray above with white underparts. Females are brown with white underparts marked with brown streaks. Their wingtips are black, and they have white rumps, see Figure 12-11. Northern harriers are widely distributed in North America, see Figure 12-12.

Northern harriers have become specialized in finding and catching mice that are hidden in tall vegetation. They do this through their acute hearing. They are able to locate mice without being able to see them because they hear the mice when they squeak.

Females lay 4–6 eggs in ground nests located in a marsh. Both parents help rear the young, and family groups gather to roost together on the ground each night.

Accipiters, or woodland hawks, range widely through North America. In addition to the goshawk this genus includes the Cooper's hawk and the sharp-shinned hawk. All three of these birds are commonly called chicken hawks, because they are occasional predators of domestic birds including chickens.

These are medium-sized hawks that are very adept at flying. They have short wings and long tails, and they are well equipped to weave swiftly through their woodland and forest habitats. These three hawks occupy much of North America, and their territories overlap significantly.

**AVIAN
PROFILE**

Goshawk
(Accipiter gentilis)

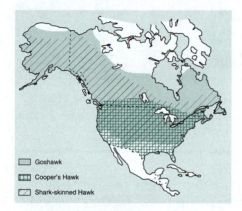

Figure 12-14 Distribution map of the goshawk, Cooper's hawk, and shark-skinned hawk.

The goshawk is gray colored on the upper body, with barred markings of light gray and white on the underside, see Figure 12-13. Young birds are more brown than gray in color. This species has a distinctive white stripe above each eye, and females are larger than males. See Figure 12-14 for their range.

These hawks fly by flapping their wings to gain speed followed by a long glide. They can maneuver swiftly through trees and branches in pursuit of prey. The preferred prey of goshawks are medium-sized birds, and it makes no distinction between domestic chickens and game birds. Forest-inhabiting grouse are a favorite prey of goshawks.

Goshawks nest high in coniferous trees. Females lay 3–4 eggs, and the male supplies food to the female during the incubation period. Both parents hunt when the demand for food becomes a burden to the male.

Figure 12-13 An immature goshawk. *Photo courtesy of Leonard Lee Rue III.*

Falcons

Accipiters and falcons are similar in their preference to feed heavily on birds. The goshawk preys on large game birds such as pheasants or ruffed grouse. Some falcons also prey on these species. Other falcons, such as the American kestrel, are small birds of prey and they feed on small birds such as sparrows and finches. The goshawks and falcons also eat small mammals.

**AVIAN
PROFILE**

American Kestrel or
Sparrow Hawk
(Falco sparverius)

The American kestrel is commonly known as a sparrow hawk. Its diet consists of mice, voles, small birds, and insects such as grasshoppers. It is a small falcon that can maintain a fixed position above the ground by hovering on rapidly beating wings. Hawks that are able to hover in this manner are called **kestrels.** Hovering allows the kestrel to keep small prey in sight before diving down to capture it.

This kestrel is reddish colored on the back, crown, and tail with blue gray head and wings. The breast and underparts are lighter in

Figure 12-16 Distribution map of the American kestrel.

color than the upper body, and a black marking is present just behind the ear, see Figure 12-15. The sparrow hawk is widely distributed in North America from Alaska and southern Canada to Mexico, see Figure 12-16.

Females lay 3–5 eggs in tree holes or in other protected areas. They do not build nests.

Figure 12-15 The American kestrel. *Courtesy of U.S. Fish and Wildlife Service. Photo by Tom Smylie.*

Among the raptors, females are often larger in size than the males. There appears to be a relationship between these size differences and how fast the prey is. Among raptors that catch birds in flight, the females are sometimes twice as large as males. The size differences are not nearly so great among raptors that eat slow-moving prey such as snails.

AVIAN PROFILE

Peregrine Falcon (*Falco peregrinus*)

Figure 12-18 Distribution map of the peregrine falcon.

The peregrine falcon is gray in color on the upper body, with a nearly white throat and distinctive bar markings on the light-colored underparts, wings, and tail, see Figure 12-17. Females are more brown than gray. This falcon occupies habitat in open prairie regions, and along sea cliffs and rivers. It ranges from Alaska and northern Canada along the Rocky Mountain range to Mexico, see Figure 12-18. It is listed as an endangered species under the Endangered Species Act.

The peregrine falcon is an extremely fast flier, and it often captures its prey in flight. Once it has spotted its victim, it makes a fast dive called a **stoop** that ends when it makes contact with its target. It has been used in falconry as a trained hunting bird to hunt game birds.

The preferred prey of this bird consists of pigeons, game birds, water birds, and ducks. It is sometimes found living among

Figure 12-17 The peregrine falcon in flight. *Courtesy of Leonard Lee Rue Enterprises. Photo by Mark Wilson.*

the tall buildings of cities where it preys on pigeons. It nests on cliffs and building ledges where females lay 2–4 reddish-colored eggs.

Falcon populations dropped dramatically with the increased use of organic pesticides such as DDT. The use of DDT caused the shells of raptor eggs to become too thin to prevent them from being crushed during incubation. Many of these chemicals were used to kill insects on farm crops and in home gardens. Some chemicals build up in the bodies of birds until they reached toxic levels. Chemicals such as DDT often persist even after their use has been discontinued. This is due to the accumulation of the chemical in the tissues of plants and animals. Such chemicals pass from plants to primary consumers to secondary consumers, and a toxic chemical can persist in the food web for a long time.

It is not necessary to eliminate all of the chemicals that are used around our yards, gardens, roads, golf courses, farms, and industrial sites to protect wildlife, but we must use such materials prudently. Chemicals that are demonstrated to have long-term negative effects on wildlife or humans must be withdrawn from use.

Vultures

The vultures discussed in this chapter are New World vultures found on the American continents. They are scavenging birds that eat carrion. The North American species of vultures are the black vulture, turkey vulture, and California condor.

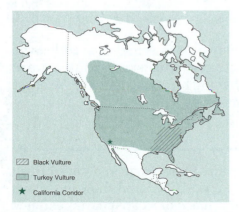

Figure 12-19 Distribution map of the black vulture, turkey vulture, and California condor.

The turkey vulture is one of the largest birds in North America. It is a scavenging bird that ranges throughout much of the United States and Canada, see Figure 12-19. It inhabits open areas where it feeds on carrion and can be seen circling at high altitudes searching the landscape below for dead animals. They have superb eyesight.

The turkey vulture is brown to black in color with a naked, red head, and broad wings that span up to 6 feet, see Figure 12-20. Its naked head and neck allow it to thrust its head into the body cavities of dead animals as it feeds. This is a sanitary adaptation that helps to keep this bird much more clean than it would be if the head and

Figure 12-20 The turkey vulture. *Courtesy of U.S. Fish and Wildlife Service. Photo by Luther C. Goldman.*

neck were covered with feathers. Females lay 1–3 eggs in a sheltered place on the ground, in rock cavities, or in hollow logs.

Vultures are awkward, ugly birds when they are on the ground, but they are graceful in flight as they ride the air currents to high altitudes. They perform an important role in the environment by eating the rotting flesh of dead animals.

The California condor is the largest bird of prey in North America. It is black in color with a naked, reddish orange head, see Figure 12-21. Only one breeding flock is known to remain in the hills of southern California.

These birds were rare even before North America was settled by Europeans, but loss of habitat for human use, pesticides, and disturbance by humans have all contributed to their decline. Their rate of reproduction is also very slow. Under natural conditions, a breeding pair raises only one offspring every two years. These birds do not breed until they are 5–6 years old.

Some captive flocks of these birds have been established in an effort to prevent them from becoming extinct. One of these flocks is located at the National Birds of Prey Center in Idaho, where they are isolated from all human contact to preserve their wild instincts and behaviors.

Figure 12-21 The California condor. *Courtesy of U.S. Fish and Wildlife Service. Photo by Glen Smart.*

Osprey

Only one species of osprey is known. It is sometimes called the fish hawk or fish eagle because its diet consists of fish that are taken alive. This bird is a superb fishing bird that plunges feet-first into the surface of the water to snag fish.

The osprey is a large bird with white underparts and dark brown upper body. The underside of the tail is striped and the wings have dark edges. The head is white, and it has a dark stripe that extends across its eyes, see Figure 12-22.

Female ospreys lay 2–4 eggs in platform nests located on trees, cliffs, bridges, or utility poles. Ospreys occupy habitats along

Figure 12-23 Distribution map of the osprey.

Figure 12-22 The osprey. *Courtesy of U.S. Fish and Wildlife Service. Photo by Bart Foster.*

rivers, lakes, and seacoasts where fish are abundant. They range through parts of the northern and coastal regions of the United States and Canada except for the far northern regions, see Figure 12-23.

OWLS

The two owl families are the typical owls and the barn owls. Owls have some unique characteristics unlike those of most other birds. They have down on the margins and tips of their wing feathers that deaden the sounds of flight, helping owls to hear potential prey and preventing prey from hearing the approaching owl. They have excellent vision, and they are able to see quite well at night. They also have a highly developed sense of hearing, and they are capable of hunting by sound alone.

Owls can be found in any habitat that has a suitable supply of food. They are nocturnal birds that hunt mostly at night or during twilight hours. They prey on rodents, rabbits, birds, reptiles, fish, and insects. Some small owls also eat worms. North American owls are not very specialized in the kinds of food they eat.

Career Option

WILDLIFE MITIGATION SPECIALIST

A wildlife mitigation specialist must be well-versed in ecology, life cycles, and population trends of wildlife, and their habitat requirements. This career requires the development and implementation of plans to reduce the negative impacts of human activities on wildlife. This is done by identifying and evaluating habitat losses, and modifying tracts of land to create critical wildlife habitat. A university degree in ecology or a related science is required.

Typical Owls

Typical owls belong to a family that is also known as "true" owls. They have large, round heads, large eyes, hooked bills and fluffy plumage that muffles the sound of flight. Seventeen species of typical owls have been identified in North America.

Figure 12-25 Distribution map of the great horned owl.

The great horned owl is a very large owl with large, yellow eyes, and ear tufts or "horns" on its round head. It is mottled brown in color with gray barring on the underside, see Figure 12-24. It hunts rabbits, rodents, and birds. Sometimes it gets into the habit of preying on domestic chickens. This owl is forced to do some of its hunting in daylight because part of its range is located in a region that does not always get dark at night.

The great horned owl may be the best-known owl in North America since its territory includes most of the continent, see Figure 12-25. It calls with a deep voice with a rhythmic sound like "hoo, hoo, hoo." The female lays 2–3 eggs in the old nest of another bird.

Figure 12-24 The great horned owl.
Courtesy of U.S. Fish and Wildlife Service. Photo by Robert Drieslein.

All of the North American owls except the barn owl belong to the family of typical owls. These owls are widely dispersed across the continent. They range in size from birds the size of sparrows to some that are as large as ducks.

The elf owl is the smallest owl in North America, comparable in size with the sparrow, see Figure 12-26. Its habitat is the Saquro desert region in the southwestern part of the continent, see Figure 12-27. It is also known to live in dry wooded canyons and arid regions covered with scrub.

Figure 12-26 The elf owl.

This owl lives on a diet of large insects. It nests in abandoned woodpecker holes in large cactus plants, where females lay 3–4 eggs.

Figure 12-27 Distribution map of the elf owl and the snowy owl.

AVIAN PROFILE

**Snowy Owl
(Nyctea scandiaca)**

The snowy owl is a large white owl with dark spots, and a wingspan up to 55 inches, see Figure 12-28. Its range extends throughout Canada and into the northern arctic region. It is a migratory bird whose summer range includes the tundra of Alaska and northern Canada. During the winter season, it moves southward.

Figure 12-28 The snowy owl. *Courtesy of U.S. Fish and Wildlife Service. Photo by Kent N. Olson.*

Lemmings make up a major part of the diet of the snowy owl. In years when lemmings are not very abundant, this owl migrates south in search of food. It nests on the ground in shallow depressions lined with moss and grass. Females produce 5–7 eggs each nesting season.

Many other species of typical owls are found in North America. They include the great gray owl, the largest of the North American owls. Another well-known owl is the spotted owl of the Pacific Northwest and British Columbia, see Figure 12-29. Efforts to protect this endangered species have led to the closure of much of the logging and timber industry in the disputed region.

Figure 12-29 A pair of spotted owls. *Courtesy of U.S. Fish and Wildlife Service. Photo by Randy Wilk.*

The screech owl is widely distributed in the United States and in parts of Canada. It feeds mostly on mice and moles. Another interesting owl is the burrowing owl. It nests in the burrows of rodents. Its range includes much of the western United States and parts of Florida.

Barn Owls

Members of the barn owl family are distinctive in their appearance because they have heart-shaped faces. Only one member of this family is found in North America.

AVIAN PROFILE

Barn Owl
(Tyto alba)

The barn owl is a medium-sized, tan-colored owl, with gray streaks in its plumage and a white face rimmed with tan, see Figure 12-30. Its habitat includes woodlands, prairies, farms, and suburban areas, and it ranges throughout much of the United States. It perches on trees and feeds mostly on rodents.

Figure 12-30 A pair of barn owls. *Courtesy of U.S. Fish and Wildlife Service. Photo by Perry Reynolds.*

The barn owl nests in protected areas of many kinds, including barns where 5–11 eggs are laid on a bare surface. It is from its nesting behavior that this bird gets its name. The barn owl is considered to be a friend of man because it preys on rodents, and is willing to live in close proximity with humans

LOOKING BACK

Birds of prey include six families of raptors. Raptor families include one group composed of eagles, hawks, and kites. Three other families are the vultures, ospreys, and falcons, and the two owl families are the typical or true owls, and the barn owl.

Birds of prey fill roles in the environment as secondary consumers. By preying on birds, reptiles, fish, and mammals, they help maintain a balance in nature between each of these animals and its food supply. Some raptors also eat carrion which helps maintain a healthy environment by recycling the flesh of dead animals.

REVIEW QUESTIONS

1. Name the major differences between birds of prey and other kinds of birds.
2. List the differences between hawk-like raptors and owls and other birds.
3. Identify the six families of raptors, and list each of the North American raptor species mentioned in this chapter with the family to which it belongs.
4. Describe some similar characteristics of all raptors, and identify ways that raptors differ from one another.
5. Define the roles that raptors fill in the ecosystems of North America.
6. Identify the two families of owls, and list some representative species of each.
7. Distinguish between typical owls and barn owls.
8. Discuss the positive and negative impacts on birds of prey of chemicals that are manufactured for uses in gardens, yards, highway weed control, golf courses, industry, and agriculture.

LEARNING ACTIVITIES

1. Identify a falconer in your area and invite this person to bring his/her bird to class. Schedule plenty of time for a discussion about falcons and other birds of prey, and request an outdoor demonstration of the bird's training.
2. Obtain permission from your state fish and game agency to prepare skull specimens of different birds of prey for educational use. Request the agency to provide the skulls from wounded birds that die in their care.
3. Establish an aviary at the school that is approved by the state fish and game agency for the care of wounded birds of prey.

13

Songbirds and Other Perching Birds

broods
conical bill
dipper
gregarious
parasitic bird
promiscuous
shrike
tanager
territorial

T HE SONGBIRDS are part of a larger group of birds identified by the Audubon Society as perching birds. This group also includes the kingfishers, hummingbirds, and cuckoos. As with all attempts to organize birds in groups, they do not fit neatly into classification. The groups discussed here have similar characteristics but their placement in groups is arbitrary. Perching birds are adapted to a wide range of habitats, and they are found in most locations in North America. As the name implies, these birds are adapted to perching upright on branches and other surfaces.

OBJECTIVES

After completing this chapter, you should be able to

■ name the different families of perching birds that are found in North America
■ describe the characteristics and habits of several kinds of songbirds.
■ speculate about the tendency for males of many species of birds to be more brightly colored than females
■ discuss the importance of the calls and songs of territorial birds
■ analyze the characteristics of house sparrows and starlings that have helped them to adapt so well to North America since they were introduced from Europe.
■ differentiate among the following terms that describe mating behaviors of birds: polygynous, polyandry, polygamous, monogamous, and promiscuous
■ name some popular members of the thrush family of birds
■ describe the reproductive behaviors of parasitic birds
■ explain why gregarious birds frequently cause extensive damage to crops while solitary birds with the same eating habits are seldom blamed for crop damage
■ describe some adaptations of hummingbirds that allow them to gather nectar for food.

SONGBIRDS

Songbirds use their calls and songs to define the boundaries of their territories. Singing and calling makes other birds aware of their presence in the area. These birds are called **territorial birds** because they defend their living areas against other birds. The calls of each bird species are unique, and an experienced bird-watcher can often identify a bird by its call. Only a few of the songbirds will be described in this chapter due to the large number of birds in this group.

The largest bird family in the world (**Fringillidae**) includes sparrows, finches, grosbeaks, and buntings. Seventy-two species are found in North America. An introduced bird species that has similar characteristics but which belongs to the Ploceidae family is the English or house sparrow. It was introduced to North America in 1850 when a few birds from England were released in New York City. This sparrow is a weaver finch, and it is not related to native sparrows. It is included in this group because it is similar to other birds that are described here.

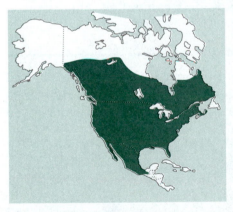

Figure 13-2 Distribution map of the house sparrow.

The male house sparrow or English sparrow is a small brown bird with black stripes on the wings and back, a dark breast, a white face with chestnut markings, and a gray head, see Figure 13-1. Females are more drab in their colors. Since introduction to New York City, it has expanded its range to include most of North and South America, see Figure 13-2.

This sparrow inhabits cities, towns, and farms, where it nests in buildings and other sheltered areas. Females lay 5–6 eggs in large nests of grass, feathers, and other soft materials. Up to three broods per year may be raised. They eat grain, seeds, fruits, insects, and buds. They are preyed upon by raptors, cats and other carnivorous mammals. Like most birds, they are most vulnerable to predators immediately after fledging.

Figure 13-1 A male house sparrow. *Photo courtesy of Leonard Lee Rue III.*

Members of the Fringillidae family have **conical bills** (cone-shaped) with sharp edges that are adapted to cracking seeds. They are technically omnivores, but they feed mostly on seeds of grasses and weeds. They live in terrestrial as well as wetland habitats, and they are found throughout North America. Many finches, grosbeaks, buntings and sparrows are **gregarious** birds that live together in flocks. Some of them are brightly colored, and many of them migrate between summer and winter ranges.

House Finch
(Carpodacus mexicanus)

Figure 13-4 Distribution map of the house finch.

The house finch is a sparrow-sized bird with males exhibiting shiny red crowns, rumps, and breasts, see Figure 13-3. It was introduced to the east coast from the west coast around 1940. Females are striped brown. The clutch consists of 3–5 eggs. This finch sings from a high perch with a song similar to that of a canary. Its habitat is diverse, ranging from deserts to coastal regions, see Figure 13-4. It often lives in close proximity to humans. The diet includes grass and weed seeds during the winter along with insects in the summer.

Figure 13-3 A male and female house finch. *Courtesy of Leonard Lee Rue Enterprises. Photo by Irene Vandermolen.*

There is a pattern among many birds for males to be more brightly colored than females of the same species. There may be several reasons for this. Males use their colors in mating rituals to bluff other males in defense of their territories. They also use color to attract potential mates. One explanation for the dull colors of most female birds is that drab colors help camouflage them during the nesting season.

Cardinal
(Cardinalis cardinalis)

Figure 13-6 Distribution map of the cardinal.

The male cardinal is a bright red bird with a high crest and black face, see Figure 13-5. Females are buff brown with red wings and tail. It is a familiar bird ranging from the eastern United

Figure 13-5 The cardinal. *Courtesy of U.S. Fish and Wildlife Service. Photo by Jon Nickles.*

States to Mexico, see Figure 13-6. Females lay 3–4 eggs in cup nests located in shrubs or bushes. It is considered a garden bird in much of this territory because it is attracted to garden plantings with berries, etc. Winter feeding by homeowners has extended its range northward. The cardinal is a nonmigratory bird that eats fruits, insects, weed seeds, and grains.

The **tanagers** of the Traupidae family have notched bills. They are colorful forest birds that prefer diets of small fruits and insects. These songbirds are migrating birds that spend their winters in Central and South America. Five species are found in North America. It is important to maintain habitat for migratory birds such as tanagers in their winter range. Loss of winter habitat in South or Central America would seriously harm these bird populations.

The summer tanager is one of the most colorful of all the songbirds. Males are red and females are olive above and yellow below. Young males acquiring adult plumage often have patches of all three colors, see Figure 13-7. They prefer oak woodlands, but they also inhabit cottonwood and willow thickets along streams, in the eastern and southern regions of the United States, see Figure 13-8. The western tanager is a close relative that occupies habitat in coniferous forests and willow thickets in the western regions of the United States and Canada. Diets include insects and small fruits that are gathered in the canopy of trees. Clutch size is 3–4 offspring that are raised in shallow cup nests attached to tree branches.

Figure 13-7 The summer tanager.

AVIAN PROFILE

Summer Tanager (Piranga rubra)

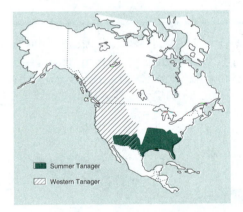

Summer Tanager
Western Tanager

Figure 13-8 Distribution map of the summer and western tanager.

Wrens are tiny brown birds with up-turned tails and loud singing voices. Ten species of wrens are found in North America. They eat mostly insects and spiders and are found from southern Canada to Mexico. They are adapted to a wide range of environments from humid coastal regions to dry deserts. The cactus wren even builds its nest among the spines of a cactus.

The males of some species of wrens build elaborate nests that are used for sleeping as well as nesting. During the mating season they sing loudly and perform elaborate rituals to lure mates to their nests. The males of some species are known to construct as many as 25–30 nests in their attempts to attract mates. Females inspect the nests as part of the mating ritual. Once a female has chosen a mate, a completely new nest may be constructed for nesting.

AVIAN PROFILE

House Wren (Troglodytes aedon)

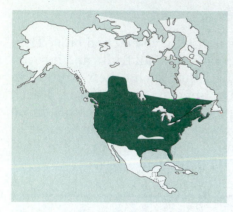

Figure 13-10 Distribution map of the house wren.

The house wren is a small brownish-gray bird with a short up-turned tail and alternating light-colored crossbar markings on its back, see Figure 13-9. It prefers open woods or brushy habitat and ranges from southern Canada to Mexico, see Figure 13-10. The diet of the wren consists mostly of insects and spiders.

House wrens nest in simple nests that are constructed in tree cavities or birdboxes. Females lay 5–8 eggs, and they may nest two or more times in a season. These birds are attracted to humans and gardens where it lives in close proximity with humans.

Figure 13-9 The house wren. *Photo by Leonard Lee Rue III.*

AVIAN PROFILE

Dipper (Cinclus mexicanus)

Figure 13-12 Distribution map of the dipper.

The dipper is a small gray bird with a short tail that closely resembles a wren, see Figure 13-11. It is found mostly in the western region of the United States between the Pacific coast and the eastern slope of the Rocky Mountains, see Figure 13-12.

A wrenlike bird that is found in and near streams of water is the **dipper.** It feeds on aquatic insects that it gathers from stream bottoms. This small bird has a habit of bobbing up and down continuously as it perches on rocks near the water. It walks and flies in and out of the swift current, and it appears to walk along the bottoms of fast-moving streams as it searches for food.

Figure 13-11 A dipper.

Dippers nest in protected areas along streambanks. Nests are insulated with dry moss, and females lay 3–6 eggs. Young dippers are able to dive for food even before they learn to fly.

Shrikes are birds with shrill voices, hooked bills and the hawk-like behaviors. They have the habit of hunting insects, mice, small birds and lizards, and storing them for later meals. They often impale their prey on thorns or barbed wire, or force them into cracks and crevices. Only two species, the loggerhead and the northern shrikes are known in North America. Both are migratory birds.

Figure 13-13 Distribution map of the loggerhead shrike.

The loggerhead shrike lives south of the coniferous forests of the United States with a range that extends into Mexico, see Figure 13-13. It has a gray upper body with white underparts and a black mask on its face, see Figure 13-14. The population of this bird is declining, and it is gone from much of its northern range east of the Mississippi River. These birds build nests of twigs in which they raise **broods** or families of 4–8 nestlings. They sit on high perches as they hunt for food. Their prey consists of large insects, lizards, small birds, and mice.

Figure 13-14 The loggerhead shrike. *Photo courtesy of Leonard Lee Rue III.*

The thrushes are members of the Turdidae family of songbirds. They are migratory birds, and thirteen species are known in North America. The best known of these birds are the robins and the bluebirds. The young birds of this family have spotted breasts, as do the adult thrushes other than robins and bluebirds.

Figure 13-15 Distribution map of the American robin.

The American robin is widely distributed, and its range spans the continent, see Figure 13-15. It is a colorful bird with a reddish-orange breast, brownish-gray upper body, and dark-colored head and tail, see Figure 13-16.

The male of this species has a strong singing voice that is often heard in the evenings as he sings from a high perch. Females lay 3–4 eggs in a nest made of grasses, roots, string, mud, and other materials. Two broods are often produced during the summer, and the young birds have speckled breasts. Robins eat mostly insects and

Figure 13-16 An American robin with her young. *Photo courtesy of Leonard Lee Rue III.*

worms, but they also become garden pests during some seasons because they eat berries and other fruits.

The thrushes are monogamous birds that form pairs each breeding season. Some species are quite aggressive in defending their nesting territories. They are found in a variety of habitats; some inhabit woodlands while others prefer meadows and pastures interspersed with trees.

AVIAN PROFILE

Mountain Bluebird (*Sialia currucoides*)

Figure 13-18 Distribution map of the mountain bluebird.

The mountain bluebird male is bright turquoise blue on the upper body with light-colored underparts, see Figure 13-17. Females tend to be drab gray, but blue coloring can be seen as they fly. This species is found mostly in the Rocky Mountain region from Alaska to Mexico, see Figure 13-18. Its preferred habitat is high mountain forests of pine or aspen interspersed with open meadows. It nests sometimes in farm buildings or nesting boxes that are within its territory.

These birds are hole nesters, and they lay 4–6 eggs in grass nests located in tree cavities or in holes in the eaves of buildings. Their diet consists mostly of insects.

Figure 13-17 The mountain bluebird.

Avocational Option

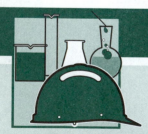

A favorite hobby for many people is to observe and identify birds in their natural environments. Field notes of bird observations by knowledgeable and interested hobbyists have contributed immensely to the body of scientific knowledge about birds and their habits. The thousands of amateur birders from around the world comprise a valuable volunteer field staff whose sightings of birds have contributed to the accuracy of range maps and have led to other scientific findings.

For example, the American robin alerted residents of Madison, Wisconsin to the dangers of DDT as an insecticide to reduce the spread of Dutch Elm disease in trees. Because the effects of this pesticide are magnified in the food chain, the contaminated earthworms that were eaten by the robins resulted in far greater concentrations of DDT in the robins. This killed the robins in large numbers, and alerted scientists to this problem with DDT.

The Mimidae family is one of the best known songbird families. It includes the mockingbirds, catbirds, and thrashers. These birds have strong voices, and their songs are quite complex. They have demonstrated the ability to mimic the sounds of other birds and animals, and this is the characteristic for which the family is named. Their food consists mostly of insects and spiders, but they also eat fruits and berries.

AVIAN PROFILE

Mockingbird (Mimus polyglottos)

Figure 13-20 Distribution map of the mockingbird.

The mockingbird is a plain-looking bird with gray coloring on the upper body and white underparts, see Figure 13-19. There is nothing plain about this bird, however, when it begins to sing. It can copy many of the sounds it hears, including other birds and mammals, and musical instruments. For this reason, it was named the mockingbird.

Figure 13-19 The northern mockingbird. *Photo courtesy of Leonard Lee Rue III.*

This bird is found in many parts of the United States, see Figure 13-20. Its habitat includes woodlands, gardens, and open grassy areas. It nests in thick shrubs which provide protective cover. Females lay 3–6 eggs in nests lined with fine plant materials. Males require high perches from which they defend their territories and sing their unusual songs.

The Icteridae family of perching birds includes the blackbirds, meadowlarks and New World orioles such as the northern oriole. These birds have powerful bills, and they feed on insects, fruits, seeds, and nectar. Males are usually larger and more brightly colored than females.

Blackbirds and closely related birds of the Icteridae family include the yellow-headed blackbird, red-winged blackbird, Brewer's blackbird, grackle, cowbird and others. Some blackbirds are polygamous, meaning that a bird of either sex may take several mates. Sometimes several males may tend a single nest with a female. In other instances, several females may mate with one male and nest nearby.

Figure 13-22 Distribution map of the great-tailed and common grackle.

The great-tailed grackle is a shiny blackbird with a long tail and wings, and a green sheen to its plumage, see Figure 13-21. It is found in parts of Mexico and Texas, and along the southern border of the United States from Texas to California, see Figure 13-22. The common grackle is a related species that ranges east of the Rocky Mountains from Canada to the Atlantic coast.

Figure 13-21 The great-tailed grackle.

Grackles and other members of the Icteridae family are omnivorous birds that feed on seeds, small fruits, and grain as well as on insects. They also eat the eggs and young of other birds. Males are polygynous, and tend to attract several mates. Females lay 3–5 eggs in large grass-lined nests anchored in trees, bushes, or reeds.

The birds in the Icteridae family tend to gather in large flocks before migrating in the fall season. Flocks of several million blackbirds of various species have been observed in their wintering areas. These flocks of birds sometimes cause extensive damage to grain crops due to the large numbers of individuals that congregate to eat in small areas.

Some species of cowbirds do not take care of their own young. They are **parasitic birds,** more specifically nest parasites, that lay their eggs in the nests of other birds. When a female lays an egg in another bird's nest, she may also remove one of the eggs belonging to the host bird. Some species of cowbirds simply take over the nests of other birds, and raise their broods in stolen nests. Cowbirds are **promiscuous** birds that mate with numerous members of the opposite sex. They do not form pair bonds.

The brown-headed cowbird is the smallest blackbird in North America. Males are shiny greenish black with dark brown heads, see Figure 13-23, and females are dull brownish gray. They gather in wooded areas along streams during the breeding season, and they inhabit pastures and fields during other times of the year. The

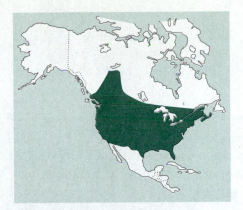

Figure 13-24 Distribution map of the brown-headed cowbird.

Figure 13-23 A male brown-headed cowbird. *Photo courtesy of Leonard Lee Rue III.*

diet of the cowbird consists mostly of seeds and insects. Their range extends from southern Canada to Mexico, see Figure 13-24. They lay 4–5 eggs in the nests of other songbirds.

The birds that make up the Icteridae family are found in a variety of habitats. Some of the blackbirds inhabit wetland areas. The meadowlarks prefer open fields and grasslands. The birds in this family eat such varied diets that they are able to live under a wide range of environmental conditions.

AVIAN PROFILE

Eastern Meadowlark (Sturnella magna)

Western Meadowlark (Sturnella neglecta)

■ Eastern Meadowlark
▨ Western Meadowlark

Figure 13-26 Distribution map of the Eastern and Western meadowlarks.

These two species of meadowlarks are nearly identical in appearance and habits. The songs of these two species are distinctly different, however, although they are able to mimic each other where their ranges overlap. These birds have yellow breasts crossed with a black v-shaped bib, and mottled brown backs, necks, and wings, see Figure 13-25. The ranges of these two birds overlap, but they occupy similar habitats, see Figure 13-26.

Meadowlarks nest on the ground in a domed-over nest constructed of dried grass and concealed in vegetation. Females lay 4–6 eggs. They are helpful birds to gardeners and farmers because they eat large amounts of harmful insects and weed seeds.

Figure 13-25 The meadowlark. *Courtesy of U.S. Fish and Wildlife Service. Photo by Kent N. Olson.*

The blackbird family includes the New World orioles. This leads to some confusion because they belong to a different family than many of the birds known as orioles. Orioles have black and orange- or yellow-colored plumage.

AVIAN PROFILE

Northern Oriole
(*Icterus galbula*)

Figure 13-28 Distribution map of the Northern oriole.

The northern oriole is a medium-sized bird with a black upper body, neck, and head. The underparts are orange, see Figure 13-27. Both the Baltimore and Bullock's orioles are races of northern orioles. These birds interbreed, and are widespread in North America, ranging throughout most of the United States from Canada to Mexico, see Figure 13-28. The diet of the Northern Oriole consists mainly of insects, seeds, and small fruit. Females lay 3–6 eggs in woven nests that hang from tree branches. The preferred habitat for the Northern Oriole is a wooded area along a stream. They also inhabit areas located near humans where tall trees are plentiful.

Figure 13-27 A male northern oriole at his nest. *Photo courtesy of Leonard Lee Rue III.*

AVIAN PROFILE

Starling
(*Sturnus vulgaris*)

Figure 13-30 Distribution map of the starling.

Starlings are gregarious birds that eat and roost in large flocks. They are mostly iridescent black with light-colored speckles especially in winter, see Figure 13-29. Females lay 5–6 eggs in cavities, and they often raise two broods each year.

The starling was introduced to North America from Europe in 1890. Since that time it has become established throughout much of the continent, see Figure 13-30.

The starling is one of the most adaptable birds in the world due to its ability to use many different kinds of food. It has become a serious pest in some places because it causes damage to vegetables and fruits. Starlings also eat or pollute grain that is intended for livestock feed.

Figure 13-29 A starling at a dump site. *Courtesy of U.S. Fish and Wildlife Service. Photo by Luther C. Goldman.*

CROWS AND JAYS

The crows and jays belong to the Corvidae family. This family includes the common crow, the raven, the magpie and the blue jay. Fifteen North American species are members of the corvidae. They are omnivorous birds that feed on vegetables, fruits, insects, and meat. Some of these birds are scavengers that eat carrion.

<table>
<tr>
<td>

AVIAN PROFILE

Common Crow
(Corvus brachyrhynchos)

Figure 13-32 Distribution map of the common crow.

</td>
<td>

The common or American crow is sometimes confused with the raven. Both of these birds are black, but the crow is the smaller of the two species, see Figure 13-31. The range of the crow includes much of the United States and Canada, see Figure 13-32.

The preferred habitat for these birds is deciduous forest and bushes close to rivers and streams, farmlands and agricultural fields. Nesting occurs in large stick nests lined with dried vegetation. Four to six eggs are common, and families stay together after the young have fledged. Crows are thought to be the most intelligent of all birds.

</td>
<td>

Figure 13-31 The common crow. *Photo courtesy of Leonard Lee Rue Enterprises. Photo by Irene Vandermolen.*

</td>
</tr>
</table>

Crows and their close relatives have been observed to perform simple tasks that demonstrate an ability to learn unusual skills. Experiments with ravens have shown that they can count to as high as five or six. Crows have been known to drop shellfish on hard surfaces to break them open. Some of these birds are also able to mimic the calls of other birds.

<table>
<tr>
<td>

AVIAN PROFILE

Black-billed Magpie
(Pica pica)

</td>
<td>

The magpie is a moderate-sized black-and-white bird with a long black tail, see Figure 13-33. It ranges along the Rocky Mountains from Alaska to New Mexico, see Figure 13-34. Magpies are sometimes quite noisy birds when they are gathered in flocks.

Nests are made of sticks and mud, and they are lined with

</td>
<td>

Figure 13-33 The black-billed magpie. *Photo courtesy of Leonard Lee Rue III.*

</td>
</tr>
</table>

Figure 13-34 Distribution map of the Black-billed magpie.

**Blue Jay
(Cyanocitta cristata)**

Figure 13-36 Distribution map of the bluejay.

fine roots and grasses. Females lay 6–9 eggs, and the young are cared for by both parents even after they leave the nest. They eat almost any kind of food, including insects, rodents, eggs, fruits, berries, and small mammals and birds.

The blue jay is a blue bird with a crested head, gray underparts, and a black necklace, see Figure 13-35. It ranges east of the Rocky Mountains from southern Canada to the Gulf of Mexico, see Figure 13-36. This bird inhabits oak and pine woods, and adapts well to gardens and human neighbors.

Blue jays build nests in trees where they raise 3–6 offspring. They use nearly all kinds of building materials. They are noisy birds that eat seeds, fruits, invertebrates, and the eggs and nestlings of other birds.

Figure 13-35 The bluejay. *Photo courtesy of Leonard Lee Rue III.*

Other birds in the corvidae family include the Clark's nutcracker, which differs from other family members in that it eats mostly pine nuts. Several other species of jays are found in North America, including the "camp robber", or gray jay that is most often observed by campers in the northern coniferous forests. The raven is another important member of this family.

CUCKOOS

The family Cuculidae includes the North American cuckoos, the anis and the roadrunner. The roadrunner is discussed in Chapter 14 as an upland ground bird.

Many of the Old World cuckoos are parasitic birds, but the New World cuckoos care for their own young. These birds eat insects, spiders, centipedes, and worms, and some cuckoos' diets consist of hairy caterpillars that are avoided by many kinds of birds. They live in wooded habitats, thickets, and orchards.

AVIAN PROFILE

Yellow-billed Cuckoo (*Coccyzus americanus*)

Figure 13-38 Distribution map of the yellow-billed cuckoo.

The yellow-billed cuckoo is a long-tailed bird that is olive brown on the upper body with white underparts. It has a long, curved bill with the lower mandible yellow in color, see Figure 13-37. This bird ranges from the East Coast to the Rocky Mountains, and from southern Canada to Mexico, see Figure 13-38. The preferred habitat for this bird is deciduous woods with large trees, usually located near rivers.

Females lay 2–4 eggs in nests made of twigs that are located in trees or bushes. They are perching birds that sit quietly as they watch and listen for the insects that make up their diets.

Figure 13-37 The yellow-billed cuckoo.

KINGFISHERS

Kingfishers are solitary birds that spend most of their time alone looking for fish. Once they see one, they plunge into the water to catch it. These birds have strong bills, and some members of this family have large crested heads. The belted kingfisher is the more common of the two North American species of kingfishers.

AVIAN PROFILE

Belted Kingfisher (*Megaceryle alcyon*)

Figure 13-40 Distribution map of the belted kingfisher.

The belted kingfisher is bluish-gray on the upper body and wings, with a white neck and underparts. The female of this species is more brightly colored than the male. The kingfisher has a large crested head and a long, heavy bill, see Figure 13-39. This bird ranges from Alaska and Canada to Central America, see Figure 13-40.

These birds prefer wooded habitats along rivers, lakes, and coastlines. They perch on limbs overlooking the water where they watch for fish. They catch their prey by hovering above a fish, and then diving into the water to catch it.

Figure 13-39 A belted kingfisher.

Females lay 5–8 eggs in tunnels that have been excavated in streambanks.

HUMMINGBIRDS

Hummingbirds are unusual in that much of their nourishment comes from nectar, and they beat their wings so fast that they can hover in place during flight. This group of birds also includes the smallest birds in the world with one member of the family Trochilidae being only about the size of a bumblebee. Fifteen species of these colorful birds breed in North America. They have long bills and tubelike tongues with which they take nectar from flowers. They also eat soft insects.

The wings of hummingbirds beat at 15–79 times per second depending on the species, and the familiar whirring sound that is created as they fly has earned them their name. These birds are able to rotate their wings in a manner similar to the way humans rotate their wrists. This allows them to fly forwards or backwards. Their spectacular flying skills allow them to perform amazing display flights during the mating season as they attempt to attract and win mates.

AVIAN PROFILE

Black-chinned Hummingbird (Archilochus alexandri)

Figure 13-42 Distribution map of the black-chinned hummingbird.

The black-chinned hummingbird is one of the most common hummingbirds in North America. It is green in color with a black chin and a purple throat, see Figure 13-41. It is less than 4 inches long, and it is found mostly in the western United States, see Figure 13-42.

Nests are constructed of wool and/or lichens, and they are usually woven together and hung in a shrub or tree using spider webs. Two eggs make up a clutch, and the males defend their territories. Nectar is their main food, but they will also eat spiders and insects. These birds migrate south for the winter to warm locations where food is available.

Figure 13-41 The black-chinned hummingbird.

There are many perching birds in North America that have not been discussed in this book. They include but are not limited to the following: wood warblers, flycatchers, titmice, wagtails, pipits, trogons, wrentits, flycatchers, gnatcatchers, vireos, waxwings, and larks.

All of these birds fill important niches in the ecosystems of North America. Their roles frequently overlap those of many of the other birds that have been discussed in more detail. Five chapters of this text have been devoted to birds, and in such limited space, it is impossible to discuss in detail the roles of all the North American birds.

LOOKING BACK

The perching birds include songbirds of many kinds as well as cuckoos, kingfishers, hummingbirds, crows, and jays. All of these birds perch upright on tree branches or other surfaces. Songbirds use songs to help define the boundaries of the territories that they defend.

Perching birds occupy many different habitats and eat many different kinds of food. Some are solitary birds while others are gregarious birds. They have a variety of mating and nesting behaviors. Some are colorful birds and others are plain with males generally being more colorful than females. The only thing that is common to all of them is their perching behavior.

REVIEW QUESTIONS

1. Identify the major families of North American perching birds using their Latin names, and list the birds that belong to these families using their common names.
2. Make a chart listing the major songbirds discussed in this chapter, describe the physical appearances of both males and females.
3. List some reasons why the males of many species of birds are brightly colored while the females often have dull and plain coloring.
4. Discuss the importance of the calls and songs of territorial birds.
5. Identify some characteristics of house sparrows and starlings that helped them adapt quickly to the ecosystems of North America.
6. Define the following terms that describe mating habits of birds: polygynous, polygamous, monogamous, polyandry, and promiscuous.
7. Name and describe some well-known members of the thrush family of birds.
8. Describe how parasitic birds are able to reproduce without building their own nests.
9. Explain why gregarious birds tend to cause greater damage to crops than do solitary birds that have similar eating habits.
10. List some unique physical characteristics of hummingbirds that make it possible for them to gather and eat nectar.

1. Call on members of the Audobon Society or a local birdwatchers group to make a presentation to your class. Ask them to concentrate on birds that are found locally and to illustrate their talk with specimens or pictures. Where possible, go on a field trip to observe and identify birds that are found in the local community.

2. Obtain and show a video or slides illustrating the appearance and habits of the songbirds and other perching birds. Sources of these materials are PBS television, biological supply houses, and public libraries. Follow up this activity by having each student prepare an illustrated written report about a particular bird species.

3. Assign each class member to make a birdhouse for a particular species of bird (bluebird, wren, etc.). Plan a field day to install them in an appropriate area where they are likely to be used. Conduct an annual inventory of bird usage of the birdhouses.

14

Other Birds of North America

$\mathbf{M}$ANY SPECIES OF BIRDS are found in North American ecosystems. Although they are too numerous to include all of them in a textbook that focuses on ecology, this chapter discusses representative species of several large families of birds.

The waterbirds include swimming or wading birds that depend on water environments for food or protection. The swallow-like birds are insectivores that often live close to human habitation. Upland ground birds spend their lives living on the ground, while the tree-clinging birds are insectivores that depend on trees for food and shelter.

OBJECTIVES

After completing this chapter, you should be able to

- list distinguishing characteristics of the different families of waterbirds
- define the roles of the long-legged wading birds in the ecosystems of North America
- evaluate the roles of gull-like birds in North American ecosystems
- define the roles of the upright perching waterbirds in the ecosystems of North America
- evaluate the roles of duck-like birds in the ecosystems of North America
- consider the roles of the sandpiper-like birds in North American ecosystems
- explain the roles of upland ground birds in the ecosystems of North America
- appraise the roles of swallow-like birds in North American ecosystems
- describe how the torpid state that affects some kinds of birds contributes to their survival during cold weather conditions
- evaluate the roles of the tree-clinging birds in the ecosystems of North America.

The waterbirds included in this chapter are grouped as long-legged wading birds, gull-like birds, upright perching waterbirds, duck-like birds, sandpiper-like birds, and chicken-like marsh birds. All of these birds depend on water habitats to survive.

LONG-LEGGED WADING BIRDS

The long-legged wading birds include the herons, egrets, bitterns, storks, ibises, and cranes. These are all wading birds that enter shallow water in pursuit of frogs, crayfish, and small fish. Some of these birds stand still and wait for their prey to pass nearby. Others actively seek out and pursue their prey.

Figure 14-2 Distribution map of the great blue heron.

The great blue heron is a large bird that stands 4 feet tall. Its back and wings are blue gray, and its light-colored underparts are streaked with black. It has a white head with a black stripe and plumes behind its eyes, see Figure 14-1.

This heron is distributed throughout the U.S. and along the southern border of Canada, see Figure 14-2. It is a migratory bird that breeds in the northern part of its range, and spends the winter in the southern states. It nests in colonies where females lay 3–7 eggs in stick nests. Its preferred habitats are marshlands, swamps and shorelines.

Figure 14-1 The great blue heron. *Photo courtesy U.S. Fish and Wildlife Service.*

Herons are adapted to their environments better than most birds. They eat several different kinds of food such as fish, voles, frogs, and insects. They choose safe nesting sites on piles of reeds or on man-made structures. The versatility of these birds helps them to survive where more specialized birds might have difficulty finding suitable food or nesting sites.

The American bittern is a moderate-sized heron that is brown with light-colored streaks that help it blend in with the marsh reeds in its habitat, see Figure 14-3. Its nesting range includes much of the United States and Canada with northern populations migrating south in the winter, see Figure 14-4. Its preferred habitats are marshy lakes and meadows, where it eats frogs, crayfish, and grasshoppers.

Figure 14-4 Distribution map of the American bittern.

Figure 14-3 The American bittern. *Courtesy of U.S. Fish and Wildlife Service. Photo by Gary Zahm.*

Bitterns are territorial, and males defend their home territories against other birds. Females lay 3–7 eggs in platform nests that are protected by heavy plant cover. This bird has a distinctive call that is heard in the morning and evening hours. During the nesting season, the young are especially vulnerable to predators such as mink, snakes, and birds of prey.

Some members of the heron family nest together in large colonies. Birds and animals that live together in large groups are described as being **gregarious**. Other members of this family are territorial, and a single nesting pair is found in a given area.

AVIAN PROFILE

Snow Egret (Egretta thula)

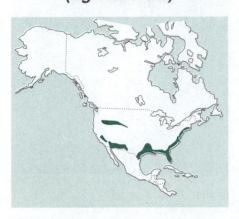

Figure 14-6 Distribution of the snowy egret.

The snowy egret is a gregarious medium-sized white heron with a black bill, dark-colored legs, and yellowish feet, see Figure 14-5.

Figure 14-5 The snowy egret. *Courtesy of U.S. Fish and Wildlife Service. Photo by Jim Leupold.*

It is found in marshy habitats along estuaries and inland lakes in widely dispersed regions of the United States, see Figure 14-6.

Males and females look very much alike, and they share the duties of incubating the eggs and caring for the young. Females lay 3–6 eggs in platform nests constructed in willows or other trees. They hunt in the marshes for small fish and other aquatic animals and insects.

The heron family contains several members that are not discussed in this chapter. They include the Louisiana heron, little blue heron, black-crowned night heron, yellow-crowned night heron, green heron, reddish egret, great egret, cattle egret, and least bittern. Other large wading birds include the stork, limpkin, ibis, and flamingo.

AVIAN PROFILE

Sandhill Crane
(Grus canadensis)

Figure 14-8 Distribution map of the sandhill crane.

The sandhill crane is a large bird with long legs and a long neck. It stands over 3 feet tall, and it is bluish-gray in color with a bald red forehead, see Figure 14-7. It is an omnivorous marsh bird that eats mostly plants and seeds, but will also eat frogs, snakes, lizards, and mice.

Breeding pairs build mounds of reeds and other plant materials in marshes where the females lay two eggs. Both parents incubate the eggs and care for the young. These birds range in marshlands from the far northern regions to Texas, and another population of these

Figure 14-7 A young sandhill crane. *Courtesy of U.S. Fish and Wildlife Service. Photo by William Radke.*

birds lives year-round in southern Florida, see Figure 14-8.

An endangered crane known as the whooping crane occupies similar habitats and has similar habits. The breeding flock of these birds has seriously declined despite the efforts of scientists to save the birds from extinction, see Figure 14-9. The

Figure 14-9 The whooping crane. *Courtesy of U.S. Fish and Wildlife Service. Photo by Steve Van Riper.*

decline is due in part to loss of habitat, shooting, and disturbance by humans. Several whooping cranes that were reared in the recovery effort have died when they flew into overhead power lines.

Some of the wading birds are adorned with beautiful plumage and/or exotic coloring. In some species these bright colors and plumage are only evident during the mating season. Other birds such as the scarlet ibis, roseate spoonbill, and flamingos have spectacular appearances during all seasons of the year.

Figure 14-11 Distribution map of the American flamingo.

The American flamingo is a large, slender wading bird with pink plumage, black wing feathers and salmon pink legs. It also has a dark-colored bill that curves sharply downward, see Figure 14-10. In North America it is only found along the coast of Florida, see Figure 14-11. It inhabits saltwater flats and lagoons. Flamingos feed by filtering small organisms from the water through the comb-like bristles in their bills.

Figure 14-10 The American flamingo. *Photo courtesy of Wendy Troeger.*

Females lay one or two eggs in cone-shaped mud nests on remote offshore islands. They nest in colonies with several hundred birds crowded together in close quarters.

GULL-LIKE BIRDS

Gulls and gull-like birds are widely distributed in North America. These birds are strong flyers, and they sometimes travel long distances each day in search of food. They feed mostly on fish and other animal matter, and they nest in colonies.

Gulls of numerous species are found along the seacoasts and inland waters of North America. They have similar habitats and behaviors, and most gulls are **scavengers**, meaning that they will eat any food they can find. They have adapted to modern agricultural practices, and some gulls feed on worms and larvae that have been exposed by tillage, or that have come to the surface during the irrigation of crops. Many of them also use garbage dumps as feeding sites. They also prey on the eggs and young of other shorebirds.

The California gull has a dark-colored eye, white head, neck, and body, with a gray back, and red and black spot on its lower bill, and black-tipped wings, see Figure 14-12. This gull ranges along the California coast and inland from northeastern California to the prairie regions of central Canada, see Figure 14-13. It is the state bird of Utah because it saved the early settlers from famine by eating the grasshoppers that were destroying their crops.

Figure 14-12 The California gull. *Photo courtesy of Leonard Lee Rue III.*

This gull nests in colonies on islands surrounded by shallow inland lakes. Females lay 2–3 eggs in nests made of grass and sticks. They feed on carrion, small rodents, large insects, and aquatic animals of various kinds.

There are over thirty-one species of gulls and terns found in North America. Gulls vary greatly in size and in their geographic ranges. They are swimming birds with webbed feet and hooked beaks. Most gulls are gregarious birds. While several kinds of gulls are often found together in nesting areas, they mate only with members of their own species.

Terns are closely related to gulls but they eat mostly fish, and they are not scavengers. They resemble gulls in their body structure, but they have sharp-pointed, not hooked bills. The tails of gulls are usually square on the ends, while the tails of most terns are forked.

AVIAN PROFILE

California Gull
(Larus californicus)

Figure 14-13 Distribution map of the California gull.

AVIAN PROFILE

Arctic Tern
(Sterna paradisaea)

Figure 14-15 Distribution map of the arctic tern.

The arctic tern is identified by its white face and underparts, black cap, red bill, and gray back and wings, see Figure 14-14. It inhabits coastal regions and tundra lakes in northern Canada and Alaska, and it migrates to the Antarctic regions during the winter, see Figure 14-15. Some arctic terns migrate close to 23,000 miles round-trip each year.

Figure 14-14 The arctic tern. *Photo courtesy of Leonard Lee Rue III.*

Females lay two eggs in scrapes on the ground. They nest either in colonies or as pairs. When the young are able to fly, flocks of terns begin their migrations south. Like most terns, this bird often hovers before diving beneath the surface of the water to catch its food.

There are many gulls and gull-like birds in North America besides those that are discussed in this chapter. Among them are the petrels, skuas, jaegers, boobies, tropicbirds, frigatebirds, fulmars, shearwaters and albatrosses.

UPRIGHT PERCHING WATERBIRDS

The upright perching waterbirds are unusual in that their legs are placed far back on their bodies, and they stand in an upright position. The auks, murres, and puffins belong to the Alcidae family. They are arctic birds that are found in the greatest numbers in the far north. They are awkward on land, but they are expert swimmers. They are capable of pursuing and catching fish that comprise of a large percent of their diet.

**Tufted Puffin
(Lunda cirrhata)**

Figure 14-17 Distribution map of the tufted puffin.

The tufted puffin is a short seabird that is black on the upper body and white on the lower body, with a beak like that of a parrot, see Figure 14-16. Puffins are sometimes called sea parrots. The range of these birds is restricted to the arctic coast and to the Pacific coast from Alaska to California, see Figure 14-17.

The puffins' diet consists mostly of small fish which they gather at sea. They nest in colonies, and females lay a single egg in burrows or rock crevices.

Figure 14-16 The tufted puffin.
*Courtesy of U.S. Fish and Wildlife Service.
Photo by Susan Steinacher.*

The alcids are unusual birds. In many respects they are like the penguins of the Antarctic except that they can fly. Small members of the auk family are called **auklets**. Some of these birds venture quite far out to sea as they search for food.

**Cassin's Auklet
(Ptychoramphus aleuticus)**

Figure 14-19 Distribution map of Cassin's auklet.

The Cassin's auklet is a seabird the size of a robin. It is slate colored on the upper body, with light gray coloring underneath except for its white belly, see Figure 14-18. Its range extends from Alaska to Baja California, see Figure 14-19. These birds produce a single egg, and they nest in burrows or rock cavities. Parents incubate the chick in shifts.

This auk sometimes goes far out to sea in search of tiny shrimplike plankton for food. Its preferred habitat is open ocean, but it nests on sea cliffs and promintory points of land that extend out into the sea. Natural enemies include seagulls, foxes, weasels, and other mammals.

Figure 14-18 Cassin's auklet.

The upright perching water birds use their webbed feet and their short wings to propel themselves as they swim. They are well adapted to living in marine biomes.

Cormorants are long-necked seabirds that dive for food from the surface of the water. They swim using both their feet and their wings, and they have been known to dive to depths in excess of 100 feet. Their diets consist entirely of fish.

Figure 14-21 Distribution map of Brandt's cormorant.

The Brandt's cormorant is a large black seabird with a long neck and body, see Figure 14-20. Its range extends along the Pacific coast from Alaska to Mexico, see Figure 14-21. This bird eats fish that it pursues and captures beneath the surface of the ocean. Females lay 3–6 eggs in nests located on cliffs and rocky island coastlines.

These birds are gregarious, and several hundred of them often dive together for food. This improves their chances of catching the fish upon which they feed, because they are able to confine their prey to an area beneath the surface making them easier to catch.

Figure 14-20 A cormorant with full-wing spread. *Courtesy of U.S. Fish and Wildlife Service. Photo by Luther C. Goldman.*

DUCKLIKE BIRDS

Ducks were discussed along with other waterfowl in Chapter 10. There are a number of other duck-like birds such as pelicans, loons and grebes. These are waterbirds that inhabit fresh- and saltwater environments at different seasons of the year. All of these birds depend on diets of fish.

Two kinds of pelicans are found in North America. They are the white pelican and the brown pelican. The pouched bills of these birds are used as nets to catch fish. They are also used as containers to carry captured fish to young pelicans still in the nest.

The brown pelican sometimes plunges into the water during flight to capture fish. White pelicans are known to form lines to drive fish to shallow water where they can be scooped up more easily. Both of these pelicans are quite vulnerable to predators during the nesting season.

The white pelican is a large white bird measuring over 5 feet in height. It has a large, pouched bill and black wing feathers, see Figure 14-22. It breeds along inland lakes and rivers from British Columbia to Utah, and it migrates to the coastal regions of California and to the Gulf Coast during the winter, see Figure 14-23.

Figure 14-23 Distribution map of the white pelican.

Pelicans nest in large colonies, and females usually lay two eggs in large mound nests located in marsh and prairie regions. Young pelicans are naked, and their parents have to shield them from the sun to prevent sunburn. Pelicans fly together in long lines to reduce wind resistance on the birds behind the leader.

Figure 14-22 The white pelican. *Courtesy of U.S. Fish and Wildlife Service. Photo by John Foster.*

Loons are swimming birds that capture fish by spearing them with their sharp bills. They are **solitary** birds, meaning they spend much of their lives alone or in pairs. They are well known in the northern woods of Canada and Alaska for their strange nocturnal calls.

AVIAN PROFILE

Common Loon (Gavia immer)

Figure 14-25 Distribution map of the common loon.

The common loon is a large dark-colored bird with white markings on its back. It has a glossy, greenish black head and neck, with a white marking around its neck, and white underparts, see Figure 14-24. The summer range of this bird includes the wooded areas of Alaska, Canada, and the Great Lakes region of the United States. It migrates to both the Pacific and Atlantic coasts during the winter, see Figure 14-25.

Females nest on piles of vegetation along the shores of freshwater ponds and lakes. This loon is a diving bird that is known to reach depths of over 200 feet in pursuit of fish. They sometimes escape their enemies by gradually sinking in the water to avoid detection.

Figure 14-24 The common loon. *Photo courtesy of Leonard Lee Rue III.*

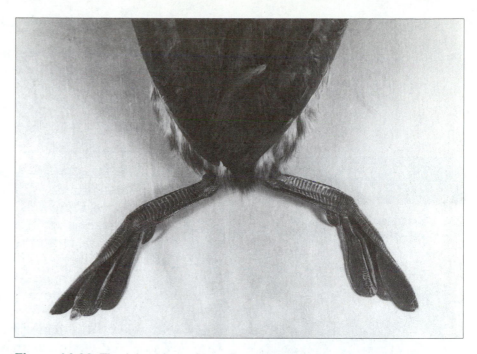

Figure 14-26 The lobed feet of a grebe. *Photo courtesy of Leonard Lee Rue III.*

Grebes are aquatic diving birds that eat mostly insects and small aquatic animals. They also eat small amounts of vegetation. Unlike many waterbirds, they do not have webbed feet. They have **lobes** on their toes to help them swim easier, see Figure 14-26. Ten species of grebes are found in North America.

**Pied-billed Grebe
(Podilymbus podiceps)**

Figure 14-27 Distribution map of the pied-billed grebe.

The pied-billed grebe is found in most regions of the United States and southern Canada, see Figure 14-27. It is a small bird with a bill similar to that of a chicken. A black mark around the bill accounts for its name, see Figure 14-28.

This grebe prefers marshland habitat, and it can be found most readily where the water is shallow and water plants are abundant. Winter habitat is usually open ocean and saltwater bays. The females lay 5–7 eggs in floating nests constructed of marsh vegetation. Nests are anchored to plants to keep them from floating into open water. Young grebes ride on the backs of their parents, or hold onto their tails to be towed.

Figure 14-28 A pied-billed grebe with a catfish. *Photo courtesy of Leonard Lee Rue Enterprises. Photo by Irene Vandermolen.*

SANDPIPERLIKE BIRDS

The sandpiperlike birds are small- to medium-sized birds whose coloring matches the habitats in which they live. They are long-legged wading birds with long slender bills, and feed mostly on insects and small aquatic animals. They nest on the ground near ponds and marshes.

Figure 14-29 Distribution map of the spotted sandpiper.

The spotted sandpiper is the most widespread of the sandpipers. It ranges throughout most of Alaska, Canada, and the continental United States, see Figure 14-29. It is a migratory bird that flies south to warm climates each winter. It is one of the smallest sandpipers and has a straight bill of moderate length. It has a brown upper body and white underparts that are covered with brown spots in the summer, see Figure 14-30.

Females lay four eggs near water in depressions on the ground. Spotted sandpipers can be identified by their habit of teetering as if they are unbalanced.

Figure 14-30 The spotted sandpiper. *Photo courtesy of Leonard Lee Rue III.*

Sandpipers, snipes, and other closely related birds are described as **precocial** birds. This means that the young are covered with down, and they become active immediately after they have hatched. They do not require care in their nests.

Figure 14-32 Distribution map of the long-billed curlew.

The long-billed curlew is a large shorebird that is mottled brown on the back and wings, with buff-colored underparts. It has a long neck and long, curved bill, see Figure 14-31. It inhabits marshes, mud flats, and open plains where it eats insects, worms, fish, shellfish and other small aquatic animals. During the winter it feeds in fields or along beaches and salt marshes. The range of this bird extends from southern Canada to the Great Basin, and eastward to Oklahoma and Texas, see Figure 14-32.

Figure 14-31 The long-billed curlew. *Photo courtesy of Leonard Lee Rue III.*

Females lay four eggs in grass-lined ground nests. They defend territories during the nesting season, but they are social birds during the rest of the year.

The **plovers** are a family of small- to medium-sized shorebirds. Ten species are found in North America. These birds have loud voices, and they nest in scrapes on the ground. Females usually lay four speckled eggs that are incubated by both parents. The chicks are active almost as soon as they hatch. These birds feed on insects and small aquatic animals. They gather food in the wet areas along the shorelines, but they don't normally wade very far from the shore.

The killdeer is about the size of a robin. It is white underneath and grayish brown above, with a long, tan-colored tail and two black bands on its breast, see Figure 14-33. This bird is found throughout much of North America, see Figure 14-34. It is easily recognized by its shrill, "kill-deeah" call from which it gets its name.

AVIAN PROFILE

Killdeer Plover (*Charadrius vociferus*)

Figure 14-34 Distribution map of the killdeer plover.

Figure 14-33 A killdeer plover at her nest. *Courtesy of U.S. Fish and Wildlife Service. Photo by Perry Reynolds.*

The adult killdeer often draws predators away from its young by dragging one wing on the ground and pretending to be injured. The parent will stay just far enough ahead to encourage the enemy to follow as it leads it away from the young birds.

Other birds that belong in this grouping are the phalaropes, oystercatchers, and avocets. The **phalaropes** are unusual birds in that females are brightly colored and the drab-colored males incubate the eggs. Oystercatchers live on coastal beaches and tidal flats, where they feed on shellfish and other marine animal life in the intertidal zone. These areas are covered with water during high tide, but they are above the water line at low tide. The bills of these birds are adapted for opening the shells of mollusks and other shellfish. The **avocets** are unusual-looking shorebirds whose bills curve upwards.

CHICKENLIKE MARSH BIRDS

The family Rallidae includes rails, gallinules, and coots. Most members of this family have short wings and tails. They are usually dull-colored birds that feed on vegetable and animal matter that they find by swimming or diving along seashores and in shallow marshland areas. Their narrow bodies are adapted to moving easily through the reeds and other aquatic plants found in their natural environments.

An interesting characteristic observed in some of the birds discussed in this section is the **frontal shield.** This is a fleshy growth that extends from the top of the bill to the forehead. In some species it is white, while in others it is dark red. The frontal shield is used by strong birds to intimidate weak birds. It is known to increase in size in males that are defending their territories.

The common gallinule is a slender bird with brown wings and back; the rest of the body is dark gray. It has a red frontal shield that extends from its red bill to its forehead. The tip of its bill is bright yellow, see Figure 14-35.

Figure 14-36 Distribution map of the common gallinule.

Figure 14-35 A common gallinule. *Courtesy of Leonard Lee Rue Enterprises. Photo by Irene Vandermolen.*

Gallinules are found in marshes and open water habitats in many areas of North America, see Figure 14-36. They are prolific birds that lay 9–12 eggs in platform nests near the edge of the water. They are omnivorous, and their diets include insects, tadpoles, seeds, and fruits.

Rails are divided into three main groups that include the long-billed rails, the crakes and gallinules, and the coots. These birds live in the reeds that line the shores of many wetlands. They nest on platform nests that float on the water. Some rails were unable to fly, and this contributed to the extinction of some species that were native to offshore islands. The remaining North American species are birds that can fly.

**American Coot
(*Fulica americana*)**

Figure 14-38 Distribution map of the American coot.

The American coot is a dark gray waterbird with a black neck and head. Its lobed toes aid in swimming or walking in soft mud, and it has a distinct white frontal shield above the bill, see Figure 14-37. Coots are gregarious birds that are found from Canada to California and the Gulf Coast, see Figure 14-38. Their preferred habitats include marshes, ponds, and lakes. These birds dabble but they will also dive to the bottoms of marshes and small bodies of water to feed on submerged vegetation.

Figure 14-37 The American coot. *Courtesy of U.S. Fish and Wildlife Service. Photo by Luther Goldman.*

Females lay up to a dozen eggs in platform nests constructed of plant materials. Nests are located offshore, and they are used as resting areas after the young have hatched.

Some of the chickenlike marsh birds, including coots and gallinules, are gamebirds that are hunted during certain seasons of the year. Many of them are similar in body structure to partridges, except that they are more slender. They are noisy birds, and it is believed that they call to one another to warn other birds away from their territories.

UPLAND GROUND BIRDS

The upland ground birds most commonly found in North America are nightjars, grouse, roadrunners, pheasants, quails, partridges and turkeys. All of these birds except the nightjars and roadrunners were discussed in Chapter 11.

The **nightjars** are nocturnal birds that eat insects, and for this reason are called insectivores. Their mouths are large and their bills are lined with bristles that aid in capturing insects. Their harsh or jarring calls after dark earned these birds the name of nightjars. The whip-poor-wills and poor-wills are common nightjars found in North America.

Figure 14-39 Distribution map of the whip-poor-will.

The whip-poor-will is a forest dweller whose range includes the eastern half of the United States and the border region of southern Canada, see Figure 14-39. This species is also found along the Gulf Coast and in some areas of the southwest and northern Mexico. Its name comes from the sound of its call which is repeated throughout the night. Its color is a mixture of mottled tan, gray, brown, and black, see Figure 14-40. It blends well with the forest floor where it lays its two eggs.

These birds were once known as goatsuckers because goat herders believed that they were sucking the milk from their animals. Nightjars call to each other during the night at about the same time of the year that the milk production of goats begins a natural decline. The actual diet of the bird consists of insects.

Figure 14-40 The whip-poor-will.
Photo courtesy of Kent A. Vliet.

The plumage of nightjars is mottled to help them blend with their surroundings. Nightjars are not known as social birds. However, during the nesting season, a single male may have several mates nesting close together on the ground. This nesting habit makes them vulnerable to most **terrestrial** or land-based predators.

The roadrunner has become famous in the cartoon that matches the bird against the wily coyote. It is brown in color, and streaked with black and white on the upper body, with a greenish sheen to its plumage. The lower body is buff colored and streaked with brown. The bill is large, and the bird has a long tail and a crested head, see Figure 14-41. The roadrunner ranges throughout the desert regions of the southwestern United States, see Figure 14-42.

Roadrunners prefer to run rather than to fly, and they seek shelter in the brush and scrub found in desert habitats. Females lay

Figure 14-42 Distribution map of the roadrunner.

Figure 14-41 The roadrunner.

3–6 eggs in nests located in mesquite bushes, large cacti, or shrubs. They are predatory birds that eat snakes, rodents, lizards, scorpions, insects, and young birds.

SWALLOWS AND SWIFTS

Swallows are insectivores that often live in complete harmony with humans, and their mud nests are often found attached to man-made structures. Sometimes they build nests in banks and cliffs by making holes and lining them with grass. These birds are easily recognized by their long wings and tails and by their large mouths with which they catch insects in flight. Eight species of swallows are found in North America.

AVIAN PROFILE

Barn Swallow (Hirundo rustica)

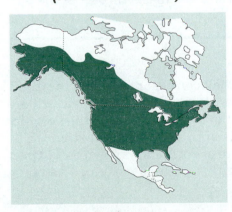

Figure 14-43 Distribution map of the barn swallow.

The barn swallow ranges throughout much of North America, see Figure 14-43. It is a bluish black bird on the upper body, with rust-colored underparts. The wings are long and streamlined, and the tail is forked, see Figure 14-44. Barn swallows inhabit open country near water where insects are abundant.

They nest in colonies and they hunt together. They also join together to drive hawks and other predatory birds away from their nesting sites. Females lay 4–6 eggs in feather-lined nests made of mud and straw.

Swallows are migratory birds, and they fly to Central or South America for the winter.

Figure 14-44 The barn swallow. *Courtesy of U.S. Fish and Wildlife Service. Photo by J.C. Leupold.*

They return to the same nesting sites year after year, with the older birds arriving first. They are monogamous birds, and both parents are involved in feeding the young. Two clutches of eggs are often produced in a year, and young birds from the first brood have been observed helping their parents gather food for the second brood. They are somewhat territorial even though they often nest in colonies. They tend to defend small territories around their nests against other members of the colony.

One group of birds often mistaken for swallows is the **swifts.** These migratory birds often have slightly forked tails, and an appearance similar to that of a swallow or martin. They have weak legs and feet, and they seldom land except during the nesting period. They cling to vertical surfaces to avoid landing on the ground. These birds are insectivores that sometimes spend several weeks in flight without landing. They even mate in flight.

Swifts are monogamous birds that return to the same nesting sites each year. They prefer to nest on cliffs and canyon walls in sheltered areas. Their nests are constructed mostly of sticks, vegetation, grass, and algae glued together with sticky saliva.

AVIAN PROFILE

White-throated Swift (*Aeronautes saxatalis*)

Figure 14-46 Distribution map of the white-throated swift.

The white-throated swift is the size of a barn swallow with a slightly forked tail and long, narrow wings. Its wings, sides and tail are black, and the throat and breast areas are marked with white, see Figure 14-45. This bird ranges from British Columbia to Central America, although much of its habitat is in the Rocky Mountains and the southwestern United States, see Figure 14-46.

These birds nest in colonies on cliffs. A typical clutch consists of three to six eggs. Immature birds or **nestling**s are fed pellets of insects that are collected in the throats of the parents during flight.

Figure 14-45 The white-throated swift.

Some birds survive cold weather conditions by slowing their metabolism until they appear near death. Their body heat is temporarily lost, all of the bodily functions slow down, and they require little or no food. This condition is similar to that experienced by animals that hibernate. It is a state of dormancy known as **Torpor.** A bird or animal that is found in this state is described as being **torpid.**

Swifts have been known to leave their nesting areas during cold, rainy weather when insect food is hard to find. They fly long distances to feed until the storm is over. While they are gone, the nestlings become cold and torpid. In this state of torpor, they are able to survive until their parents return to feed them.

Figure 14-47 Distribution map of the chimney swift.

The chimney swift is an inhabitant of the eastern half of the United States and southern Canada, and it is slowly expanding its range westward, see Figure 14-47. It is charcoal brown with a light-colored throat. It has a short, square tail and long, narrow wings, see Figure 14-48. In flight, it tends to alternate between rapid flight and gliding. This bird used to be found mostly in woodland areas, but it now inhabits cities and towns where females lay 4–5 eggs in chimney nests. They also nest in hollow trees or caves. A nest consists of a few twigs glued together with the bird's saliva.

Figure 14-48 The chimney swift.

TREE-CLINGING BIRDS

The tree-clinging birds include the woodpeckers, nuthatches, and creepers. All of these birds are insectivores, and they spend most of their time climbing tree trunks in search of insects. These birds play important roles in forest and desert ecosystems by controlling insects that damage trees and cacti. In addition, woodpeckers make holes in trees and other woody plants that are used for nesting sites and shelter by many birds and animals.

Twenty species of woodpeckers live in North America. These birds are usually distinctively colored, and they are equipped with strong skulls and straight bills. They use their bills like chisels to dig into the surfaces of trees as they hunt for insects. Their long tongues are used to get insects out of the holes in the wood. They are solitary birds that come together only to nest.

The common flicker is a woodpecker that is widely distributed in North America, see Figure 14-49. Three different color variations are known for this bird depending on its race. The wings and tail of the race known as the red-shafted flicker are lined with pink that is easily noticed as the bird flies. The lower body is black and white,

Figure 14-49 Distribution map of the common flicker.

the back is barred with cinnamon and brown, the rump is white, the neck and face are gray, and the head is brown. Males have red mustaches, see Figure 14-50.

Common flickers prefer deciduous or mixed woodland habitats, but they are adaptable to a wide variety of environmental conditions. A nesting female lays 6–10 eggs in a hole drilled into a tree, cactus, or post. The diet of the flicker consists of ants and other insects, with berries making up part of the diet during the winter. Unlike many other woodpeckers, the flicker is frequently seen on the ground pursuing ants.

Figure 14-50 A female yellow-shafted flicker with her young. *Photo courtesy of Leonard Lee Rue III.*

The diets of woodpeckers vary quite a bit, with insects and spiders making up a large part of the food supply. They also are known to eat seeds, berries, and other fruit. One woodpecker eats only acorns, and another known as the sapsucker eats the sap of trees and the insects that the sap attracts.

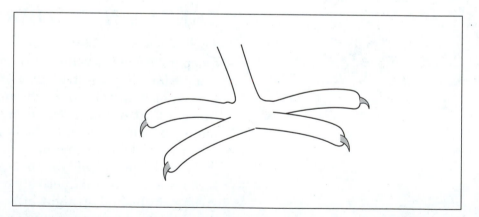

Figure 14-51 The toe arrangement of the climbing birds are two toes forward and two toes behind.

The toes of many of the climbing birds are arranged with two toes forward and two behind, see Figure 14-51. The toes are long for gripping the bark of trees as they search for insects. The rear toes also provide leverage as these birds peck holes in woody plants and trees.

Pileated Woodpecker
(Dryocopus pileatus)

Figure 14-53 Distribution map of the pileated woodpecker.

The pileated woodpecker is a large bird about the size of a crow. It is mostly black in color with white markings on the flank, throat, face and underwing areas. The male has a red crest and mustache, see Figure 14-52, and the female is similar but without as much red coloring. This woodpecker is found in the eastern United States and in the Pacific Northwest, and it ranges across much of southern and central Canada, see Figure 14-53.

This woodpecker especially favors carpenter ants that it finds in dead trees and stumps. It pecks out a large oval or oblong hole opening in decaying wood, and captures ants by extending its long sticky tongue into their nest. A breeding pair works together to excavate a hole for a nesting site, and 3–5 eggs are incubated in the cavity.

Figure 14-52 A male pileated woodpecker at his nest hole. *Photo courtesy of Leonard Lee Rue III.*

Most species of woodpeckers go through a courting ritual that includes inspecting or excavating nesting holes, and drumming loudly with their bills to attract prospective mates to likely nesting sites. Males and females usually share the duties of incubating the eggs and caring for the nestlings. After the young woodpeckers leave the nest, the parents of some species divide the brood. Each parent assumes the care of some of the young birds until they are mature enough to care for themselves.

Four species of nuthatches are native to North America. They are migratory birds that depend mostly on insects, nuts, and some berries for food. They range in size from 4.5–6 inches. They spend most of their time climbing up and down the trunks of trees as they gather food. They position themselves on the tree trunk with their heads facing down as they seek food.

Red-breasted Nuthatch
(Sitta canadensis)

The red-breasted nuthatch is a common bird in many parts of North America, see Figure 14-54. It is a small blue-gray bird with a rust-colored breast and a black cap, see Figure 14-55. It inhabits coniferous forests and mixed woods where it gathers insects, spiders, and seeds for food.

Figure 14-54 Distribution map of the red-breasted nuthatch.

These birds nest in holes located in dead trees, and the entrance areas are often protected with sticky pitch. This helps to defend the nest from larger birds and other predators. These birds are monogamous and they usually produce a single brood of 4–7 young each year.

Figure 14-55 Red-breasted nuthatch.

Creepers are small brown birds that are usually observed climbing up tree trunks in search of insects. They have long bills that curve slightly downward which they use to probe for insects. Only the brown creeper is found in North America. It can easily be distinguished from the nuthatches by the way its head is positioned to face up the tree trunk in contrast with nuthatches that face downward.

AVIAN PROFILE

Brown Creeper (*Certhia familiaris*)

Figure 14-57 Distribution map of the brown creeper.

The brown creeper is a small bird with mottled brown coloring on the upper body and white underparts, see Figure 14-56. It is distinguished by its long, curved bill, and by its stiff tail feathers which brace it on the trunk as it works to gather food from beneath the bark of the tree. Its preferred habitat is coniferous or mixed forests, see Figure 14-57.

This bird builds a nest of twigs, moss and bark behind loosened chunks of tree bark in which it incubates 4–8 eggs. Breeding pairs are sometimes assisted with the duties of gathering food by

Figure 14-56 The brown creeper.

their male offspring from earlier broods. These family groups sometimes tend more than one nest at any given time. Females tend to scatter earlier than males, and they do not assist other females with parenting duties.

LOOKING BACK

A large number and variety of bird species occupy the ecosystems of North America. Each performs an important role in the food chain, and each species interacts with other living and nonliving resources in the environment. The birds that were discussed here occupy niches in many different habitats. Each has adapted to its environment in different ways. This is evident when their differences in diet, coloring, body structure, metabolic functions, and habits are considered. The groups of birds included in the chapter are arbitrarily grouped by commonly shared characteristics. They include long-legged wading birds, gull-like birds, upright perching birds, waterbirds, duck-like birds, sandpiperlike birds, chicken-like marsh birds, upland ground birds, and tree-clinging birds.

REVIEW QUESTIONS

1. Make a chart listing the characteristics of the different kinds of waterbirds discussed in this chapter. The information on the chart should include identifying traits, habitats and range, food, and reproductive traits.
2. Name several species of long-legged wading birds discussed in this chapter, and describe their diets and habitats.
3. Name the major species of gull-like birds, and describe their preferred habitats and diet in the North American ecosystems.
4. Explain the roles of several upright perching waterbirds named in this chapter.
5. Name some roles of the ducklike birds described in this chapter and list the diet and preferred habitat of each.
6. Describe the habitats of the sandpiperlike birds considered in this chapter, and name some major species.
7. Explain the role of upland ground birds and list the major species discussed here.
8. Describe the similarities and differences between swallows and swifts.
9. Describe how torpor in birds and hibernation in animals contributes to the ability of an organism to survive.
10. Name several kinds of tree-clinging birds and describe their diets and preferred habitats.

LEARNING ACTIVITIES

1. Select an observation area near your school, and lead the students in conducting a survey to identify the species of birds that use the area. Evaluate the role of each bird species that occupies the habitat. Prepare a group presentation, and solicit opportunities for class members to deliver it to other classes, civic clubs, and community groups.

2. Conduct a magazine picture contest in your school with categories for several different kinds of pictures obtained from magazines or printed materials. Each photo that is submitted should be attached to a data sheet listing habitat, diet, range, etc. Obtain sponsors for the prizes, and display the pictures in a prominent location in the school or community. Make sure that the activity and winning entries receive plenty of publicity.

SECTION IV

Ecology of Fishes, Reptiles & Amphibians

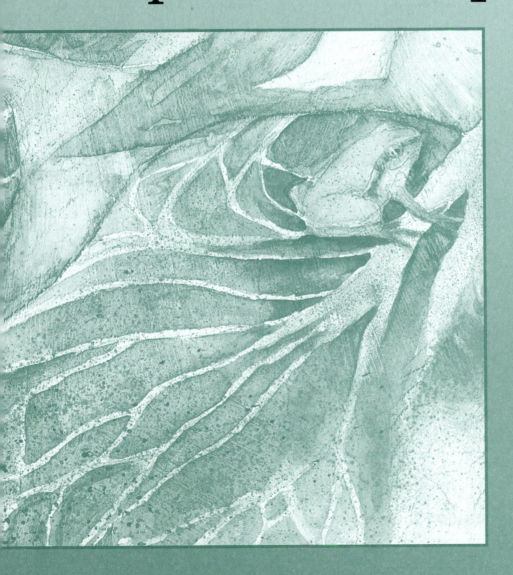

Fishes, reptiles and amphibians are cold-blooded animals that depend on their environments for body heat. Most of these animals are oviparous or ovoviviparous in their reproductive habits. Most of these animals are protected by scales, shells or plates. Some of them guard their eggs and care for their offspring, and others do not. They are widely distributed in the ecosystems of North America.

15

Freshwater Fishes

KEY TERMS

adipose fin
aerated water
anal fin
anatomy
aquaculturist
barbel
bullhead
cannibalistic
caviar
char
darters
dorsal fin
freshwater
hybrid
laterally compressed
milt
pectoral fin
pelvic fin
race
roe
school
scute
spawn
tapetum lucidum

FRESHWATER FISHES include all of the fish species living in water habitats that are not salty. The freshwater habitats of North America include springs, canals, streams, rivers, lakes, marshes, and ponds. Each of these waters provides a different living environment, and each environment is inhabited by fishes that are adapted to the conditions that are found there. The freshwater habitats of North America provide living environments for a great variety of fishes.

OBJECTIVES

After completing this chapter, you should be able to

- distinguish between freshwater and saltwater habitats
- describe the spawning process by which fish reproduce
- analyze the roles of catfishes in the freshwater habitats of North America
- define the roles of sunfishes in the freshwater environments of North America
- consider the importance of salmonids in freshwater ecosystems
- appraise the roles of perches in North American ecosystems
- explain the predatory roles of pikes in freshwater habitats
- evaluate the importance of sturgeons in the river environments of North America
- define the roles of minnows and suckers in the ecosystems of North America.

Freshwater sources include all of the waters on the continent that are not high in salt content. Most springs, streams, rivers, lakes, marshes, canals, and ponds are classed as freshwater habitats. A notable exception is the Great Salt Lake located in the Great Basin area of Utah, see Figure 15-1. This watershed has no outlet to the ocean, and the salt content of the water in this lake is much higher than that of ocean water.

Figure 15-1 The Great Salt Lake. *Photo courtesy of Salt Lake Conservation and Visitors Bureau, Utah.*

Freshwaters differ from one another in many ways. Minerals become dissolved in water as it passes over rock formations. Water also differs greatly in temperature, oxygen content, pH, rate of flow, dissolved nutrients, and degree of pollution, see Figure 15-2. All of these factors combine to create unique water environments of many kinds. Fishes have adapted over long periods of time to the water habitats in which they live. Some species are able to survive in a wide range of water environments, but others are unable to tolerate even small changes in their living conditions.

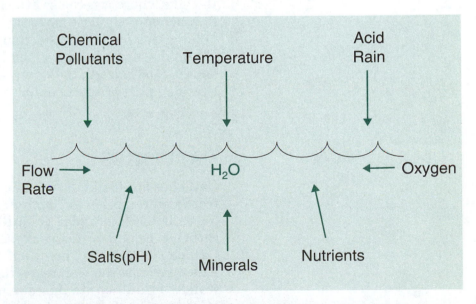

Figure 15-2 Fresh water differs from one another from levels of minerals to rate of flow.

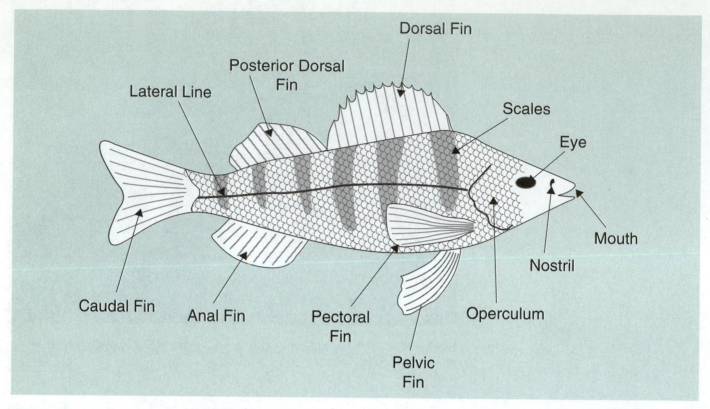

Figure 15-3 The external anatomy of the fish.

It is important to be familiar with the **anatomy** or structure of fish. While the shapes and sizes of particular body parts differ from one fish species to another, the structure is similar for most of them, see Figure 15-3. The descriptions of the fishes in this chapter will focus on similarities and differences in their anatomies. Respiration in fish is accomplished when water containing dissolved oxygen passes the gills of a fish. Oxygen enters the tissue in the gills where it becomes attached to blood cells. Waste materials flow across the gill tissue to the surrounding water.

Reproduction in fish is accomplished when they **spawn.** This is a sexual process during which the females deposit eggs or **roe** in depressions beneath the water, and the males fertilize them by discharging **milt** or sperm on the surfaces of the eggs.

CATFISHES

Catfishes are found in two different families and include forty-six freshwater species in North America. The largest family of catfishes is the Ictaluridae or **bullhead** family. Forty-five species of catfishes belong to this family, and they are found in waters from Canada to Mexico. The other family is the Clariidae family to which belongs the walking catfish. It is so named because it can walk overland on its fins during rainy weather. This fish is an exotic species that is established in Florida, introduced from South Asia to other regions of the United States.

FISH PROFILE

Channel Catfish
(Ictalurus punctatus)

Figure 15-5 Distribution map of the channel catfish (introduced in other waters).

The channel catfish has dark spots on its back and sides against light blue to greenish black coloring. The underparts are white. Fish of this species can reach 50 inches in length, see Figure 15-4. This fish is a popular game species that inhabits deep rivers and

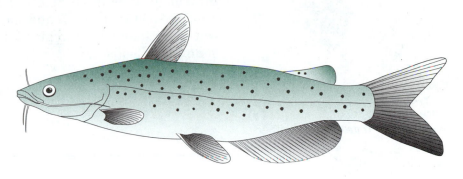

Figure 15-4 The channel catfish.

lakes, and has been introduced to waters beyond its native habitat, see Figure 15-5. It is now found throughout much of the United States. This fish is a scavenger that eats mostly animal matter. Similar species include the Yaqui, blue and headwater catfishes.

Bullhead catfishes have four pairs of **barbels**, sometimes called feelers or whiskers, surrounding their mouths. Barbels are organs that give the fish a sense of touch, and they are used to locate food. They also have bony spines located at the bases of their **dorsal fins** (located on their backs), and their **pectoral fins** (corresponding to front legs in terrestrial vertebrates.

Bullhead catfishes also have **adipose fins** that store fat, and **pelvic fins** that correspond to the rear limbs of terrestrial vertebrates. The **anal fin**, located on the underside between the anus and the tail, is very prominent in the catfishes. North American catfishes have tough skins with no scales on their bodies. Their sharp spines are used as weapons, and in some cases they are poisonous. The Madtom species of catfish are well known for their ability to injure their enemies with their poisonous spines.

FISH PROFILE

Black Bullhead
(Ameiurus melas)

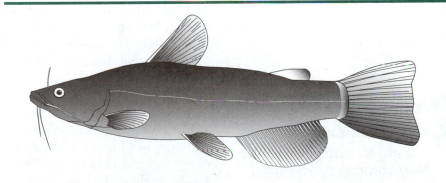

Figure 15-6 The black bullhead.

Figure 15-7 Distribution map of the black bullhead (introduced in other waters).

The black bullhead measures up to 24 inches long. It is olive to yellowish brown on the back fading to a greenish gold coloring on the sides. Its barbels are black, and its underparts range in color from white to yellow, see Figure 15-6. Habitats for these fish include sluggish rivers, pools, lakes, and ponds. They are native to interior waters in the United States, from the Great Lakes to Mexico, see Figure 15-7, and they have been introduced to suitable waters in many other regions. Similar species are the brown and yellow bullheads.

Catfishes are considered to be warm-water fishes because they are able to tolerate higher water temperatures than other species. These fishes are nocturnal bottom feeders.

Numerous other species of catfishes are found in North America, and they vary greatly in size. The flathead catfish, for example, grows to lengths of more than 5 feet, whereas the pygmy madtom matures at 1 1/2 inches.

The second catfish family includes only one North American species. It is an imported fish known as the walking catfish. It is abundant in Florida waters and it was introduced in other states as far west as California. It is capable of breathing air because a small part of each gill functions like lung tissue. This catfish uses its tail and pectoral fins to propel itself over the ground.

SUNFISHES

The sunfishes are natives of North America. They belong to a family called Centrarchidae. Included in this family are such game fish as the basses, crappies, and bluegills. These fishes are **laterally compressed,** meaning that they measure much longer from their backs to their bellies than from side to side. Sunfishes become active in direct sunlight, and their activity decreases in subdued light. This dependence on sunlight has resulted in the sunfish name.

Several distinctly different fishes belonging to different families are included among the basses. Many of them are freshwater fishes, but some basses live in the ocean. They are described as moderately compressed fishes with elongated bodies that become deeper as the fish gets older.

FISH PROFILE

Largemouth Bass (*Micropterus salmoides*)

The largemouth bass is a popular game fish in North America. Its native range extends across the eastern United States and Canada, see Figure 15-8, and it has been introduced to many other areas beyond this range. This fish sometimes grows to a length of 36–38 inches. The mouth is large, with the upper jaw extending back past the eye. The upper body is metallic green in color, the

Figure 15-8 Distribution map of the largemouth bass (introduced in other waters).

Figure 15-9 The largemouth bass. *Courtesy of U.S. Fish and Wildlife Service. Photo by Brian Montaque.*

sides are mottled, and the underside is white, see Figure 15-9.

The favored habitat for this fish is the quiet waters of lakes, sloughs, and ponds, and deep pools in rivers and streams. They prefer waters where plants are abundant. This bass is a predatory fish that feeds mostly on aquatic animals.

Males of the centrarchidae family build nests in which the females lay their eggs. Each nest is a shallow depression in the gravel or sand on the bottom that is created by the male as he fans loose material away with his fins. Each male defends a small area in the vicinity of the nest, and he guards the eggs and the young fish.

FISH PROFILE

Smallmouth Bass (*Micropterus dolomieu*)

The smallmouth bass is brownish olive on the back and mottled with dark markings. It has dark bars on yellowish-green-colored sides, and the underparts are yellow, see Figure 15-10. This fish is distributed beyond its native range, and it is found today from southern Canada to the central region of the United States, see Figure 15-11.

Figure 15-11 Distribution map of the smallmouth bass.

Figure 15-10 The small mouth bass.

The preferred habitat for this fish is clean, flowing water in streams, rivers or shallow lakes that have rock or gravel bottoms. It is a predatory fish that feeds mostly on small fish and aquatic animals.

The black crappie has a deep, laterally compressed body that is gray green on the back. Its silvery blue sides are marked with black-colored wavy lines and blotches, and it is white underneath, see Figure 15-12. This fish prefers clean, quiet water, and it lives in

Figure 15-12 The black crappie.

Figure 15-13 Distribution map of the black crappie (introduced to other waters).

lowland ponds, lakes, and sloughs. It is also sometimes found in deep, calm pools in slow-flowing streams. Its native range has been expanded from the eastern United States by introductions of the fish into other suitable waters, see Figure 15-13.

Sunfishes of different species frequently mate together. The off-spring of these matings are called **hybrids.** Hybridization is observed most often when the water in the spawning areas is polluted or dirty. It is thought that the fish may have difficulty distinguishing between the different species due to poor visibility.

One of the most common fishes in freshwater ponds and lakes is the bluegill. This popular game fish is native to the eastern United States and its current range extends through the southern states and from Canada to Mexico, see Figure 15-14. Its preferred habitat includes ponds, lakes, swamps and pools where vegetation is abundant.

This sunfish has a very compressed body that is quite deep from top to bottom. It is olive colored on its back and sides, with green and yellow dots and narrow black bars marking the sides. Underparts are colored reddish-orange (Figure 15-15). Males

Figure 15-14 Distribution map of the bluegill (introduced to other waters).

Figure 15-15 The bluegill.

engage in elaborate mating rituals to attract females to their territory. Eggs are deposited in shallow depressions near the shoreline.

PERCH

The perch and other members of the Percidae family are widely distributed in North America. The Percidae ranks second among North American fish families in the number of different species. The most numerous of these are small fish called **darters.** There are 150 species of these bottom-dwelling fish, and they are used as food by larger fish that live in the same environments.

FISH PROFILE

Yellow Perch (Perca flavescens)

The yellow perch has a deep, laterally compressed body and a large mouth. Its coloring is greenish brown on the back, with yellow sides and underparts, see Figure 15-16. This fish is found in Atlantic drainages from Central Canada to the northeastern part of the United States, see Figure 15-17. It inhabits ponds, lakes, and

Figure 15-17 Distribution map of the yellow perch (introduced to other waters).

Figure 15-16 The yellow perch.

deep pools in streams and rivers. They prefer vegetated habitats where they can hide as they stalk their prey. The diet of the perch consists mostly of aquatic insects and animals. Perch grow to lengths of about 16 inches when they are mature.

Some of the most popular game fish in this family are the sauger and the walleye. These are large fish that have the ability to see in the dark. They have special tissue in their eyes known as **tapetum lucidum** that gathers light and gives them their night vision.

The walleye is a large predatory fish up to 36 inches long, with a very large mouth that extends back beyond the middle of the eye. Coloring ranges from yellow to brown on the back, with yellowish blue sides that are marked with dark green, see Figure 15-18.

Figure 15-18 The walleye.

Figure 15-19 Distribution map of the walleye.

Walleyes inhabit rivers, lakes and pools, preferring clearwater and brushy banks. They are native to Canada and the central region of the United States, see Figure 15-19. This fish is popular with sportsmen due to its large size and its fighting instinct when it is caught.

TROUT, CHAR AND WHITEFISHES

All of the fishes in this group belong to the Salmonidae family. In addition to the freshwater species, there are also several salmonids that migrate to the ocean during certain stages of their lives. These fishes will be discussed in a later chapter. The species considered here are freshwater fishes.

The brown trout is a brightly-colored fish with an olive brown back, silvery yellowish-brown sides, and white or yellow underparts. The sides and head are covered with red and black spots, and some of the fins may be orange or red, see Figure 15-20. Males have hooked lower jaws.

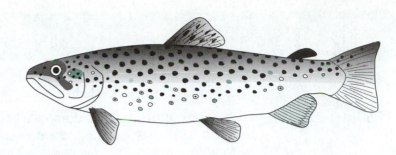

Figure 15-20 The brown trout.

Figure 15-21 A true distribution of the brown trout is not available since it has been introduced to many waters.

These predatory fish live in cold, clean streams and lakes. Their diet consists mostly of insects, small fish, frogs and other aquatic animals. They were introduced to North America from Europe in 1883, and they now occupy many regions of the United States and Canada, see Figure 15-21. These fish require cool temperatures and well-**aerated water**, containing high levels of dissolved oxygen. Its favorite habitats are deep pools in fast-moving streams. Reproduction is accomplished by laying eggs in the gravel beds of rivers and streams.

The salmonid family includes a large number of fishes that belong to the same species even though they appear to be distinctly different from one another in their color and markings. They are divided up into similar groups or **races** within the species. The difficulty of classifying these fishes is complicated by the many hybrids that have resulted from crossbreeding between species.

FISH PROFILE

**Rainbow Trout
(Oncorhynchus mykiss)**

The rainbow trout varies in color from one habitat to another. Those living in lakes are often silver over much of their bodies, while those living in streams may be nearly brown on their backs, with yellow to greenish coloring and silvery sheens on their sides, see Figure 15-22.

Figure 15-22 The rainbow trout.

Figure 15-23 Distribution map of the rainbow trout (introduced to other waters).

This fish is native to the Pacific drainage, and it is found in many of the cold-water streams, rivers, and lakes in the region, see Figure 15-23. It is also found in isolated streams in other areas. This is one of the most important fishes in North America. It is a

Career Option

AQUACULTURIST

A **aquaculturist** raises domesticated fish for human consumption or use. Such uses may include stocking fishing waters or processing the fish for food. An aquaculturist is responsible for maintaining a high quality water environment for the fish from the time they are placed in the hatchery as eggs until they are mature. A person who pursues this career will require a strong understanding of the sciences associated with aquatic environments and fish culture. Testing water quality and preventing fish diseases are only two of many applied science skills that will be required in this career. A BS degree is needed that includes strong components in business, science and fish culture.

popular game fish, and it is raised on fish farms as a commercial fish that is processed for food. Its diet consists of insects, worms, and small aquatic animals and fishes. Reproduction is accomplished by depositing eggs in shallow scrapes in gravel streambeds. Males discharge milt on the eggs to fertilize them.

Most of the fishes in the Salmonidae family migrate to spawning areas. Those living in lakes and ponds move out to the quiet water and into the flowing streams that are the sources of fresh water. They spawn in clean gravel beds where the eggs are trapped in the rocks and where plenty of well-aerated water is available. Many of these fishes retreat back to the lakes when spawning is completed. Fish living in rivers often migrate upstream or into the small streams that flow into the rivers before they spawn.

FISH PROFILE

Mountain Whitefish (*Prosopium williamsoni*)

The mountain whitefish has a small mouth, greenish brown coloring above, and silvery or white coloring on the underside, see Figure 15-24. They grow as big as 22.5 inches in length. It lives in cold water habitats in lakes and streams, and it is distributed

Figure 15-25 Distribution map of the mountain whitefish.

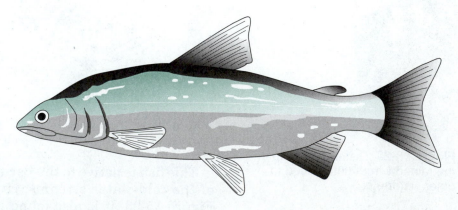

Figure 15-24 The whitefish.

throughout the United States and Canada, see Figure 15-25. These fish live together in large groups called **schools.** The diet of this fish consists of insects, worms and other small aquatic animals. Reproduction is accomplished by spawning in substrate of rivers and streams.

Brook trout and closely related fishes are also known as **chars.** During the spawning season, these fish are especially bright in their coloring and many of them have red on their fins and sides, see Figure 15-26.

FISH PROFILE

Brook Trout
(*Salvelinus fontinalis*)

The brook trout is native to the eastern waters of the continent, and it has been introduced into the waters of the Pacific drainage. These fish live in the cool, flowing waters of creeks and streams. It is a predatory fish and its diet consists of insects and other aquatic organisms. Other closely related species include the Dolly Varden

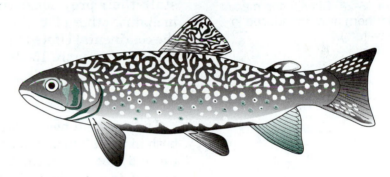

Figure 15-27 The brook trout.

Figure 15-26 Distribution map of the brook trout.

and the bull trout. Males are brightly colored during the spawning season with bright orange or red on its lower body and fins and pink or red spots surrounded by blue highlights on its sides, see Figure 15-27. Spawning occurs in flowing streams where eggs are deposited in clean gravel on the streambed.

There are many fishes in this family that are not discussed in this chapter. They include the grayling, salmon, and cisco.

PIKE

Pike are members of the Esocidae family of fishes. They are vigorous predators that attack and eat other fish, and they are capable of eating fish that are almost as large as themselves. Their jaws are long, and they swallow their prey whole. Pikes are also **cannibalistic,** meaning that they eat other fish of their own kind.

Figure 15-29 Distribution map of the northern pike (introduced to other waters).

The northern pike is a large gray fish with a huge mouth lined with sharp teeth, see Figure 15-28. It grows as long as 6 feet, and it has an appetite to match its size. This fish inhabits marshes, rivers, lakes, pools, and ponds that have heavy vegetation growing in and near the water. They use this vegetation for cover as they

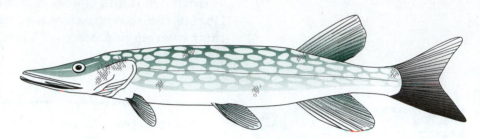

Figure 15-28 The northern pike.

stalk their prey consisting of fish and other aquatic animals, including other pike. This fish is found in the northern regions of the continental United States and throughout much of Canada and Alaska, see Figure 15-29.

The muskellunge or "musky" is very similar to the northern pike both in its habits and in its appearance. It is a large fish that grows up to 6 feet in length. A popular game fish known as the "tiger musky" is a hybrid produced by musky and northern pike parents. It is a more ferocious hunter than either of the parent species.

STURGEONS

Sturgeons are the largest fish in North America. The white sturgeon, see Figure 15-30, grows to lengths of 20 feet. Sturgeons inhabit large rivers, and they belong to the family Acipenseridae. They are long-lived fishes that are known to live up to seventy-five years. Sturgeons stir up the mud on river bottoms in their search for food. The eggs of sturgeons are used as a delicacy food called **caviar.**

Figure 15-30 The lake sturgeon.

Shovelnose Sturgeon (*Scaphirhynchus platorynchus*)

■ White Sturgeon
▨ Shovelnose Sturgeon

Figure 15-31 Distribution map of the white and shovelnose sturgeon.

The shovelnose sturgeon has a flat shovel-shaped snout and a sharklike tail. Its mouth is on the underside of the snout, and this fish has four barbels to aid in locating its food. It eats aquatic organisms and carrion that it locates in the mud or on the bottom of the river. It is protected by large bony plates called **scutes** that cover its head, back, and sides. Habitat for this sturgeon is the bottom of river channels in flowing water. Sturgeons lay eggs to reproduce, and the eggs are used as a human food called caviar. It ranges in the Mississippi, Mobile Bay, and Rio Grande drainages, see Figure 15-31.

MINNOWS AND SUCKERS

The minnows make up the largest family of fishes, the Cyprinidae. Members of the family are found on every continent, and include 231 species that occur north of Mexico. Many people incorrectly believe that all minnows are small. Some species of minnows are large in size. For example, the grass carp grows to lengths of 49 inches. The Cyprinidae includes carp, goldfishes, squawfishes, shiners, chubs, and many others.

Common Carp (*Cyprinus carpio*)

Figure 15-33 A true distribution of the common carp is not available since it has been introduced to many waters.

The common carp has a long, deep body. The back is arched with a long dorsal fin, and the belly is somewhat level. This fish has large, dark scales, and mature fish show a dark, metallic green coloring, see Figure 15-32. This carp is reported to grow to lengths of 48 inches.

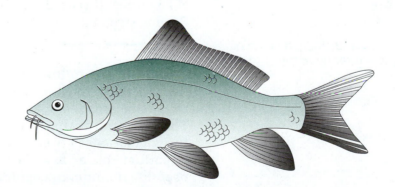

Figure 15-32 The common carp.

This fish was introduced to North America in 1831, from Europe, and it thrives in muddy pools containing organic matter. Its diet consists mostly of plants, but it also eats aquatic organisms when they are available. Its habit of stirring up mud in the water is thought to cause deaths of more desirable fish species that require clean water to live. It can be found in the quiet waters of rivers, lakes, and ponds throughout the United States and southern Canada, see Figure 15-33.

Some carp are strict vegetarians, and they eat large amounts of aquatic plants. The grass carp was introduced to North America in the 1960s for the purpose of controlling weeds and grasses in ponds, lakes, and waterways. It has become widespread since its introduction, and some biologists are concerned that some native species that depend on vegetation may have trouble surviving.

Suckers are members of the Catostomidae family. There are sixty-three species of suckers in North America. The name sucker comes from the feeding habits of these fish. Their mouths are shaped for feeding off the bottoms of streams and lakes and they eat by sucking small invertebrates into their mouths.

FISH PROFILE

White Sucker (Catostomus commersoni)

Figure 15-35 Distribution map of the white sucker (introduced to other waters).

The white sucker is black to olive brown on the back and sides, and it has light-colored underparts. The mouth is located on the underside of the head, with the upper and lower lips shaped for sucking up food, see Figure 15-34. Its diet consists of aquatic organisms and plant materials gathered from the bottoms of streams, lakes and ponds. The fish is distributed widely in North America, see Figure 15-35. It lives in habitats ranging from small creeks to large lakes, and it sometimes grows to lengths of 25 inches. It reproduces by spawning in the substrate material of flowing streams.

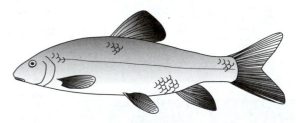

Figure 15-34 The white sucker.

Many of the minnows and suckers are considered to be trash fishes because they are not very desirable for human food, and they tend to be more competitive than some of the more useful fishes. The net result is that populations of desirable fishes decline while the populations of trash fishes increase. They do, however, serve an important role as food fishes for larger game fishes that are more useful to the human population.

LOOKING BACK

Freshwater fishes include all of the fishes that live in water environments that are not salty. Each species has adapted to the living conditions that are found in its habitat. The great variety of fishes is evidence of the diversity of the freshwater habitats that are found in North America. In this chapter we learned about the most common freshwater fishes including the sunfishes, catfishes, trouts, perches, pikes, sturgeons, minnows, and suckers. Each group of fishes plays important roles in the biomes of North America.

REVIEW QUESTIONS

1. List the characteristics that distinguish freshwater habitats from marine habitats.
2. Illustrate and label the external anatomy of a fish.
3. Describe the spawning process by which fish reproduce.
4. Name some freshwater habitats, and describe ways that each is different from the others.
5. Create a chart listing each of the fish families considered in this chapter. List their preferred habitats, diets, sizes and identifying characteristics.
6. Explain how hybrids occur among similar fishes, and describe how hybridization complicates the process of identifying fish species.
7. Describe how the anatomies of different fishes are modified to accommodate their feeding habits.
8. Suggest some factors that should be considered before non-species of fishes are introduced to a water environment.

LEARNING ACTIVITIES

1. Identify a freshwater environment near the school, and design a water quality study to determine the characteristics of the water over a period of several weeks. Contact a state agency (e.g., the Department of Environmental Quality, the Department of Water Resources, etc.) that deals with water quality, and ask for instructions on water sampling techniques. Gather weekly samples, and measure factors that affect water quality. Involve the students in every aspect of this activity. Prepare charts and written reports, and present your findings to the community. Enter the project in science student competitions such as science fairs.
2. Develop a community plan to protect and/or to improve the water quality of the site that you selected for your water assessment study. Implement your plan, and enter the project in community service competition.

16

Anadromous and Diadromous Fishes

SOME OF THE FISHES in North America hatch in freshwater streams from which they migrate to the ocean. A fish that migrates between freshwater and marine habitats is classed as a **diadromous fish**. A fish that migrates up a river from the sea to spawn is called an **anadromous fish.** After reaching maturity, these fishes leave the ocean, and migrate up rivers and streams to spawn. They return to the same streams in which their lives began. A freshwater fish that migrates to the ocean to spawn is known as a **catadromous fish.**

OBJECTIVES

After completing this chapter, you should be able to

- define the difference between anadromous, diadromous, and catadromous species of fishes
- name some common anadromous fishes in North America, and describe their life cycles
- explain the spawning process that is common among anadromous fishes
- discuss the effects that silty water might have on live fish eggs
- appraise the positive and negative effects of the drawdown plan as a method for increasing endangered populations of salmon
- evaluate the roles of fish hatcheries in maintaining populations of fish
- suggest ways that the anadromous fishes are able to find the way back to the stream where their lives began
- consider the reason why anadromous steelhead trout grow larger than freshwater rainbow trout of the same species.

An anadromous fish lives its life in two very different environments. The early part of its life cycle is spent in the freshwater

Spawning Process

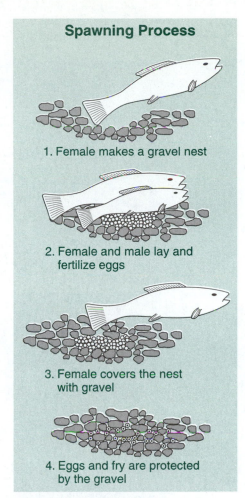

1. Female makes a gravel nest

2. Female and male lay and fertilize eggs

3. Female covers the nest with gravel

4. Eggs and fry are protected by the gravel

Figure 16-1: The spawning process of anadromous fish.

environment of a stream or river. Female fish make nests in the gravel for their eggs by rolling on their sides and churning the gravel with their tails. The female lays her eggs in the nest as the male discharges sperm or milt on them. The female then covers them with gravel to protect them while they develop into tiny fish, see Figure 16-1.

The eggs are held in position in the flowing stream by falling into the spaces between the stones in the gravel. It is important for the gravel in spawning areas to be free from silt. Silt tends to coat the egg surfaces preventing oxygen from getting into the eggs. Under such conditions, the live eggs suffocate and die.

The tiny fish is called a **fry** after it hatches from the egg. At first the fry carries a **yolk sac** for nourishment, see Figure 16-2. This is a membrane filled with a highly nutritious substance called **yolk** that is obtained from the egg from which the fish hatched. As the yolk is used to nourish the fish, the sac shrinks until it finally disappears.

As the fry grows, it is sometimes called a **fingerling.** It may be called by this name until it is about a year old. At about two years of age, when the young fish is ready to migrate to the ocean, it is called a **smolt.** During this phase of development its body adjusts to enable it to maintain its body fluids while living in a salt water environment.

The migration of smolts to the ocean is the most dangerous time in the life cycles of anadromous fishes. During this migration they are often eaten by predatory fish such as the Squawfish. They also have trouble migrating through the backwaters of dams that have been constructed on many major rivers. The smolts face upstream during their migration down the river, see Figure 16-3. In this position, they depend on water currents to carry them and experience difficulty moving through the slack water behind the dams, see Figure 16-4.

One management practice that is used to assist migrating smolts is known as **barging.** Smolts are collected above a dam and placed in large river vessels called barges, see Figure 16-5. River water flows through the interior of a barge to maintain a constant water environment. The barges are transported below the dams through the shipping locks that allow ocean-going ships to pass the dams.

Figure 16-2: Yolk-sac fry.

Stream flow

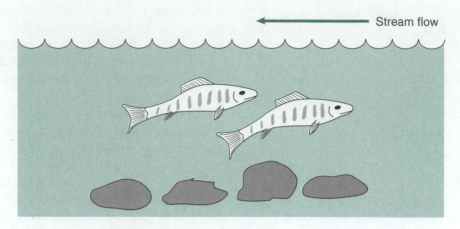

Figure 16-3: Smolt migration downstream.

Figure 16-4: Smolts depend on water currents to carry them to the ocean during their migration. Slack water behind dams create problems for the migrating smolts since the current is interrupted and their movement is impaired. *Photo courtesy Bureau of Reclamation, U.S. Department of the Interior.*

The smolts are released back into the river below the dams to continue their migration to the ocean.

Anadromous fishes usually remain in the ocean for one to six years, where they grow and mature. When they are mature, they begin the long journey back to their freshwater birthplaces to spawn, see Figure 16-6. These fish possess powerful homing instincts that guide them as they return to the streams where their lives began. Along the way a migrating fish must find its way past dams, river rapids, and fisherman. Some of them are also caught and eaten by predators such as seals, eagles, bears, humans and otters.

In addition to the hazards encountered during migration, offshore fishing fleets have contributed to reduced fish populations and to the endangered status of some anadromous species. The populations of most species of migrating fishes have declined in recent years.

Barging Smolts around dams

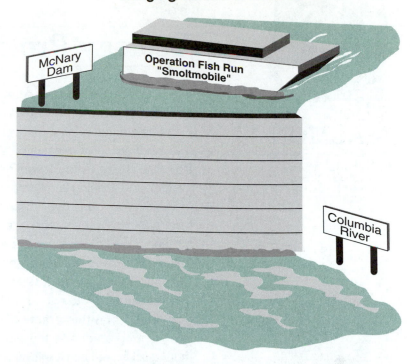

Figure 16-5: To assist migrating smolts through the slack water behind dams, they are collected above the dam, placed in barges, and transported below the dam to continue their migration.

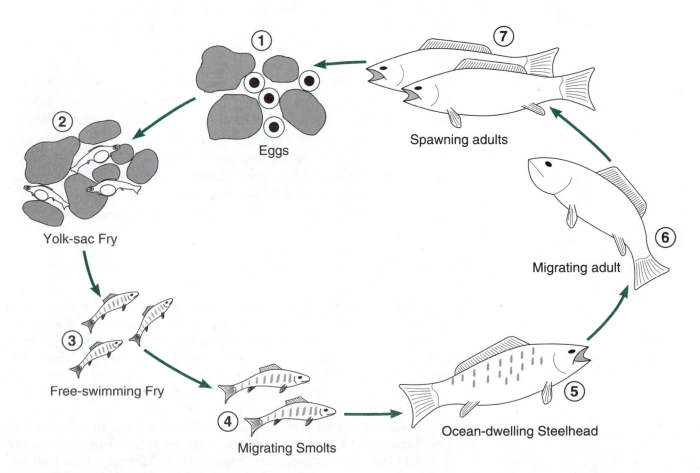

Figure 16-6: The life cycle of anadromous fishes.

SALMON RECOVERY PLAN

Ecology Profile

A management practice for sockeye salmon known as the drawdown plan has created great controversy. It is the practice of releasing water from the dams in the lower river system, and flushing large amounts of water down the river channel from water storage reservoirs on the upper end of the river system. This practice was attempted in the Columbia River system in 1994, but it is too soon to know the results. It is hoped that this practice will carry young smolts to the ocean more quickly, and that it will improve their chances of survival during migration.

The controversies over this practice arise for several reasons. One problem that occurs is that riverbanks and structures are damaged by the drawdowns. Commercial shipping on rivers is halted. Great economic losses are incurred by people in the shipping industry, and by people who use ships and barges to transport goods to the coast. Water that is stored for the purposes of generating electrical power and irrigating crops goes into the ocean instead. The cost to implement the plan is expected to be $350 million per year, and this does not include any of the losses from agriculture, commerce or the generation of electrical power. Proponents of the drawdown suggest that the economic benefits of healthy salmon will make up for the other losses to the economy of the area. Additional controversy is generated because offshore fishing fleets and sportsmen along the Columbia River are allowed to harvest substantial numbers of these fish before they reach their inland spawning areas.

Citizens are divided over the issue of whether the use of drawdowns to aid recovery of salmon runs can be justified when the economic losses incurred by the plan are so high. There is no way to know whether the plan will help the salmon recovery effort, but a team of scientists has suggested that the drawdown plan is the best alternative available at this time. If the practice works as planned, it is expected that the practice will be implemented on other rivers.

Two populations of anadromous fishes now exist. One population includes **wild stock** that spawn naturally in rivers and streams. The other is made up of **hatchery stock** that has been spawned and raised in artificial environments in fish hatcheries. These fish can be distinguished from one another by the clipped adipose fins of hatchery stocks of fish. The fins of these fish are clipped to identify them separately from wild stocks of fish.

Government agencies and power utilities have developed fish hatcheries along major rivers, and these are used to supplement the wild fish populations, see Figure 16-7. Mature adult fish are trapped as they arrive back in their spawning areas, see Figure 16-8. The eggs from these females are fertilized with milt from the

Figure 16-7: Fish hatcheries are used to supplement the wild fish populations. *Photo courtesy of Rick Parker.*

Figure 16-8: A fish trap. *Photo courtesy of George Lewis, Cooperative Extension, The University of Georgia.*

males, and the eggs are placed in trays. Fresh river water is directed over them to provide oxygen and to maintain a constant temperature until they hatch, see Figure 16-9.

The young fish are raised in long fish runs or tanks until they are about one year of age, see Figure 16-10. These hatchery-raised smolts are then released to streams and rivers to migrate to the ocean.

SALMON

Salmon belong to the Salmonidae family of fishes. They are close relatives of freshwater species of trouts and whitefishes. These fishes are found in the coastal waters and rivers along the Atlantic and Pacific coasts of North America.

When anadromous fish enter their home rivers, they quit eating, and their bodies begin the process of producing eggs or milt. As the migrations are completed, and the time for spawning approaches, the fish become brighter in their coloring. Males of most species are more colorful than the females.

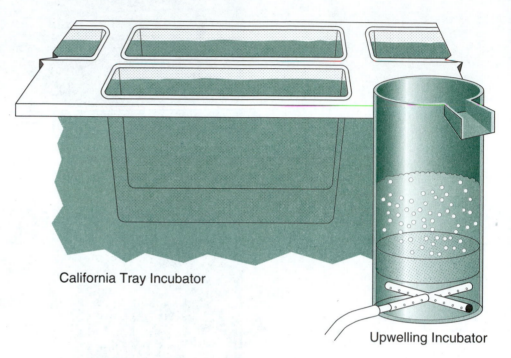

California Tray Incubator

Upwelling Incubator

Figure 16-9: Fresh river water is directed over incubating eggs to provide them the proper amount of oxygen and to maintain a constant temperature.

Figure 16-10: Young smolts are raised in tanks until they are about a year old. Then they are released into rivers and streams. *Photo courtesy of Rick Parker.*

Ecology Profile

HOMING INSTINCT OF ANADROMOUS FISH

It is believed that adult fish locate their rivers of origin by the smell or the taste of water. This theory has been tested by two scientists who captured salmon from two different branches of the Issquah River in the state of Washington. Half of the fish from each branch had their nasal sacs plugged with cotton. The fish were taken back down the river and released. The untreated fish returned to the same branches of the river in which they were originally captured. Those fish with plugged nasal sacs were not able to locate their home waters, and approximately half of them returned to the wrong branch of the river.

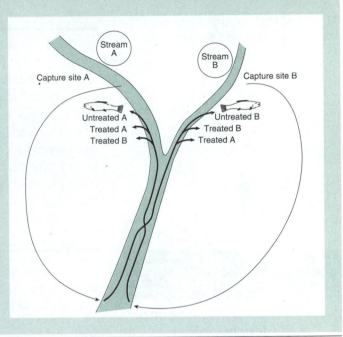

The migrations up the rivers to the spawning streams deplete the energy and strength of these fish. After they spawn they become weak. The Atlantic salmon usually recovers and returns back to the ocean after spawning. None of the Pacific salmon are known to survive after they spawn. Steelhead trout, however, frequently survive after they have spawned. Populations of salmon have been reduced due in part to pollution of streams, rivers, and coastal waters.

The Atlantic salmon is a long, streamlined fish that reaches up to 55 inches in length. It has silver sides with brown, green, or blue coloring on its back, and the males have hooked lower jaws, see Figure 16-11. Colors become darker when these fish enter freshwater, and they often develop red spots on their bodies as the time for spawning approaches.

These fish range along the Atlantic coast where they feed on shrimp and other marine

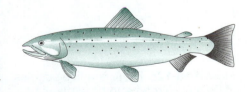

Figure 16-11: The Atlantic Salmon.

Figure 16-12: Distribution map of the Atlantic and pink salmon.

animals and enter coastal rivers and streams during the fall season to spawn, see Figure 16-12. Eggs are deposited in the gravel substrate where they are fertilized by the males. Young fish live in freshwater for two to three years before entering the ocean where they remain for a year or more. They migrate back into the rivers to spawn, but return to the ocean when spawning is complete.

The Atlantic salmon is an important commercial fish, and in some instances has been harvested beyond its capability to reproduce. Some of the rivers and streams that used to be included in the range of this species are no longer populated by them.

Five species of Pacific Salmon are known. They are the chinook, coho, pink, chum, and sockeye salmons, see Figure 16-13. These are valuable commercial fish that are harvested in coastal regions from the Arctic to California. The sockeye salmon is included in the discussion of endangered species in chapter 4. They live their first year in freshwater, and they spend from one to several years in the ocean before returning to their spawning areas. After they spawn, they die.

FISH PROFILE

Chinook or King Salmon
(*Oncorhynchus tshawytscha*)

The chinook or king salmon is the largest salmon, and it often weighs 30–50 pounds. It is blue, gray, or green with black spots on the back, reddish sides, and silver underparts, see Figure 16-14. The diet of the king salmon includes fish, shrimp, and other marine organisms. It ranges in the Pacific and Arctic Ocean drainages of

Anadromous North American Salmons
Atlantic Ocean:
Atlantic Salmon
Pacific Ocean:
Chinook
Coho
Pink
Chum
Sockeye

Figure 16-13: There are five species of Pacific salmon and one Atlantic species.

Figure 16-14: Chinook Salmon. *Photo courtesy of U.S. Fish and Wildlife Service.*

North America, see Figure 16-15, and it has been successfully stocked in the Great Lakes. It is the least abundant of the Pacific salmon. This salmon returns to its home river to spawn, and it dies soon afterward.

Figure 16-15: Distribution map of the King or Chinook salmon.

FISH PROFILE

Pink Salmon (*Oncorhynchus gorbuscha*)

The pink salmon is metallic blue to blue green, with large black spots on the back on the caudal fin. Males often have a pink or brown stripe down their sides. They are silver below, and they grow to lengths of 30 inches. Males develop hooked upper jaws and

Figure 16-16: The pink salmon. *Courtesy of U.S. Fish and Wildlife Service. Photo by G. Haknel.*

humped backs, see Figure 16-16. This salmon occupies coastal ocean and stream habitats from California to the Arctic region, see Figure 16-12. It has also been successfully introduced to Lake Superior and to Newfoundland. Its diet consists of fish, shrimp, and other marine organisms. It dies soon after it spawns.

STEELHEAD

The **steelhead** is a race of large rainbow trout that migrates to the ocean from freshwater streams at about 2–3 years of age. Steelhead usually remain in the ocean for 1–3 years before returning to spawn in the rivers and streams where they were hatched. It is believed that the abundant supply of food that is available in the ocean contributes to the large size of the steelhead.

FISH PROFILE

Steelhead
(Salmo gairdneri)

The Steelhead is an anadromous rainbow trout. It grows to lengths of 18–40 inches. It is silver colored with a bluish tint on the back, small black spots on the fins and back, and a pink or reddish side stripe, see Figure 16-17.

Figure 16-18: Distribution map of the steelhead.

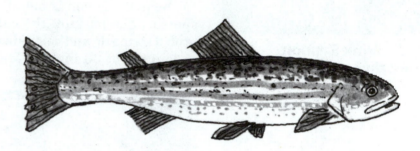

Figure 16-17: A steelhead trout.

These fish are native to the streams, rivers, and coastal waters in the Pacific coast region of North America, see Figure 16-18. They have been transplanted to other waters including inland lakes. The diets of the Steelhead consist of aquatic insects, worms, crustaceans and small fish.

Different races of steelheads are known. They spawn at different times of the year, and they differ in size from one another. Each race returns to its spawning area at a predictable time. Eggs require 4–7 weeks to hatch depending on the temperature of the water in which they develop.

The Fish and Game agencies of several states define a steelhead as any Rainbow Trout that is over 20 inches long. The name "steelhead" has also been used in the past for other kinds of ocean-run trout, such as the brook trout, cutthroat trout, and brown trout.

Career Option

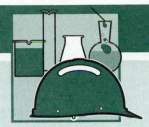

ICHTHYOLOGIST

Ichthyology is the branch of zoology that deals with the classification, structure, and life cycles of fishes. An ichthyologist is a scientist who studies fishes. Ichthyologists are employed for fish management in marine and freshwater fisheries, hatcheries, and commercial fish farms. Other ichthyologists serve as curators of scientific collections of fish in museums and educational institutions. Employment in this field requires a graduate degree in the biological sciences. A master's degree will sometimes suffice in fish management positions, but a doctorate is required of curators.

Figure 16-19: Distribution map of the striped bass.

STRIPED BASS

The striped bass belongs to the Moronidae family of fishes. It is an anadromous fish that spends most of its life in a marine environment, but it migrates far upstream to spawn in coastal rivers. This fish is a native of the Atlantic coastal waters, but it was transplanted to the Pacific coast in 1879 and 1882, see Figure 16-19, and has since reproduced and populated the Pacific coast from Vancouver Island, British Columbia to Baja, California, Mexico.

FISH PROFILE

**Striped Bass
(Morone saxatilis)**

The striped bass can be identified by its dark olive to bluish gray color on the back and 6–9 gray stripes on its silvery sides, see Figure 16-20. It is a large fish that grows as long as 79 inches, and

Figure 16-20: The striped bass. *Photo courtesy of U.S. Fish and Wildlife Service.*

it has been known to weigh over 70 pounds at maturity. They are predatory fish that feed on smaller fish and other marine organisms.

Striped bass move into freshwater to spawn in the spring of the year. Females become sexually mature at about five years of age. Males mature by two years of age, and they are about half the size of the females when they spawn. In many areas where striped bass have been introduced, they have mated with white bass, and hybrids of these two species are quite common.

SHAD

Shad are anadromous fishes that are native to the Atlantic seacoast of North America, where their range extends from Newfoundland to Florida. They have also been introduced to Pacific coastal waters where they now range from Alaska to California, see Figure 16-21.

Shads are members of the herring or Clupeidae family. They are deeper-bodied fish than most herrings, and they swim together in large schools. They spend most of their lives in the ocean, and they only enter freshwater habitats to spawn. Their diets consist mostly of plankton that is strained from the water.

Figure 16-21: Distribution map of the American shad.

FISH PROFILE

American Shad (Alosa sapidissima)

The American shad ranges in color from green to blue on the back, with silver sides tinged with yellow. Its fins are light green or clear. A bluish black spot is evident in the upper gill area aligned with one or two rows of smaller spots, see Figure 16-22. These fish enter coastal rivers to spawn in the spring, and they prefer large rivers with open water. Shad live in vast schools, and they feed on

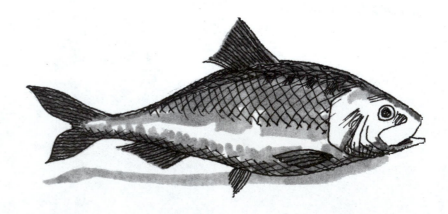

Figure 16-22: The American shad.

plankton. They grow as long as 24 inches with mature fish weighing about two pounds.

FRESHWATER EELS

Freshwater eels are members of the Anguillidae family. They are fishes that lack pelvic fins, and their scales are so small that they are seldom noticed. They are 3–5 feet in length, and they inhabit rivers and lakes in the eastern regions of North America. Females migrate up streams and rivers along the Atlantic coast where they remain for as long as fifteen years. Males remain in saltwater environments and in streams near the ocean. Some eels are capable of traveling short distances over land to other bodies of water.

FISH PROFILE

American Eel
(Anguilla rostrata)

Figure 16-24: Distribution map of the American eel.

The American eel is a long, snakelike fish with a long dorsal fin that extends around the tail and merges with the anal and caudal fins. This eel ranges in color from brownish green to yellow on the back, and it fades to light yellow or white on the belly, see Figure

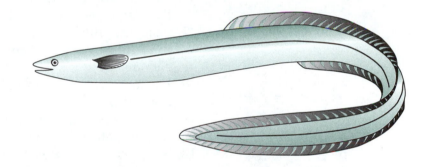

Figure 16-23: The American eel.

16-23. It inhabits the rivers and coastal waters along the Atlantic coast of North America, see Figure 16-24, and ranges as far inland as the midwestern states.

The freshwater eel is a catadromous fish because it migrates down freshwater streams to spawning areas in the Sargasso Sea. Adults die soon after spawning, and their larvalike offspring called **leptocephali** spend the next year migrating back to the streams of North America. Along the way their bodies change to look like young eels, at which time they are called **elvers.** They reach maturity shortly after this time, and the females return to freshwater rivers. The males remain in salty or brackish coastal waters.

LOOKING BACK

Anadromous fishes migrate up freshwater rivers from the ocean to spawn. Catadromous fishes spend their adult lives in freshwater environments, and they migrate down the rivers to spawn in the ocean. Diadromous fishes include fishes that migrate between freshwater and saltwater habitats in either direction. Anadromous North American fishes include members of the Salmonidae (salmon and steelhead), the Clupeidae (shad) and the Moronidae (striped bass) families. The Anguillidae (freshwater eel) family is catadromous.

REVIEW QUESTIONS

1. Define the difference between anadromous, diadromous, and catadromous species of fish.
2. Name some anadromous fishes found in North America.
3. Describe the life cycle that is typical of anadromous fishes such as the salmon, steelhead, and ocean-run basses.
4. Contrast the life cycle of a catadromous fish such as a freshwater eel with that of an anadromous fish.
5. Describe the spawning process that is common among anadromous fishes.
6. Discuss the effects that silty water conditions might have on live fish eggs that have been spawned in freshwater streams.
7. Appraise the positive and negative effects of the drawdown plan as a recovery method for endangered populations of salmon.
8. Evaluate the roles of fish hatcheries in maintaining populations of fish.
9. Describe how anadromous fishes are able to find the streams where their lives began.
10. Suggest a reason why ocean-run steelhead trout grow larger than rainbow trout of the same species that live in freshwater habitats.

LEARNING ACTIVITIES

1. Take a field trip to a fish hatchery and observe the process of raising fish in this protective environment. Have each of the students take field notes containing their observations. Make certain that the students have learned what they can expect to see before they arrive, and help them develop a set of questions that they can ask of the tour guide. Assign each student to prepare a written report on the tour activity using the field notes that each has prepared. If you do not have a fish hatchery available to you, contact your state Fish and Game agency for a guest lecturer or a video.

2. Obtain some fish scales and observe them under a microscope. Determine the age of the fish from which the scale was obtained by counting the annual growth rings on the scale.
3. At the conclusion of this unit of study, divide the class into two groups and conduct a debate on the use of barging and/or the drawdown salmon recovery plan as management tools for aiding the recovery of endangered or threatened salmon populations. Students may want to prepare fact sheets before engaging in debate.

17 Saltwater Fishes and Fauna

SALTWATER FISHES and **fauna** include the fishes and all of the other animals that live in estuaries and oceans. Their bodies are adapted to living in salty environments. The oceans are rich in nutrients in comparison with freshwater environments, and most marine animals are able to find abundant food supplies.

OBJECTIVES

After completing this chapter, you should be able to

- describe the characteristics of a marine biome
- identify factors that contribute to differences in marine environments
- evaluate the roles of sharks and rays in the marine ecosystems of North America
- define the roles of mackerel, tuna, and marlins in the ocean environments of North America
- speculate about the environmental factors that may have contributed to the body shapes and anatomies of flatfishes
- explain the roles of food fishes such as herring and codfishes in the food web
- specify the roles of smelts and similar fishes in marine ecosystems
- evaluate the roles of sea bass and groupers in saltwater habitats
- identify the characteristics that distinguish mollusks from other marine animals, and define their roles in marine environments
- name four important crustaceans and identify their distinguishing characteristics.

A marine biome is a water environment in which the salt content is 3 to 3.7 percent. It is the world's large biome. Oceans cover 71 percent of the surface of the earth, and because they are connected to one another organisms may pass from one ocean to another.

Many different environmental conditions exist in the ocean, see Figure 17-1. Water temperatures are different in arctic regions as compared to southern coasts. Water pressure is much less in shallow waters and near the surface than it is at depths of several hundred feet. Food supplies vary in different depths of the oceans due to the lack of penetration of sunlight into the deeper strata. The amount of salt varies in the upper layers of the oceans, and salt content is also affected by proximity to large rivers of freshwater. These differences provide environments for many kinds of saltwater fishes and other marine fauna.

Figure 17-1: Factors affecting marine environments.

SHARKS, SKATES AND RAYS

The sharks and rays belong to a group called **cartilaginous fishes,** meaning their skeletons are composed of hard cartilage instead of bone. Cartilage is a tough elastic tissue that allows more freedom of movement than bone. Sharks and rays are also unusual because they have 5–7 gill slits located on each side of their heads.

Most sharks are moderate in size, but the whale shark is the largest of all the fishes. It belongs to the Rhincodontidae family, and individuals can reach as long as 60 feet. It eats plankton that it sieves out of the seawater, along with small fishes and crustaceans.

FISH PROFILE

Whale Shark (Rhincodon typus)

The whale shark is the world's largest fish. It has a broad head and mouth, with tiny, hooked teeth. Three rows of ridges extend along its back on each side. This shark varies in color from dark gray to reddish or greenish brown, with light-colored spots on its upper body. Its belly is white or yellow, see Figure 17-2.

Figure 17-3: Distribution map of the whale shark.

Figure 17-2: A whale shark.

Whale sharks range throughout warm regions near the Atlantic and Pacific coasts and in the Gulf of Mexico, see Figure 17-3. They reproduce by laying eggs. Their diet consists of plankton that are gathered as they swim near the surface of the ocean.

Sharks are probably the most feared fishes known to man. This is because some sharks are ferocious predators, and although attacks on humans are rare, they have been documented. Sharks have several rows of sharp teeth that are used to kill and to cut their prey into bite-size chunks. Their teeth are really enlarged scales that are soon replaced with new ones when they break or fall out. One of the best known sharks is the white shark. It is a member of the Lamnidae family, and it is sometimes called the great white shark.

FISH PROFILE

White Shark (Carcharodon carcharias)

The white shark is a large predator of the sea that eats seals, sea lions, birds, turtles, some shellfish, other sharks, and many kinds of fishes. It is a large shark, up to 30 feet long, with coloring that blends from near black or slate on its back to white on its underparts, see Figure 17-4. It has large triangular teeth with serrated edges. White sharks are found in temperate to tropical seas on both

Figure 17-5: Distribution map of the white shark.

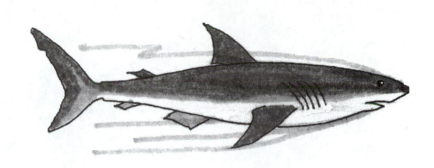

Figure 17-4: The white shark.

North American coasts, see Figure 17-5. White sharks give birth to live offspring.

Sharks range in length from less than a foot to 60 feet. They also vary greatly in their manner of reproduction, see Figure 17-6. Some sharks are oviparous, and the females expel **egg cases** from their reproductive tracts. Other species such as the sand tiger sharks are **ovoviviparous,** meaning that the eggs of the female remain inside her body as the young develop and until they have hatched. Still other sharks are viviparous, and give birth to live young. In all of these cases, reproduction requires internal fertilization of the eggs, or ova, produced by female sharks with the sperm of male sharks.

Shark Reduction		
Oviparous	**Ovoviviparous**	**Viviparous**
• Female lays eggs	• Egg hatch inside female • Young sharks expelled	• Female gives birth to live offspring

Figure 17-6: The manner of reproduction varies depending on the species of shark.

FISH PROFILE

Sand Tiger Shark (Carcharias taurus)

The sand tiger shark is an Atlantic coast species that belongs to the Carchariidae family. It ranges from Maine to Florida, see Figure 17-7, and lives in shallow waters on both sides of the Atlantic Ocean. It is a light-colored shark, with numerous dark spots on its sides, see Figure 17-8.

Figure 17-7: Distribution map of the sand tiger shark.

Figure 17-8: The sand tiger shark.

Females become sexually mature when they are about 7 feet in length. These sharks are ovoviviparous, and only two young are born at a time. After young sharks hatch in each half of the female's uterus, they become cannibalistic and eat any other eggs that have not yet hatched.

Rays and skates are fishes with flattened bodies, long thin tails, and large winglike pectoral fins. Like the sharks, they have skeletons made of hardened cartilage. They are found in North American coastal waters in both the Atlantic and Pacific Oceans. They live on the seafloor where they eat shellfish.

Some of the rays and skates are very unusual. The stingray, for example, (Family: Dasyatidae) has poison in a spine on its tail with which it defends itself. Electric rays (Family: Torpedinidae) are able to emit an electrical shock to their enemies. The sawfish (Family: Pristidae) is a shark-like ray that is equipped with a long projection on its snout that has teeth along both sides like a saw blade. Skates (Family: Rajidae) have wide pectoral fins with which they appear to "fly" through the water.

FISH PROFILE

**Pacific Electric Ray
(Torpedo californica)**

The Pacific electric ray, sometimes called a **torpedo,** is a smooth-skinned, disk-shaped fish, see Figure 17-9. Its upper body is various shades of gray, and the underparts are lighter colored. Females are larger than males, and can reach $4^1/2$ feet in length and weigh up to 90 pounds.

Figure 17-9: The Pacific electric eel.

This ray often buries itself in the sand on the seabed. It uses a powerful electric shock to discourage its enemies and to paralyze its prey. It eats halibut, herring, shellfish and other fishes.

MACKEREL, TUNA AND MARLINS

Mackerel, tuna, and marlins are large, streamlined fishes that are very fast swimmers. They are surface feeders that swim together in large schools as they chase other fishes to the surface of the water. Hunting in this way, they are able to surround and catch the fishes that they depend upon for food.

FISH PROFILE

Chub Mackerel (Scomber japonicus)

Chub mackerel are green or blue on their upper bodies, with numerous vertical bars on their sides that blend to a silvery color on their undersides, see Figure 17-10. They are small fish in comparison with related species such as the tuna. Most mature chub mackerel are 16-18 inches in length. They are caught in large numbers by commercial fishermen.

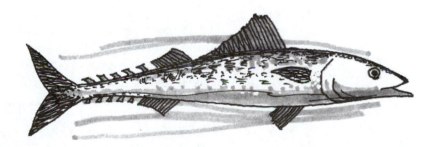

Figure 17-10: The chub mackerel.

Chub mackerel are **pelagic** fishes, meaning that they inhabit the open ocean. They feed mostly on plankton, but they also eat shrimp and small crustaceans. They are found in temperate ocean waters in many parts of the world, but they are most abundant in the Pacific Ocean from Alaska to Mexico. Chub mackerel are also known as Pacific Mackerel because of their abundance in the Pacific region.

Mackerel and tuna belong to the Scombridae family of fishes. They occupy most temperate and tropical oceans, and they are found both inshore and offshore. They are migratory fishes that travel north in the summer and south in the winter, following the migration patterns of the fishes that make up their diets. Tuna are important sources of human food. They eat plankton, crustaceans, and other fishes. The tuna are unusual among fishes because their body temperature is higher than the surrounding water.

FISH PROFILE

Albacore
(Thunnus alalunga)

The albacore is a premium quality commercial tuna. It is abundant in the Pacific Ocean, with a smaller population found in the Atlantic Ocean. It is a migratory species that prefers pelagic habitat and that lives in schools. It prefers warm tropical or subtropical waters.

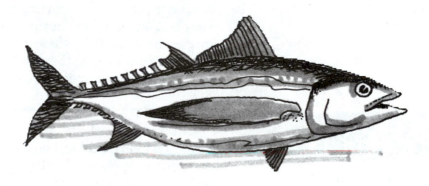

Figure 17-11: The albacore.

Albacore are bright blue in color on their upper bodies. They shade to silvery white on their underparts, and they have long pectoral fins, see Figure 17-11. At maturity they are about 4 1/2 feet in length and usually weighs 75–80 pounds.

A group of fishes that is closely related to the tuna and mackerel is the Istiophoridae family. In members of this family, including the marlins, sailfishes, spearfishes, and swordfishes, the upper jaw is long forming a swordlike bill. These fishes are also known as **billfishes,** and they are highly prized as gamefishes.

Blue Marlin
(Makaira nigricans)

The blue marlin is a large fish that can reach lengths of 11 feet or more. Large members of this species have been observed that weigh over 1,800 pounds. This fish is brown to dark blue above and white or silver below, with vertical bar markings on its sides, see Figure 17-12.

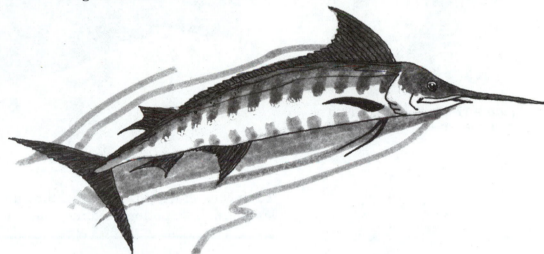

Figure 17-12: The blue marlin.

Blue marlins are found in both the Atlantic and Pacific Oceans, although the Atlantic specimens are smaller. They are surface feeders that can swim as fast as 50 miles per hour. Blue marlins eat other fishes and they use their bills to strike and kill their prey.

FLOUNDER, HALIBUT AND SOLE

The flounder, halibut, sole, and similar fishes are members of an order of **flatfishes,** the Heterosomata. These fishes spend most of their lives lying on the ocean bottoms. Their bodies are flattened to accommodate this behavior and both eyes are located on the upper side of the head.

The eggs of flatfishes float in the water until they hatch. Soon after hatching, the bodies of the young fry begin to change. The eye on the "blind side" of the head migrates across to the other side of the head, and the dorsal fin grows forward. One of the pectoral fins often grows larger than the other, and the young fishes sink to the bottom of the ocean where they lie on their blind sides

Dover Sole
(Microstomus pacificus)

The Dover sole is a flatfish that is considered to be among the best flavored fish for human consumption. It is a slender, right-eyed flatfish that is found in Pacific waters from Alaska to Mexico, see Figure 17-13. It is brown in color on the upper side, and gray on the blind side. This fish is soft and limp, with a small mouth in comparison to other species of soles, see Figure 17-14. It averages

Figure 17-13: Distribution map of the Dover sole.

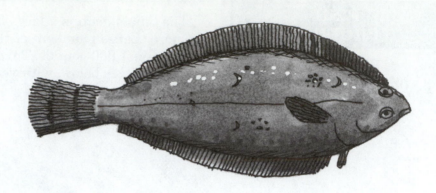

Figure 17-14: The Dover sole.

12 inches in length. Its mouth lies underneath its head, which suits it well for bottom feeding.

Some species of flatfishes are always left-eyed. They are known as **sinistral fishes**, and they belong to the Bothidae family of fishes. Others are always right-eyed. They are called **dextral fishes,** and they belong to the Pleuronectidae family. Some species, such as the starry flounder, may be either sinistral or dextral even though they are classed as right-eyed flatfishes.

FISH PROFILE

**Pacific Sanddab
(Citharichthys sordidus)**

The Pacific sanddab is the most common sinistral or left-eyed flounder. Its coloring is light brown mottled with dark brown, and some individuals have yellow or orange spots. The blind side of this fish ranges from white to tan in color, see Figure 17-15.

Figure 17-16: Distribution map of the Pacific sanddab.

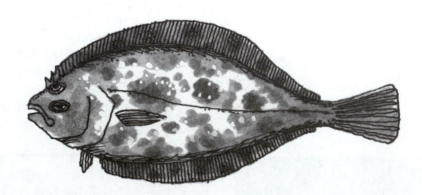

Figure 17-15: The Pacific sanddab.

The Pacific Sanddabs are found from Alaska to Mexico, see Figure 17-16, and they are popular food fishes. They often bury themselves in the sediment on the ocean floor with only their eyes uncovered. They feed on crustaceans and small fishes.

The flatfishes are carnivorous fishes that eat bottom-dwelling marine animals such as squids, crustaceans, and a variety of fishes. In turn, flatfishes are eaten by other predatory fishes, and many species are significant to commercial fisheries. Some flatfishes use a method of protection against their natural enemies known as **camouflage.** They are able to change their coloring to blend with the color of the seabed.

FISH PROFILE

**Atlantic Halibut
(Hippoglossus hippoglossus)**

**Pacific Halibut
(Hippoglossus stenolepis)**

The Atlantic and Pacific halibut are large dextral or right-eyed fishes that have been known to reach lengths of 8 feet and weights of 400 pounds. The Atlantic halibut tends to live longer and to reach larger average sizes than the Pacific species. Both of these halibut have flattened shapes, but they are more plump than most other flatfishes. They are generally dark colored on their upper surfaces and white on their blind sides, see Figure 17-17.

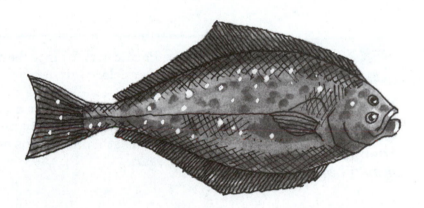

Figure 17-17: The Atlantic or Pacific halibut.

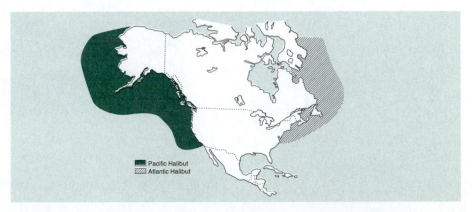

Figure 17-18: Distribution map of the Atlantic and Pacific halibut.

Like other flatfishes, both eyes of a halibut are on the same side of its head. The halibut is different from other flatfishes, however, because it has a normal-shaped mouth. The jaws of halibut are not twisted like those of the flounders. Halibut thrive in cold and temperate oceans, see Figure 17-18.

HERRING

The herring and their relatives are the most important food fishes in the world. They are members of an order of fishes known as **isospondylous fishes.** These fishes are similar in that the vertebrae near the head are the same as the vertebrae located in the tail. This group includes such well known fishes as herring, salmonids, tarpon, sardines, anchovies, and shad.

Herring belong to the Clupeidae family of fishes, and 175 different species of herring are found in the oceans of the world. Different races of herring are also identified within some of the species. They are surface feeders that eat floating plankton.

Herring form vast schools of fish that tend to move in a counterclockwise rotation in the Northern Hemisphere. The movements within these schools of fish are coordinated, and when they change the direction in which they are swimming, they do it together.

FISH PROFILE

Pacific Herring
(Clupea pallasi)

Atlantic Herring
(Clupea harengus)

The Pacific and Atlantic herring are olive to bluish green above, with silver-colored underparts and forked tails, see Figure 17-19. As a group, these herring may be the most important fishes on earth because of the vast amounts of food that they provide for humans and for predatory animals. They average only about 12 inches in length, but billions of them are harvested or preyed upon every year.

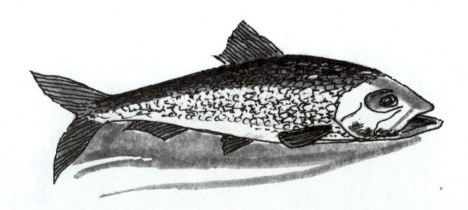

Figure 17-19: The Atlantic and Pacific herring.

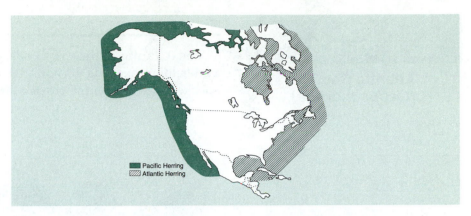

Figure 17-20: Distribution map of the Atlantic and Pacific herring.

The Pacific herring is considered by some scientists to be a population of Atlantic herring located in the Pacific Ocean, see Figure 17-20. Their diets consist of plankton obtained from the ocean surface.

Herring are known as **demersal spawners** because they attach their eggs to weeds and other materials on the ocean floor. Herring that spawn in deep water are protected by this behavior, but when spawning activity occurs in shallow water, large numbers of predators assemble in the area to eat the eggs and catch the fish.

Some of the other important species of North American herring include the skipjack herring that is found in the Gulf of Mexico and in freshwater habitats of the Mississippi River and its eastern tributaries. The alewife is a herring that inhabits the Atlantic coast from Newfoundland to North Carolina.

CODFISHES

Codfishes are second in importance to herring as food fishes. They are members of the Gadidae family, and they inhabit the cold and temperate waters of all of the oceans. These fishes live on the ocean bottom where they feed on small fishes and squid.

Career Option

MARINE BIOLOGIST

A career in marine biology involves the study of animals and plants that live in saltwater environments. It includes studies of the effects that environmental conditions such as light intensity, salinity, temperature, pollutants, and other factors have on marine organisms.

This career requires an advanced science degree with emphasis on the biological sciences. Field work is often an important activity in this career.

The Atlantic codfish is a codfish that usually averages 5 pounds and seldom exceeds 60 pounds. It is a plump, elongated fish with specked olive green to brown coloring. This codfish has a single bar-

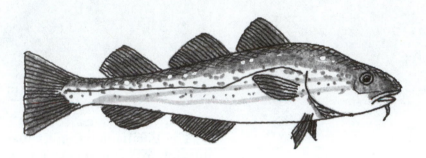

Figure 17-21: The Atlantic codfish.

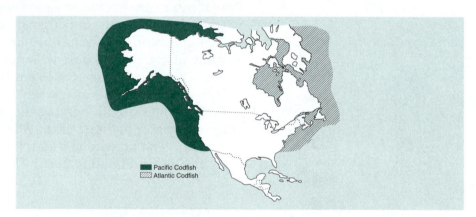

■ Pacific Codfish
▨ Atlantic Codfish

Figure 17-22: Distribution map of the Atlantic and Pacific codfish.

bel under its chin that is used as a feeler, see Figure 17-21.

This codfish is found on both sides of the North Atlantic Ocean, see Figure 17-22. It grows rapidly, and many fish of this species are 15 inches long by the time they are two years of age. They spawn at about five years of age.

Codfishes usually spawn during the winter months, and females are capable of laying several million eggs. The eggs float until hatching occurs. The newly hatched, immature fry are called **larva,** and they float near the surface of the water with the plankton deposits. About the time they are one inch long, they sink to the bottom.

**FISH
PROFILE**

**Pacific Codfish
(Gadus macrocephalus)**

The Pacific codfish has a gray to brown upper body, with pale areas and spots of brown on its sides and back. The underparts are light colored, and it has a small barbel on its chin, see Figure 17-23.

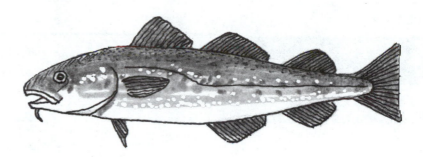

Figure 17-23: The Pacific codfish.

It is also known as the gray cod.

This fish is distributed throughout the cold waters of the Pacific Ocean from Alaska to California, see Figure 17-22. It is an important commercial fish that seldom exceeds 35–40 inches in length.

A close relative of the codfish that is important in the Atlantic region is the haddock. This fish occupies many of the same areas as the codfish, and is also an important food fish.

SMELTS

Smelts are small, slender fishes that live together in schools. They are members of the Osmeridae fish family. They are important in North American ecosystems because they provide a dependable food supply for larger fishes, seals, and cormorants. Most smelts are marine, but some are anadromous. These fishes are found on both coasts of the North American continent and in much of the cold Arctic region, see Figures 17-24 and 17-25.

Figure 17-24: Distribution map of smelts.

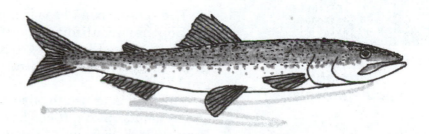

Figure 17-25: Smelts.

SEA BASS AND GROUPERS

The sea bass and groupers belong to the Serranidae family. They are important food fishes, as well as carnivores that feed on crustaceans and small fishes. Much of their time is spent on the ocean floor looking for food. These fishes have large mouths and sharp teeth that equip them well for their predatory roles.

FISH PROFILE

Snowy Grouper
(Epinephelus niveatus)

The snowy grouper is a reddish brown fish with white spots and a large mouth, see Figure 17-26. It inhabits both the Atlantic and

Figure 17-27: The snowy grouper.

Figure 17-26: Distribution map of the snowy grouper.

Pacific Oceans, and it is found in warm and temperate seas, see Figure 17-27. This fish grows up to 4 feet long in the Atlantic Ocean, but it seldom reaches more than 31 inches in the Pacific Ocean.

Most of the members of the Serranidae family are less than a foot long, although a few of them attain lengths of 4 feet. Most of these fishes are **hermaphrodites,** meaning that they have the sex organs of males and females at the same time. Some species function as females when they are small and as males when they become larger.

FISH PROFILE

Spotted Sand Bass
(Paralabrax maculatofasciatus)

The spotted sand bass is a popular sport fish found off the coast of Southern California southward to Mexico and the Gulf of California, see Figure 17-28. It is a bottom fish that occupies water from the shoreline to depths of 200 feet. It is often found in bays and harbors.

This fish is brown to olive colored on the upper body, with light-colored underparts. It has bar markings on the back and sides, and round black spots on its fins and body, see Figure 17-29.

Figure 17-28: Distribution map of the spotted sand bass.

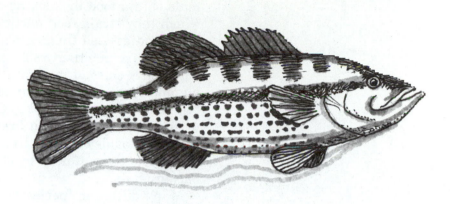

Figure 17-29: The spotted sand bass.

MOLLUSKS

The mollusks make up one of the largest phyla of organisms in the animal kingdom. This group of invertebrates include the slugs, snails, bivalves, squids, octopuses, and cuttlefishes. All of these animals have soft bodies, but many of them are protected by shells. Slugs and snails belong to a class known as Gastropoda. Each of these organisms moves about on a muscular foot.

Animals known as **bivalves** are organisms with a shell of two parts, or valves, that are hinged together. A well-known member of this group is the clam, see Figure 17-30. Other members of this

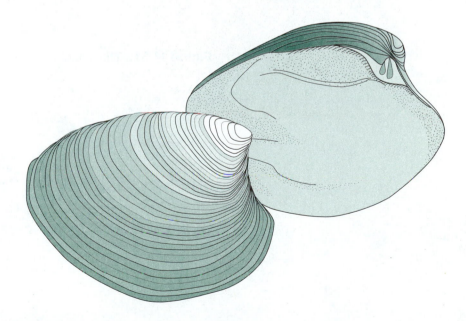

Figure 17-30: The clam.

group of mollusks are the mussels, scallops, and oysters. These animals obtain food by sucking seawater through their bodies, from which they strain food particles. The movement of water through their bodies also flushes wastes from their systems. All of the bivalves live in either fresh- or seawater.

Another class of mollusks is the Cephalopoda. Its best-known members are squids, octopuses, and cuttlefishes. The head of a cephalopod is surrounded by 8–10 arms. Each of which is equipped with sucking organs that are used to capture prey for food. They are also capable of hiding from their enemies by discharging clouds of ink into the water, or by changing their coloring to match their backgrounds.

Many different species of squids have been identified. Squids have eight arms like those of the octopus, and two additional tentacles that are longer than the other eight, see Figure 17-31. Their bodies are supported by an internal plate. These animals move about by drawing water into their bodies and forcing it out at the back of the head. This propels them backwards through the water. The squid's diet consists of fishes and crustaceans. Among the unusual characteristics of squids are their internal shells, the ability of some to produce light from special cells, and the unusual blue color of their blood.

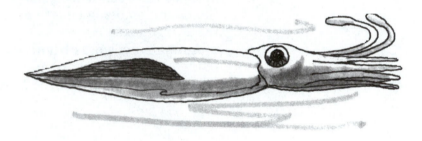

Figure 17-31: The squid.

Figure 17-32: The octopus.

The octopus is a cephalopod with eight arms and a round, short body, see Figure 17-32. Octopuses range widely in size. The Pacific octopus is the largest, with a span of 33 feet. Some species of octopuses live in deep water, while others live among the rocks in shallow coastal waters. They eat crustaceans and small fishes. Unlike the squids, octopuses have no internal shell.

The cuttlefish is similar to both the octopus and the squid. It has two long tentacles and eight arms, and its body contains a chalk-like plate containing gas-filled cells that increase its buoyancy. The largest of these animals spans more than 5 feet, the smallest less than 2 inches. Like the squid and octopus, it catches prey with the suckers on its tentacles.

CRUSTACEANS

Crustaceans are marine animals that have exoskeletons and several pairs of jointed limbs. They belong to the phylum Arthropoda, which also includes insects, spiders, centipedes, millipedes, and scorpions. The best known marine crustaceans are the shrimps, crabs, and lobsters, but the most numerous are tiny animals called plankton. Crustaceans provide a food supply for many of the fishes and other animals that are found in the oceans. Respiration occurs using gills that take oxygen from the water.

Crustaceans molt or shed their hard outer shells as they grow. These exoskeletons are replaced with larger ones which provide their bodies more room. Some crustaceans may lose legs or pincers during escapes from natural enemies or in battles with others of

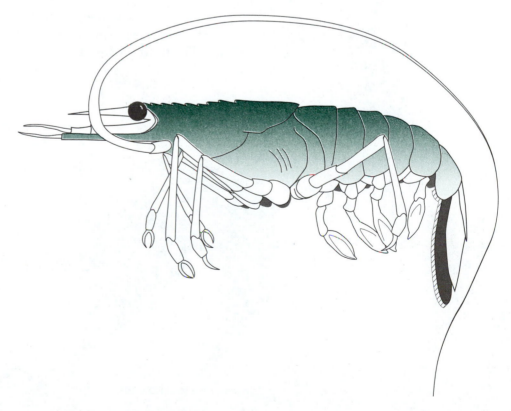

Figure 17-33: The shrimp.

their species. These limbs are replaced with new ones when they molt.

Shrimps include a number of small crustaceans that are similar in appearance and in their habits. They have two pairs of **antenna** located on their heads that serve as sensing organs or feelers. They also have five pairs of legs, see Figure 17-33. Some species spend daylight hours buried in the sand or mud, and they feed at night. Shrimp eat many kinds of small aquatic animals, as well as seaweed.

Crabs are animals with four pairs of legs, a pair of pincers, a flattened shell, and a broad abdomen, see Figure 17-34. Some of them live in seawater; others live in freshwater. Most crabs breed in marine habitats even when they must migrate from freshwater areas to do so. Some crabs eat only plant materials, while others eat only meat. Still other crabs are scavengers that eat nearly anything consisting of organic material.

Lobsters are among the large crustaceans. They are strange-looking animals, with **compound eyes** made up of many simple eyes that function together. They live on rocky seabeds near the shore and they eat several kinds of prey, including other lobsters, mollusks, crabs, and small fishes.

Lobsters have four pairs of walking legs and one pair of pincers. Their shells are dark green or blue, but they turn red when they are cooked. Lobsters also have other appendages called **swimmerets** that are located on their undersides, see Figure 17-35. These are used to propel the lobsters forward. The powerful tail is used to propel the lobster backward. Freshwater relatives of lobsters with similar anatomy are the crayfishes. They are found in many freshwater habitats in North America.

Numerous tiny crustaceans live in the oceans of the world and float on the surface of the water with other microscopic plants and animals. These life-forms are grouped together and called plank-

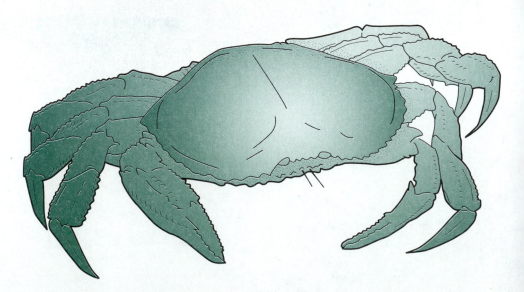

Figure 17-34: The crab.

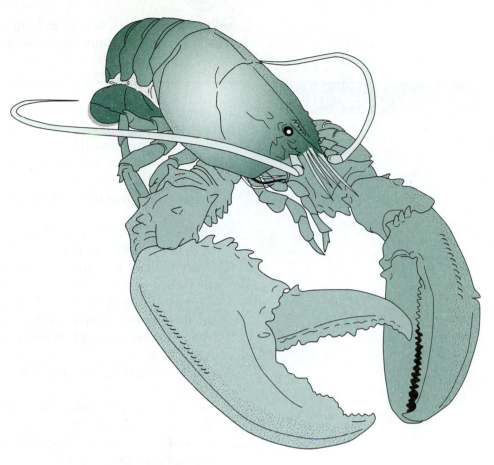

Figure 17-35: The lobster.

ton. Many of the fishes and other marine animals depend on plankton for food. Some of the whales are able to strain enough plankton from the water to survive and to grow at extremely rapid rates. The tiny planktonic crustaceans may be the most important of all the crustaceans because they provide such a massive volume of food to other living creatures.

There are many marine organisms that are found in the ecosystems of North America that have not been included in this book. The best-known fishes and marine animals have been used here as representatives of the many other kinds of organisms that are found in marine environments.

LOOKING BACK

Saltwater fishes and fauna include the fishes and all of the other animals that live in estuaries and oceans. The sharks and rays are predatory cartilaginous fishes. Several important food fishes that live in ocean environments include mackerel, tuna, marlins, and herring. All of these are surface feeders. Food fishes that are bottom feeders include flounder, halibut, soles, codfishes, sea bass, and groupers. Smelts are important fishes in the marine food chain because they provide food to many of the larger fishes. Mollusks and crustaceans include marine animals such as octopuses, squids, shellfishes, crabs, lobster, and shrimps. Some mollusks and crustaceans are microscopic animals called plankton.

REVIEW QUESTIONS

1. Describe the characteristics of a marine biome.
2. List the major factors that contribute to differences among marine environments.
3. What function do sharks and rays perform in marine environments?
4. Define the roles of mackerel, tuna, and marlins in the ocean environments.
5. Speculate about the kinds of environmental factors that may have contributed to the body shape and anatomy of flatfishes.
6. Describe how the unusual anatomy of a flatfish contributes to its ability to compete in its ocean floor environment.
7. Explain the role of food fishes such as herring and codfishes in the food webs of marine ecosystems.
8. Specify the roles of smelts and similar fishes in marine ecosystems.
9. Evaluate the roles of sea bass and groupers in saltwater habitats.
10. List the characteristics that distinguish mollusks from crustaceans and fishes.
11. Name four important crustaceans and list their distinguishing characteristics.

LEARNING ACTIVITIES

1. Divide the class into groups and have each of the groups prepare a wall chart that illustrates how each of the marine fishes and animals fits into the food web. Display the charts on the walls of the classroom.
2. Provide opportunities for class members to earn class credit by visiting the fish counter at a local supermarket. Assign students to observe the kinds of seafoods that are available, and prepare a list of these foods. Make a chart listing important characteristics of the organisms and their habitats from which the foods were obtained.
3. Visit a public aquarium if such a facility is available, or identify video materials on marine environments for viewing by students.

18

Reptiles and Amphibians

REPTILES AND AMPHIBIANS are two distinct classes of organisms, see Figure 18-1. They are **ectotherms,** or cold-blooded animals, that depend on their surrounding environment for body heat. **Endotherms,** or warm-blooded animals, obtain heat from metabolism of their food. **Reptiles** are distinguished by the scales, plates, or shields covering their bodies, eggs with leathery skins, and by the claws on their feet. Young reptiles look like their parents except for their smaller size. **Amphibians** have moist skin with no visible scales and their toes are never clawed. Their eggs must be deposited in the water to keep them from drying out. Young amphibians develop through an aquatic larval stage before they become adults. The change from juvenile to adult amphibian occurs through a process known as **metamorphosis.**

OBJECTIVES

After completing this chapter, you should be able to

- distinguish between ectotherms and endotherms
- describe differences between reptiles and amphibians
- compare the different kinds of reproduction that occur in reptiles
- illustrate the structure of an amniote egg, and describe the functions of the four embryonic membranes
- evaluate the roles of alligators and crocodiles in North American ecosystems
- distinguish between vipers and elapids
- analyze the roles of lizards and snakes in the ecosystems of North America
- explain how the tongue of a snake aids its sense of smell
- discuss the differences between turtles and tortoises, and define their roles in the ecosystems in which they live
- identify the major steps in the metamorphosis of an amphibian

319

■ compare the similarities and differences between frogs and
toads
■ appraise the roles of salamanders and newts in North American
ecosystems.

Distinguishing between Reptiles & Amphibians		
Distinguishing Feature	**Reptile**	**Amphibian**
Blood supply:	Cold blooded Ectotherm	Cold blooded Ectotherm
Reproduction:	Internal fertilization Oviparous, Ovoviviparus	External fertilization Oviparous
Skin:	Scales, plates or shields	Smooth, soft, moist
Offspring:	Miniatures of parents	Larval stage with gills
Feet:	Claws	No claws

Figure 18-1: Distinguishing features of reptiles and amphibians.

REPTILES

Reptiles usually spend their entire lives within a short distance of
the place where they hatched. They are not very mobile, and sometimes they are unable to survive environmental conditions that are
different from their natural habitats. North American reptiles
include two species of alligators, one species of crocodile, turtles,
lizards, and snakes.

Reptiles are found in a wide range of environments. They use
heat from the sun to warm their bodies, and water and shade to
cool them. They spend much of their time basking in the sunshine
or floating just beneath the surface of the water with only their
eyes showing above the water. The rise and fall in the body temperature of a reptile increases or decreases the rate of its metabolism, see Figure 18-2.

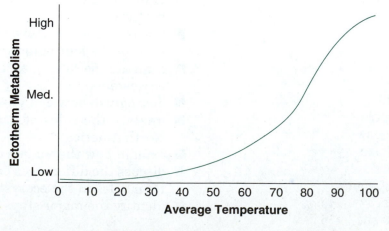

Metabolism and Temperature

Figure 18-2: The rise and fall in the body temperature of a reptile
increases or decreases the rate of its metabolism.

THE AMNIOTE EGG

Ecology Profile

An **amniote egg** contains a protective membrane and a porous shell that surrounds the developing embryo. Such eggs are key elements in the reproduction of reptiles and birds. Four different membranes develop as a fertile egg is incubated. They arise from the developing embryo and are called **embryonic membranes.** Each of them is needed for the survival of a growing embryo.

The **chorion** is a membrane that grows out of the embryo and surrounds the embryo and the other three embryonic membranes. Gasses are able to pass through it quite freely. The **allantois** is a membrane that is filled with blood vessels. This membrane joins to the chorion, and is the structure through which respiration occurs. The blood carries wastes and dissolved gasses away from the embryo and dissolved oxygen into it.

The **amnion** is the innermost protective membrane that surrounds the embryo. It contains a salty liquid called **amniotic fluid** in which the embryo floats. This liquid cushion also prevents the embryo from dehydrating.

The **yolk sac** is an embryonic membrane that surrounds the yolk of the egg. It arises from the digestive tract of the embryo, and it releases digestive juices into the yolk. Blood vessels form in the yolk, and they carry the dissolved nutrients obtained from the yolk to the body of the embryo.

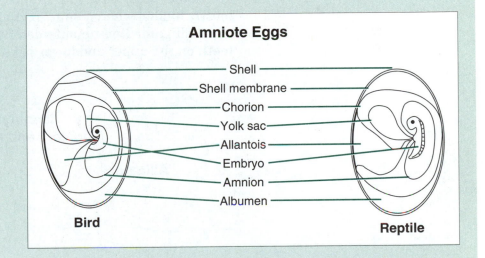

Amniote Eggs

Shell
Shell membrane
Chorion
Yolk sac
Allantois
Embryo
Amnion
Albumen

Bird

Reptile

Reproduction in reptiles requires internal fertilization of the eggs. Once the shell is formed, fertilization is impossible. Most reptiles are oviparous. They lay their eggs in nests that have been dug in earth, sand, or decaying vegetation. Some reptiles are ovoviviparous, and their young develop within the eggs inside the body of the female. After the eggs hatch, these females give birth to living offspring.

Reptile eggs are similar in many ways to the eggs of birds. Gases are able to move in and out of them allowing the growing embryos to obtain oxygen and to eliminate carbon dioxide as a waste product. The outer shells of both reptile and bird eggs are impermeable to water. This prevents the embryos from drying out and dying before they hatch. The most obvious difference between reptile and bird eggs is that reptile eggs have leathery shells, and the eggs of birds have rigid, brittle shells.

Alligators and Crocodiles

The alligators, caimans, and crocodiles are similar in appearance to large lizards, and they belong to the order Crocodilia. These animals are covered with plates and scales that protect their bodies, and they live in warm climates. This order of animals has existed since the age of the dinosaurs, yet their four-chambered hearts are more like those of mammals than those of other reptiles.

Crocodiles are represented by a single species in North America. They can be distinguished from alligators by their pointed snouts, by the arrangement of their teeth, and by their aggressive behaviors, see Figure 18-3. The snout of an alligator is broader and more rounded than that of a crocodile.

The fourth lower tooth and many of the upper teeth of a crocodile are visible when its mouth is shut, but the teeth of an alligator are mostly hidden when it shuts its mouth. The teeth on the upper jaw of an alligator line up outside of the teeth on the lower jaw. The teeth on the upper and lower jaws of a crocodile are in line.

Identifying
Alligators and Crocodiles

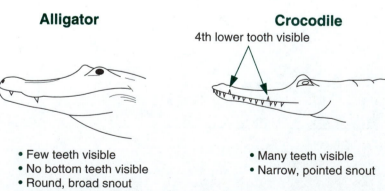

Alligator

- Few teeth visible
- No bottom teeth visible
- Round, broad snout

Crocodile
4th lower tooth visible

- Many teeth visible
- Narrow, pointed snout

Figure 18-3: Identifying features of alligators and crocodiles.

Alligators are represented in North America by two species. The natural range of the **caiman** extends from southern Mexico to many parts of South America, but it has also been introduced into some parts of southern Florida. The American crocodile is found only in southern Florida and in parts of Mexico and Central America.

REPTILE PROFILE

American Alligator (*Alligator mississippiensis*)

Figure 18-5: Distribution map of the American alligator.

The American alligator is a large reptile that averages 7.5–12 feet in length. It is black in color with yellow crossbands that tend to fade away with age, see Figure 18-4. Its body is covered with large scales, and its snout is broad and round.

Figure 18-4: The American alligator. *Photo courtesy of Isabelle Francais.*

Alligators are predatory animals that eat birds, fishes, and small mammals. They inhabit swamps, marshes, rivers, and southern coastal waters of the Atlantic Ocean and the Gulf of Mexico, see Figure 18-5. Alligators call to one another with loud bellowing roars that can be heard for great distances. Females lay 15–80 eggs in nest mounds made of mud and rotting vegetation. They communicate with their young using grunts. Alligators also hiss when they are threatened.

Alligators and crocodiles build large nests of vegetation in which they lay as many as 80 eggs. Female crocodiles return regularly to their nests, but they do not guard them as do the alligators. Both of these animals open their nests after the young have hatched. They carry their young to the water in their mouths, and protect them from predators for several months.

Lizards

Lizards, along with snakes, belong to the order of animals known as Squamata. This order of reptiles includes more species than any other reptilian order. These animals are adapted to many habitats, and they can be found in most regions of the world. Most lizards are terrestrial animals, but some are adapted to marine environments.

Most of the lizards are small carnivores that eat spiders, insects, and other small animals. Some lizards grow to a large size and eat bigger prey; a few species are herbivores. Lizards usually reproduce by laying eggs, but some species give birth to live young.

Many lizards have long slender tails that break off easily. This unusual trait helps lizards to escape their enemies by leaving their tails behind; the tail piece continues to wiggle, thereby distracting the predator. The tail of a lizard is regenerated when this happens.

Many North American lizards are **iguanas.** These lizards belong to eight separate families. They live in both temperate and tropical regions. The best known of these are probably the horned toads and the green anole, often called the American chameleon.

REPTILE PROFILE

Texas Horned Lizard (Phrynosoma cornutum)

Figure 18-7: The Texas-horned lizard.

The Texas horned lizard is usually called a horned toad. It has a spiny covering, with two head spines that are larger than those on the rest of its body. Most mature lizards are brown in color, and they blend in well with the environment in which they live, see Figure 18-6. These lizards live in areas of flat terrain with sparse plant cover. Their diet consists of ants and other insects, and they drink dew to obtain water. They escape their enemies by running for cover in holes or under rocks and vegetation. It sometimes squirts blood from its eyes when it is grasped. This unusual behavior is thought to be related to high blood pressure. The native range of this lizard is from the plains states to the desert regions of the southwestern United States, see Figure 18-7. This lizard reproduces by laying eggs.

Figure 18-6: The Texas-horned lizard. *Photo courtesy of Leonard Lee Rue III.*

REPTILE PROFILE

Green Anole (Anolis carolinensis)

The green anole is a small lizard with a colored area on its throat. Males have a colored flap of skin on their throats called a fan; it may vary from pink to purple among individual lizards. The green anole has the ability to change its body color to match its background. Its coloring ranges from green to brown and may include a mixture of colors, see Figure 18-8.

Figure 18-9: Distribution map of the green anole.

This lizard is not a true chameleon even though it is often called by that name. It is a common sight on fences, walls, and vegetation in the southern region of the United States, see Figure 18-9. Many of these small lizards are sold as pets.

Figure 18-8: The green anole. *Photo courtesy of Kent A. Vliet.*

There are two species of poisonous lizards in North America: the Gila monster and the beaded lizard. Both are easily recognized by the bead-shaped scales on their bodies. The Gila monster is native to the southwestern United States and Mexico, see Figure 18-10. The beaded lizard lives in Mexico. These lizards kill their victims by biting them and chewing the wounded area. This draws poison to the wounds through grooves in their teeth. The venom paralyzes the nerves that control breathing.

REPTILE PROFILE

Gila Monster (*Heloderma suspectum*)

Figure 18-10: Distribution map of the gila monster.

The Gila monster is a poisonous lizard. Its body is covered with black bead-shaped scales interspersed with patches of yellow and pink scales. They are short-legged animals up to 2 feet in length,

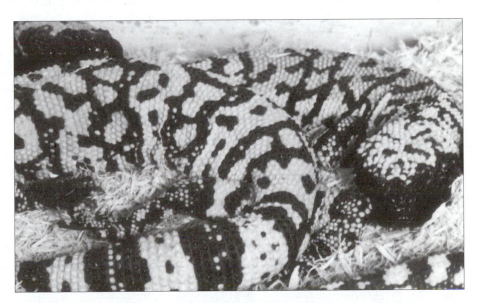

Figure 18-11: The gila monster. *Photo courtesy of Isabelle Francais.*

with stout bodies, large heads, and muscular tails, see Figure 18-11. Gilas are slow animals that eat insects, young birds, eggs, rodents, and other lizards. They live in burrows in the sand or under rocks, and they hunt mostly at night. The gila monster is oviparous, and the females lay up to a dozen tough shelled eggs in a hole located in moist sand. They prefer desert habitat.

Skinks are the most widespread species of lizards found in the United States. They are ground lizards whose smooth, flat scales give their bodies a shiny, silky appearance. They have short legs but are swift runners. Most of them are insectivores, but a few skinks are vegetarians. They are also known to eat spiders, worms and small vertebrates. They are diurnal animals that are most active during the daytime, and they usually seek shelter at night in moist areas under rocks, debris, or decaying logs. Some skinks are known to live in trees. Skinks hibernate during the winter in ground burrows or cinder logs.

During the mating season, the males usually display colored spots of red or orange on their heads. Males of some species are known to retain these colored spots throughout the year, but with some skinks they disappear after the breeding season is over. Skinks are oviparous, and mating occurs before the females lay their eggs. The female guards the eggs, which are laid in sandy nests, during the incubation period.

REPTILE PROFILE

**Ground Skink
(Scincella lateralis)**

Figure 18-13: Distribution map of the ground skink.

The ground skink is a small, shiny lizard that is brown in color, with a dark lateral stripe extending along its back, see Figure 18-12. Its belly is white or yellow. Adults average 3–5.75 inches in length from their heads to their tails. The young are about 1.75 inches long when they hatch. Females lay several eggs in the sand and guard them while they incubate.

This lizard lives in wooded habitats where it hunts among the fallen leaves and other decaying plant materials for the insects that make up its diet. It hides when it is disturbed, so it is seldom seen. It inhab-

Figure 18-12: The ground skink.

its the central and southeastern regions of the United States and isolated locations in Mexico, see Figure 18-13.

Two different families of lizards are called **geckos.** They are able to climb with their claws, or hold to surfaces like ceilings and walls with tiny suction cups located on the ends of tiny bristles on their feet. They are the only lizards that make loud calls. Some of the geckos have moveable eyelids, while others have lidless eyes that remain open all of the time.

The diets of geckos are mostly composed of insects and spiders but some geckos eat other lizards. Their preferred habitat is around and on buildings or other structures. Most geckos reproduce by laying eggs.

Some lizards look very much like snakes or worms because they have no legs. Unlike snakes, however, they have eyelids that can be closed over their eyes, and they have external ears. The most common of these lizards are worm lizards and glass lizards.

Snakes

The true snakes are the most numerous creatures in the reptile world, and they are all carnivores. They live in most environments in North American with 126 species found in the United States, and 22 species found in Canada. The diets of many snakes make them valuable to humans because they eat insects, mice, rats, and other rodents.

REPTILE PROFILE

**Eastern Garter Snake
(*Thamnophis sirtalis sirtalis*)**

Figure 18-14: Distribution map of the Eastern garter snake.

The Eastern garter snake is widespread in the eastern half of North America, ranging from Canada to the Gulf of Mexico, see Figure 18-14. It is a small snake between 18 and 26 inches in length, and it occurs in a range of colors including black, brown, olive, and green. Three yellow stripes and a series of dark spots extend along the length of its body, see Figure 18-15.

This snake lives in marshes, woodlands, streambanks, meadows, and parks. It preys upon fish, salamanders, frogs, toads, worms, and small birds and mammals. Garter snakes are sometimes called garden

Figure 18-15: The common garter snake. *Photo courtesy of Isabelle Francais.*

snakes. Many species have been identified, and they are found in a wide variety of locations. Young are born alive, twenty or more at a time.

Water snakes are the most numerous snakes in North America and are found in abundance in many parts of the continent. Water snakes prefer habitats that are near water from where they obtain most of their food. They prey upon frogs, salamanders, crayfish, fish, and other small aquatic animals. They have solid teeth, and will strike and bite when they are threatened.

REPTILE PROFILE

Northern Water Snake
(Nerodia sipedon sipedon)

Figure 18-16: Distribution map of the Northern water snake.

Northern water snakes include several subspecies that occupy a range that includes much of the eastern United States, see Figure 18-16. These snakes vary in length from 24–42 inches, and they are known to range widely in color from light gray to dark brown. Markings are often visible on light-colored snakes, but they may be completely obscured in dark-colored ones, see Figure 18-17. These

Figure 18-17: The Northern water snake. *Photo courtesy of Len Rue, Jr.*

snakes inhabit swamps, marshes, streams, lakes, and other waterways. Females are usually larger than males, and they give birth to live young.

The bodies of snakes are covered with smooth, dry scales. As a snake grows, it molts by shedding its old skin. This process is also called **ecdysis.** Once the new skin has formed, the old skin is rubbed loose. Rattlesnakes add new rattle segments each time that ecdysis occurs. Each new segment is larger in size than the previous segment, and the rattle grows as the snake matures.

Two distinct kinds of poisonous snakes are found in North America. They are the vipers and the elapids, see Figure 18-18.

Poisonous Snakes		
	Vipers	**Elapids**
Features:	Long, hinged fangs	Short, permanently fixed fangs
Common names:	Copperhead Cottonmouth Rattlesnake	Cobra Sea snake Coral snake

Figure 18-18: Features and common names of poisonous snakes.

Vipers include copperheads, cottonmouths, and rattlesnakes. These snakes have long fangs in the front of their mouths that are hinged. They fold back out of the way when the snake is not using them. The fangs of **elapids** are permanently fixed in place. These snakes include cobras, sea snakes, and coral snakes; only the coral snake is found in North America.

Figure 18-20: Distribution map of the Eastern coral snake with Texas and Mexican subspecies.

The eastern coral snake is a brightly colored snake with red, yellow, and black alternating bands that completely encircle its body, see Figure 18-19. The snout is black followed by a yellow band on the head and a black band on the neck. After that a pattern of yellow, red, yellow, and black bands is repeated along the length of its body.

The eastern coral snake inhabits a wide range of environments from pine woods to the lands bordering waterways. It is also found in hardwood forests with dense undergrowth. Another coral snake, known as the western coral snake, occupies similar habitats in the western United States. A Texas subspecies

Figure 18-19: The Eastern coral snake. *Photo courtesy of Leonard Lee Rue III.*

inhabits cedar brakes, rocky canyons, and hillsides, see Figure 18-20. These are diurnal snakes that hunt mostly during the daytime. Their diets are restricted to slender prey such as lizards and other snakes because their jaws do not open wide. The coral snakes are secretive in their habits, and they are seldom seen. They reproduce by laying 3–14 eggs.

Thirteen different species of rattlesnakes live in North America. Each of them is equipped with a rattle on the end of the tail. A young rattlesnake only has a button, but by the time it is mature, it will have several segments on its rattle. A new segment is added on the rattle each time the snake sheds its skin. Two to four segments are added each year, so the number of rattles is an indicator of the age of the snake. Rattles are made of material that is similar to that in horns or claws. Segments are loosely attached together, and when they are shaken rapidly they create a buzzing sound similar to the sound of a cicada.

Prairie rattlesnakes and their western subspecies are brownish green to yellowish brown in color. They have blotches of dark brown rimmed with white extending along their backs and sides, see Figure 18-21. They eat rodents, birds, and other small animals.

Figure 18-22: Distribution map of the prairie rattlesnake.

Figure 18-21: The prairie rattlesnake. *Courtesy of Leonard Lee Rue Enterprises. Photo by Charles G. Summers Jr.*

These snakes are abundant in the grasslands of the Great Plains and in the arid desert regions of the western United States. Their range extends from southern Canada to northern Mexico, and from Iowa to the Pacific coast, see Figure 18-22. The western variety is often found in rocky canyons and rock outcroppings at elevations below 8,000 feet. These snakes move into winter dens located in rocks and ledges when the cold nights of the fall season begin. This snake gives birth to live young, usually about 12 per litter.

Several species of rattlesnakes besides the prairie rattler are distributed across North America. They include the eastern and western diamondbacks, the pygmy rattlesnake, the timber rattler, the sidewinder rattlesnake, and several other species that are not very well known. All of them are poisonous and must be treated with respect.

Snakes hear through inner ears that are not visible externally. They do not hear the same way humans do, but rather their entire body picks up vibrations through the ground. They also have a good sense of smell that is enhanced by their forked tongues. A snake's tongue gathers molecules from the air, and transports them to a sensory organ in its mouth called **Jacobson's organ.** This organ is used for smelling.

A unique characteristic of many snakes is the flexible attachment of their lower jaws to their skulls. This allows them to swallow large prey and food particles that are bigger than their mouths, see Figure 18-23. To do this, the mouth and the body of a snake must expand to accommodate the size of the meal that is swallowed.

Reproduction in snakes requires internal fertilization. Oviparous snakes lay their eggs before they hatch, and the eggs of ovoviviparous snakes hatch internally before the young snakes are born or immediately after they are released from the mother. The

Hinged Jaws of Snakes

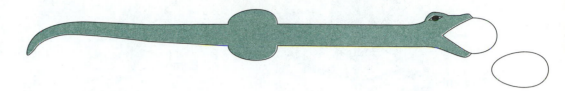

Figure 18-23: The lower jaw of the snake is flexible so that it can expand to accomodate large prey and food particles.

embryos of all snakes are nourished by the yolks in the eggs from which they hatch. They are not nourished by the body of the mother snake.

Rat snakes are usually large non-venomous snakes ranging from 24–96 inches in length. They eat mice, rats, small birds, lizards and frogs. They are good climbers, and they have adapted to a fairly wide range of habitats. They have stout bodies, and they kill their prey by constricting or squeezing them in their coils. For this reason they are classed as **constrictors.**

REPTILE PROFILE

Black Rat Snake (Elaphe obsoleta obsoleta)

The black rat snake is the largest American rat snake averaging 42–72 inches long. Its range extends through much of the northeast quadrant of the United States and in eastern Ontario, Canada. The preferred habitat includes rocky, timbered hills to farmlands with some of these snakes living in the cavities of hollow trees. This snake reproduces by laying eggs. It is shiny black in color with some evidence of a spotted pattern, especially in young snakes. Its chin and throat are cream-colored or white.

Turtles and Tortoises

The turtles and tortoises are slow-moving reptiles whose bodies are surrounded by upper and lower shells. The upper shell is called a carapace, and the lower shell is called a **plastron,** see Figure 18-24. The shell of a turtle is fused to its body, and the shells of most turtles are made of a hornlike material that is very hard. Some turtles have shells that are made of a softer material that is similar to leather.

The **tortoise** is a turtle that lives on land. It is equipped with strong feet for walking, and its toes have claws that are used for digging. Nearly all of the tortoises are vegetarians, and several species live in desert environments.

Turtles that live in saltwater marshes are called **terrapins,** and those that live in freshwater habitats are called **turtles.** There are a number of instances in the southern regions of the United States, however, where freshwater turtles have been incorrectly called terrapins.

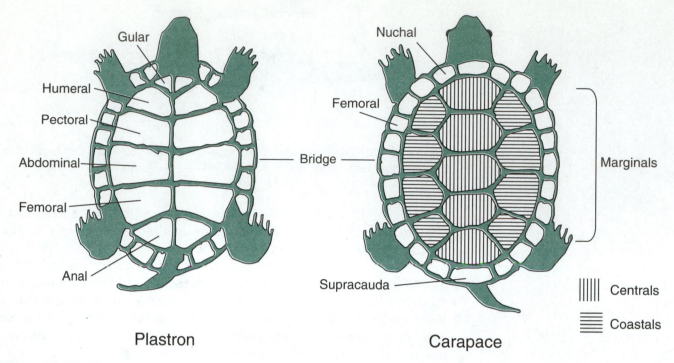

Plastron

Carapace

||||| Centrals

≡ Coastals

Figure 18-24: The scutes on a turtle's shell.

**Snapping Turtle
(*Chelydra serpentina*)**

Figure 18-26: Distribution map of the snapping turtle.

The snapping turtle is a freshwater reptile that usually measures 8–14 inches in diameter when it is mature. The carapace is usually brown to black in color with a rough surface on which three distinct ridges are evident. These ridges are called **keels,** and they become less prominent as the turtle ages. The tail of the snapping turtle is as long as the carapace, and the head is large, see Figure 18-25.

Figure 18-25: The common snapping turtle. *Photo courtesy of Isabelle Francais.*

These turtles are omnivorous animals that eat small invertebrates, fishes, birds, small mammals, plants, carrion, and small reptiles. They are feisty animals that bite when they are threatened. They can live in almost any body of water where food is abundant, and they range from Canada to the Gulf of Mexico and from the Atlantic Ocean to the Rocky Mountains, see Figure 18-26. They reproduce by laying eggs.

Turtles have scales on their heads, necks, tails, and legs just as do other reptiles. The mouth of a turtle is much like the beak of a predatory bird, and a turtle has no teeth. The jaws of turtles are strong, and they are able to cut plants and meat into bite-size pieces.

REPTILE PROFILE

Texas Tortoise (*Gopherus berlandieri*)

Figure 18-28: Distribution map of Texas tortoise.

The Texas tortoise is a moderate-sized turtle that grows up to 8 inches in diameter when it is mature. The color of this tortoise varies from tan to brown. The carapace is broad and rounded over the top, and the plastron is quite long with an upward curve, see Figure 18-27.

These tortoises are found in a small section of southern Texas and northern Mexico, see Figure 18-28. They live in a desert habitat where they feed on grasses and prickly pear cacti, in addition to other vegetation. They reproduce by laying eggs.

Figure 18-27: The Texas tortoise.
Photo courtesy of Kent A. Vliet.

A few large sea turtles are found in warmer waters of both the Atlantic and Pacific coasts, see Figure 18-29. These animals spend most of their time in the water and seldom come ashore. The

Figure 18-29: A sea turtle.

female comes ashore to lay her large clutch of eggs. Sea turtles differ from land and pond species in several ways. For example, they have flippers instead of legs. The flippers are well adapted for swimming, but they don't work very well on land. Sea turtle populations have seriously declined, because they have been hunted heavily or drowned in fishing nets. They are now protected by law in many countries, and efforts are being made to help their populations recover.

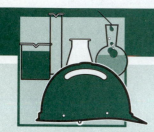

Career Option

HERPETOLOGIST

A **herpetologist** is a scientist who specializes in the study of reptiles and amphibians. They are studied in their natural settings, as well as in the laboratory. A person who engages in this career sometimes devotes an entire lifetime to learning about the interrelationships between reptiles and/or amphibians and other living and nonliving elements found in their environments. This career usually requires a graduate degree in the zoological or biological sciences.

AMPHIBIANS

Two classes of amphibians are native to North America. One of these groups includes the frogs and toads. The other includes the newts and salamanders. All of these animals require moist living conditions, because their skins are permeable to water, and they will dehydrate in dry environments. A characteristic of amphibians is that they mate and lay their eggs in water. Part of the life cycle of an amphibian is spent in larval form in water habitats, and some species spend their entire lives in water. Amphibians hibernate during cold weather.

Frogs and Toads

Frogs and toads are the best-known amphibians in North America. They are widely distributed on the continent, with frogs living in or near water. Frogs are able to obtain some of the oxygen that they require by absorbing it through their moist skins from the water in which they spend much of their time. They also have lungs for respiration. Toads and some species of frogs live in terrestrial environments. The diets of frogs consist mostly of insects, worms, and spiders, and they have long sticky tongues that flick out to capture their prey. Frogs have round external eardrums located behind each eye. This organ is called a **tympanum,** and it is quite sensitive to sound. Frogs have two breathing holes called **nares** located just in front of their eyes, and they often hide beneath the surface of the water with only the eyes and nares visible.

**Northern Leopard Frog
(Rana pipiens)**

**Southern Leopard Frog
(Rana utricularia)**

Figure 18-31: Distribution map of
the leopard frog.

The northern and southern leopard frogs are closely related species. They are brown or green in color with irregular rows of dark spots along their backs and sides, see Figure 18-30. Their rear legs are long and powerful, enabling them to leap long distances. The toes on the hind feet are webbed. The southern species has a light colored spot in the center of the tympanum that distinguishes it from the northern species.

Figure 18-30: The leopard frog.
Photo courtesy of Leonard Lee Rue III.

Leopard frogs live in many different kinds of freshwater habitats, see Figure 18-31. Many of them move into meadows and fields when sufficient vegetation exists to provide shelter. Mating calls are heard during the spring months in northern habitats and throughout much of the year in southern habitats. Reproduction is accomplished by laying eggs.

Toads are amphibians with short, compacted bodies and heads. They also have wart-covered skins. These warts help to protect them by secreting a white, milky substance that predators dislike, causing them to avoid toads in later encounters.

Toads live in areas that are too dry for many frogs, and they venture further from water than most frogs. They have short hind legs that are used for hopping, but they are unable to leap as do frogs. In arid regions toads mate following seasonal rains, and eggs are deposited in shallow standing water. These toads reproduce when there is enough water available to keep their offspring alive.

**Woodhouse Toad
(Bufo woodhousii
woodhousii)**

A common North American toad is the Woodhouse Toad. This species and several subspecies range across much of the United States and into parts of northern Mexico, see Figure 18-32. It is a large toad that grows up to 4 inches in length, with yellowish brown to gray coloring on its back and sides. Its belly is usually white or yellow. Several dark spots are evident on its back with one or more warts located in each spot, see Figure 18-33. The southwestern Woodhouse Toad and the Fowler's Toad are subspecies.

Figure 18-32: Distribution map of the Woodhouse toad and related subspecies.

Figure 18-33: The Woodhouse toad.

The diet of this toad is made up mostly of insects. It is able to survive in a variety of habitats ranging from grasslands to deserts.

Many species of frogs and toads live in North America. They range from tiny tree frogs to large bullfrogs. Chorus frogs sing constantly during some seasons, but they are seldom seen. Gulf Coast toads are large toads that are found in fields, prairies, and towns where they live. Frogs and toads fill important roles in the North American habitats in which they live. They feed on insects, and they are preyed upon by many kinds of birds, reptiles, fishes, and mammals.

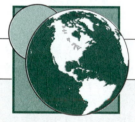

Ecology Profile

TADPOLES

A distinguishing characteristic of amphibians is the larval stage of development. It occurs in nearly all amphibian species. Adult amphibians seek water during the mating season because their offspring must pass through the first part of the amphibian life cycle in an aquatic environment.

The eggs of amphibians are fertilized externally. Male frogs and toads deposit sperm on the eggs as the females lay them. Male salamanders deposit small packets of sperm that the females collect and apply to their eggs. Eggs are incu-bated in the water, and they hatch into immature amphibians called **larvae.** The larvae of frogs and toads are called **tadpoles** or **polliwogs,** and they look nothing like their parents. The larvae of salamanders look like small adults with gills.

The process of metamorphosis occurs during the larval stage of development, when a number of changes occurs in tadpoles' bodies as they grow into adults. Tadpoles are equipped with gills for respiration, but as the larvae grow they develop lungs for breathing and their gills disappear. Tadpoles have digestive tracts suitable for plant materials, but during metamorphosis the intestines change to accommodate meat diets. Legs develop, and the tadpoles of frogs and toads lose their tails as they mature. Eventually the tadpoles become young adults, and they leave the water to live on land.

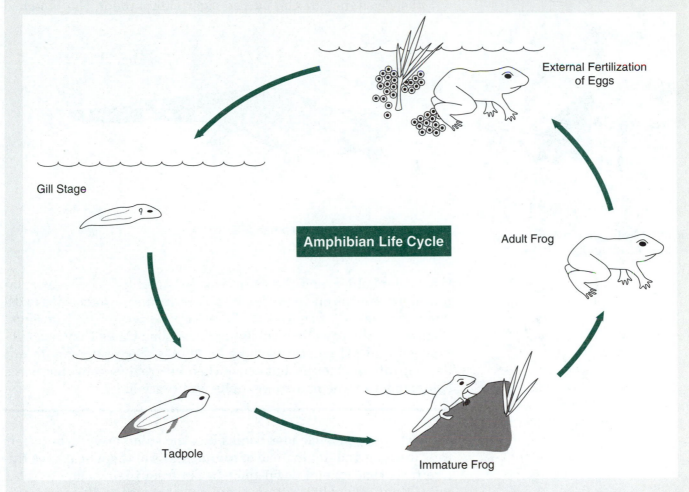

Gill Stage

External Fertilization of Eggs

Amphibian Life Cycle

Adult Frog

Tadpole

Immature Frog

Salamanders and Newts

Salamanders and newts are amphibians with long bodies and tails. They look much like lizards, but they lack scales on their bodies and claws on their toes. The skins of these small animals are moist. Some species are protected from their enemies by poison secretions that come from special glands in the skin. They molt from time to time by shedding their old skins.

Most salamanders live in damp environments, and some species live in water all of the time. Others only return to aquatic environments to breed. They breathe air, but they also absorb oxygen through their skins. Some aquatic salamanders, such as the mud puppies, never lose the gills they depended on during the larval stage of development. Most salamanders eat worms, insects, and slugs.

Figure 18-35: Distribution map of the tiger salamander.

The eastern tiger salamander is a bright-colored amphibian with irregular yellowish or olive spots on a black or brown background, see Figure 18-34. The belly is usually yellow with dark markings. Their diets consist of insects, worms, and slugs.

Subspecies include the barred tiger salamander, the blotched

Figure 18-34: The tiger salamander.

tiger salamander, and the gray tiger salamander. The tiger salamanders are found in water habitats in many parts of North America, see Figure 18-35. They reproduce by laying eggs. Sometimes the eggs float at the surface; sometimes they are submerged and attached to sticks or plants. Some larvae do not develop into land forms, but spend their entire lives in water. Some salamander larvae mature sexually and reproduce.

Newts have long slender bodies like the salamanders, but their tails are long and flat instead of round. Many of them also have fin-like frills that extend down their backs from their heads to their tails. Many newts live in terrestrial environments following metamorphosis, but they return to water environments to mate.

A male newt stimulates egg production in a female through his courtship behaviors. After displaying to the female, the male produces small packets of sperm. These are collected by the female, and her eggs are fertilized externally.

Red-spotted newts and their subspecies (central newts, peninsula newts and broken-striped newts) go through a complex life cycle. Their appearances differ from one stage to the next. In the adult stage this newt is brownish yellow to olive green, with large red spots lining its sides. Small dark spots randomly adorn its entire body, tail, and yellow-colored belly. They average 2.25–4.25 inches in length. The hind legs and frills of males sometimes disappear following the breeding season, but they reappear later. During the breeding phase of life, these newts become aquatic animals laying their eggs in the water. Eggs hatch into larvae, and

Figure 18-36: Distribution map of the red-spotted newt.

eventually the larvae metamorphase into the land-based form called **efts**.

In this phase of life, the color of the eft is red or orange and the preferred habitat becomes moist areas in mountain forests. Up to three years may pass before the eft enters the aquatic stage of its adult life thus starting the cycle over again.

In some instances, the larval form of this newt does not leave the water, but it matures sexually and changes directly to the aquatic adult stage. A newt that develops in this manner is described as a **neotenic** newt. Red-spotted newts and their subspecies are found in numerous locations in the eastern United States and Canada, see Figure 18-36.

Newts are equipped with teeth that are used to feed on small crustaceans and insect larva while they are living in water habitats. They eat mostly insects, slugs, and worms when they are living on land. Salamanders and newts are important prey for fish. Other animals that are similar in appearance to newts are the mudpuppies and waterdogs. Unlike most newts, they are aquatic animals throughout their lives.

LOOKING BACK

Reptiles and amphibians are cold-blooded animals classed as ectotherms. Reptiles are identified by the scales and plates of armor that protect their bodies, and by the claws on their feet. Young reptiles closely resemble their parents. North American reptiles include alligators, turtles, lizards, snakes, and one species of crocodile.

Amphibians usually live on land for a period of time before they return to water environments to reproduce. They have skin that is usually moist, and some amphibians are able to absorb oxygen through their skin. North American amphibians include frogs, toads, salamanders, and newts.

Major differences between reptiles and amphibians exist. Reptiles have dry skin covered with scales. Their eggs are covered with leathery shells, and their eggs are fertilized internally. Amphibians have moist skins with no scales. Their eggs are fertilized externally and have thin, soft outer membranes, and must be deposited in water to survive.

1. Describe how ectotherms and endotherms differ from one another, and give examples of each kind of animal.
2. Identify similarities and differences between reptiles and amphibians.
3. Compare the different ways that reptiles reproduce.
4. Illustrate the structures that are found in an amniote egg, and describe the functions of the four embryonic membranes.
5. Discuss the roles that alligators and crocodiles play in the ecosystems that they occupy.
6. Make a sketch that clearly illustrates the differences between vipers and elapids, and list some examples of each kind of snake.
7. Name some functions that lizards and snakes perform in the ecosystems where they live.
8. Explain how a snake uses its tongue as a sensing organ to help it smell.
9. Contrast the difference between turtles and tortoises, and describe ways that are similar.
10. Make a poster illustrating the life cycle of an amphibian such as a frog or salamander, and identify the changes that occur during metamorphosis.
11. Compare the similarities and differences between frogs and toads.
12. Identify the roles of salamanders and newts in the ecosystems they occupy.

LEARNING ACTIVITIES

1. Set up a classroom display with live tadpoles, snakes and/or other reptiles or amphibians. Study the feeding habits of these animals, and obtain instructions for their proper care.
2. Make a survey of your town or a section of your city to determine which species of reptiles and amphibians live nearby. Invite guest speakers who are interested in herpetology to talk with your class about local species.
3. Assign individual students or groups of students to prepare presentations about the life cycles of particular reptiles or amphibians. Have each group share their reports with the rest of the class.

SECTION V

Conservation and Management

Management and conservation of our wildlife resources is taking on a new urgency in North America. Private citizens, public institutions, and governments will be active participants as decisions are reached for management and conservation of these national treasures in the twenty-first century.

19

Responsible Management of Wildlife Resources

KEY TERMS

acid precipitation
biodegradable
decomposer
erosion
multiple use
nonbiodegradable
nonpoint sources
overgrazing
oversight
poaching
point sources
riparian zone
silt load
soil conservation
toxic waste

AMONG our most valuable natural resources are the wild animals with which we share our world. They are valuable because they are part of a living environment in which living organisms are dependent on one another. The food chain links many different life forms together, and the rise or fall of a particular species impacts the other species with which it interacts. Responsible management of our wildlife resources will be critical in the effort to preserve them for the benefit of future generations.

OBJECTIVES

After completing this chapter, you should be able to

■ describe ways that private individuals and institutions can contribute to responsible management of wildlife resources
■ list some positive and negative effects that farming has had on wildlife habitats in North America
■ suggest ways that farmers and ranchers might improve the environment for wildlife
■ evaluate the effects that modern industries have had on wildlife habitats, and suggest ways to resolve problems
■ identify some sources of urban pollution, and describe the effects on wildlife environments
■ propose ways that game farms, ranches, and preserves might profit from restoration and conservation of wildlife habitat
■ define the roles of environmental organizations in preserving wildlife populations and habitats
■ explain how some recreational activities contribute to damaged environments, and cause stress to wildlife populations
■ appraise the roles of government agencies in managing and protecting wildlife habitats and populations
■ consider the value of national parks, monuments, and preserves in the effort to preserve wildlife populations and habitats
■ define the role of national and international law in preserving wild animals and the environments in which they live

■ explain the multiple-use concept of management for public lands
■ discuss the effects of soil erosion on wildlife habitats and populations.

Some people believe that the extinction of a living organism is of little consequence unless it is directly linked to the loss of privileges or benefits formerly enjoyed by humans. Other people believe that the human race is morally obligated to protect all living things against population declines, irregardless of the costs.

Responsible management of a habitat, a biome or the world at large, is based on the premise that all living things have value to human society. It is probably unrealistic to expect that most people will voluntarily give up a lifestyle, or make great personal sacrifices for the preservation of any species other than their own. Human societies are most likely to practice conservation when personal benefits are expected, see Figure 19-1.

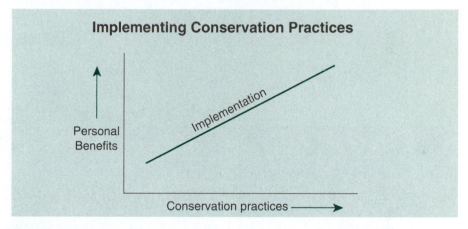

Implementing Conservation Practices

Personal Benefits

Implementation

Conservation practices ⟶

Chapter 19-1: When personal benefits increase so does the desire to implement conservation practices.

ROLE OF PRIVATE INSTITUTIONS

Individual landowners have a greater opportunity to improve the management of habitats and wildlife resources than does any other segment of society, see Figure 19-2. They make the management decisions concerning the uses of the property they control. It is also true that property owners have a greater potential for damaging habitats through poor management than other members of society.

Farmers and Ranchers

There is a public perception in some areas that modern agricultural practices are not compatible with healthy wildlife populations, see Figure 19-3. It is true that some conflict does exist between wildlife and agriculture, but most farmers and ranchers are as dedicated to protecting wildlife as anyone else. Agricultural

Figure 19-2: It is the responsibility of landowners to properly manage the habitat and wildlife resource on the property that they control. *Photo courtesy of USDA.*

Figure 19-3: Spraying of pesticides and other agriculture chemicals must be properly managed to avoid damage to public lands and wildlife habitats. *Photo courtesy of USDA.*

Figure 19-4: Extreme overgrazing such as this can present major erosion and other hazards to the land. *Photo courtesy of Dr. Jay McKendrick, University of Alaska, Fairbanks.*

practices that bring them into conflict with wildlife include tillage practices that lead to soil damage, see Figure 19-4, improper uses of pesticides and other agricultural chemicals, poor grazing practices on public lands, and conversion of wildlife habitats to fields.

Erosion is a destructive process that occurs when land is unprotected against the forces of flowing water or strong winds. Soil erosion is the number one source of water pollution in North America. It damages wildlife populations by polluting water supplies, killing young fish and aquatic animals, and filling in reservoirs and lakes. It also destroys terrestrial habitats, hindering plant growth which ultimately has negative impacts on the food chain. **Soil conserva-**

Career Option

SOIL CONSERVATIONIST

A soil conservationist develops plans and recommends practices for controlling soil erosion. Other duties include land-use planning activities, developing soil management plans such as crop rotations, and reforestation projects, establishing permanent vegetation and developing other practices that are related to soil and water conservation.

A career as a soil conservationist will require a bachelor of science degree in soil science, agronomy, forestry, or agriculture. A majority of time is often spent doing outdoor fieldwork.

Courtesy of USDA.

tion is the practice of protecting soil from the destructive forces of wind and water, see Figure 19-5. Soil conservation practices are important to farmers and ranchers because abused land eventually loses the ability to produce profitable yields of crops.

Figure 19-5: No-till corn coming up through wheat stubble in Maryland.
Photo courtesy of USDA-Soil Conservation Service.

Figure 19-6: It seems unreasonable to believe that farmers would endanger their own water supplies by being careless with agriculture chemicals. Their own children would be the first to suffer from such abuses. *Photo courtesy of USDA.*

Farmers and ranchers are stewards of most of the privately owned land and much of the public land in North America. Much of the potential farmland in North America is now tilled, and it has been converted from wildlife habitat to the production of crops and livestock. As these changes in land use have occurred, animals such as waterfowl and bison have been displaced, while other wild species such as songbirds have expanded into these regions. Fields of growing crops provide excellent habitats for many birds and some small animals.

Most agricultural producers are sensitive to the need to manage their property in a responsible manner because they are likely to lose more than anyone else if the soil is damaged or lost due to poor management practices. Farmers operate farm businesses that are just as sensitive to economic losses as the businesses in towns and cities.

Improper uses of agricultural chemicals pose greater dangers to farm families than to anyone else. Abuses of toxic materials are carefully avoided by most farmers. They live on the land with their families and drink the water from farm wells. Their own children would be the first to suffer from chemical abuses because they work and play in the fields, see Figure 19-6. Many farm families still eat meats, fruits, and vegetables that are homegrown, and improper use of chemicals would surely affect these foods, see Figure 19-7.

It seems unreasonable to believe that they would endanger their own water supplies by being careless with agricultural chemicals. There is no great conspiracy by farmers and ranchers to get rich at the expense of wildlife and other human beings by boosting crop yields through the use of poisonous chemicals. The chemicals that are used in agriculture have been tested carefully, and they can be safely applied when used according to the manufacturer's direc-

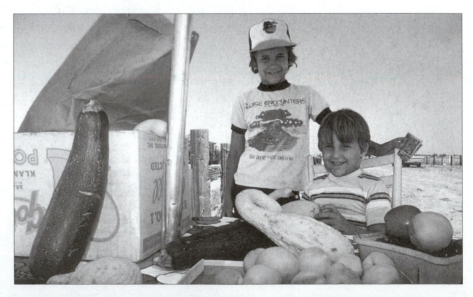

Figure 19-7: Many farm families still eat the fruits and vegetables that they grow and improper use of chemicals would affect these foods and the health of the families. *Photo courtesy of USDA.*

tions, see Figure 19-8. Many times, agricultural chemicals that are left over from a job are considered to be **toxic waste** because they are poisonous to living organisms. Laws prescribe how they should be properly handled and disposed. These materials should be safely stored until they can be delivered to a toxic waste treatment center, or until they can be properly degraded. Chemical abuses must be corrected for the safety of living plants, animals, and people.

Figure 19-8: It is important to carefully follow manufacturer's directions when applying chemicals.

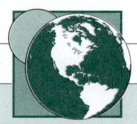

Ecology Profile

BIODEGRADABLE VERSUS NON-BIODEGRADABLE MATERIALS

Biodegradable materials are substances that can be eaten or broken down by living organisms to form nontoxic products, see Figure 19-9. Bacteria, fungi, insects, worms, and other organisms that are capable of breaking down complex materials into simple materials are called **decomposers.**

Materials that cannot be broken down by living organisms are considered to be **nonbiodegradable.** Some agricultural and industrial chemicals are toxic in addition to being nonbiodegradable. They must be handled carefully by professionals who are trained to dispose of them properly, see Figure 19-10.

Educational programs and government regulations that are designed to assure safe use of agricultural chemicals are being implemented in most regions of the United States and Canada. There is an increased awareness of the need to follow chemical labels carefully. Empty chemical containers are still a threat to the environment, but major efforts have been made to educate those who use agricultural chemicals about the best ways to dispose of used containers.

Biodegradeable
Reduced to non-toxic products by bacteria
Examples: • Detergents   • Paper products   • Aq chemicals

Non biodegradeable
Unaffected by bacteria
Examples: • Plastics   • Metals   • Nuclear waste

Figure 19-9: Biodegradable vs non-biodegradable.

Figure 19-10: A toxic waste disposal site. *Photo courtesy of USDA.*

It would be naive to believe that farmers and ranchers never abuse land and water resources, but it would be just as wrong to accuse them of destroying the environment with wanton disregard for their own families, neighbors, customers, and resident wildlife populations.

Ranchers who graze livestock on public lands are required to make improvements to the range. Water sources have been developed for use by livestock and wildlife populations, see Figure 19-11. New varieties of grasses have been seeded on many ranges. These produce much more forage than the native species they replaced, and they are equally nutritious to wild and domestic ruminant animals, see Figure 19-12.

Figure 19-11: Rangeland water must be carefully maintained and monitored. *Photo courtesy of USDA-Soil Conservation Service.*

Figure 19-12: Reseeded rangeland. *Photo courtesy of Michael Dzaman.*

Public lands are managed for both wildlife and for domestic cattle and sheep. Scientific research has demonstrated that these uses are compatible with one another. The key to proper management of rangelands is to graze an area quickly, and then remove the animals from the area to allow the plants to build up their food reserves. Plants that are not allowed to build up food reserves in their roots are more easily killed during conditions of extreme drought or cold. Range management must also consider seasonal variations in plant growth and the different impacts of grazing by horses versus cows versus sheep versus wild species. **Overgrazing** is a condition in which domestic livestock and/or wild animals destroy the plants in an area by harvesting them beyond their ability to recover.

Rangelands from which forage is harvested in the summer by livestock will generally produce new growth in the late summer and fall. This is then available as winter feed for deer and elk, and it is usually of high quality. When spring growth is not harvested, it tends to become coarse and unpalatable, and its nutrient value is usually low.

Farming practices that do damage to wildlife populations include tilling every acre of land and drainage of swamps and other wetland areas. Windbreaks, wooded areas, farm ponds, ditches, and fencerows provide habitat for birds and other small animals. Undisturbed grasses and weeds in these areas provide shelter and food for wildlife. As fields have become larger, these areas have disappeared on many farms.

Wildlife agencies and organizations are encouraging farmers to skip over small areas as they harvest their grainfields, and to leave some areas untilled over the winter season, see Figure 19-13. Small refuges such as these provide winter cover for birds and other animals, helping them to survive on agricultural lands.

Figure 19-13: Rows of grain sorghum have been used in semiarid Colorado as a windbreak and as a barrier to catch snow to add to soil moisture. *Photo courtesy of Colorado State University.*

Agricultural producers who fail to use wise management practices that are friendly to wildlife populations are placing the entire agricultural industry at risk. A single farmer who applies chemicals improperly, thereby causing injuries or death to wildlife populations, is likely to create public sentiment against all farmers, including those who use chemicals safely. One rancher who is lax in his or her stewardship over a grazing allotment on public land can place all ranchers in jeopardy of losing their grazing privileges because publicity tends to focus on problems instead of successes.

Farmers and ranchers should seriously consider the possible effects of all agricultural practices on populations of wild animals. Instructions for safe use of agricultural chemicals should be carefully followed, and empty containers disposed of properly. Agricultural organizations should police their own members to assure that abuses of public trust lead to the loss of privileges.

Industry

Industrial processes consume huge quantities of energy and raw materials, and they produce massive amounts of by-products and industrial wastes. Many species of wild animals in North America have suffered from the effects of industrial pollution. Responsible management by private industries is needed if we are to eliminate these negative impacts of pollution, see Figure 19-14.

One serious problem that has become evident in recent years is **acid precipitation.** It is caused when sulphur or nitrogen oxides are released into the atmosphere from coal burning industrial furnaces, cars, incinerators, etc. Raindrops become polluted with sulphur and nitrogen compounds, and weak acids are formed. Acid precipitation is very destructive to ecosystems, and some habitats, particularly those in areas of granite bedrock, are more vulnerable than others. Acid rain is capable of destroying both plant and animal life. Some streams that are polluted by acid precipitation no

Figure 19-14: Many of our natural resources are seriously polluted by waste.

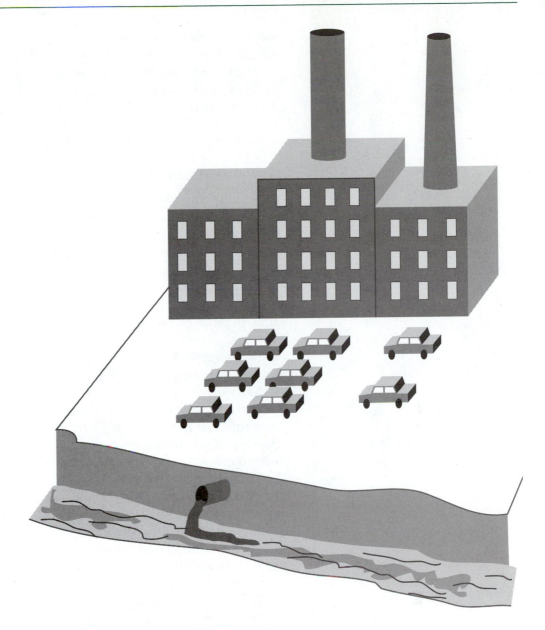

Figure 19-15: An example of point source pollution.

longer support life forms such as fish and aquatic insects and plants.

The greatest single cause of acid rain is pollution of the air by car exhaust gases. The problem is most evident near population centers where large numbers of cars are concentrated in a relatively small area. Responsible citizens should make sure that their automobile engines are properly tuned to reduce emission levels of damaging exhaust gases.

Untreated industrial by-products and waste materials are sometimes spilled or dumped into streams, rivers, and lakes. Some of these materials have also been released into the ocean. Water in some of our industrial areas is so polluted that it is incapable of supporting life. Pollution of this kind is called **point source** pollution because the pollutants can be traced to a particular source, such as a factory drain, see Figure 19-15. **Nonpoint sources** of pollution are those in which pollutants cannot be traced to a single origin.

Many industries have responded to environmental laws by reducing pollution. Some have gone far beyond the requirements of the law in efforts to demonstrate to the public that their industries are environmentally responsible. Failure to properly treat industrial waste materials or to safely dispose of them is irresponsible.

Private Citizens

Some of the chemical abuses that are blamed on agriculture can actually be traced back to urban neighborhoods. They often come from lawn and garden chemicals that are applied at excessive rates. Part of the problem results from the difficulty in accurately measuring and mixing chemicals in small amounts and containers. It is also difficult to know how fast one should walk when applying chemicals from a hand-held sprayer. Another problem that arises is how to properly dispose of chemicals that are left over at the end of the job. Too often they are poured out on the ground or flushed down sewers.

Large commercial machines that are used to apply chemicals are usually carefully calibrated to avoid overapplication. Homeowners and small landscaping firms seldom have large mixing tanks or carefully calibrated equipment. Excessive chemical applications to yards and gardens have become a problem in many urban and suburban areas.

Excess chemicals often wash into storm drains and sewage systems during heavy rains. They create pollution problems in surface water, especially when storm drains empty directly into streams without first passing through sewage treatment plants. Those who use lawn and garden chemicals should make sure that they are precise in following the instructions for application and disposal of such materials.

Wildlife populations are directly affected by the encroachment of humans on their environments. Habitat is lost to wild animals when streets are paved and natural plant cover is destroyed or removed. The presence of humans disturbs many wild animals, and construction of homes, towns and cities deprives them of their most basic needs for food and shelter. Domestic dogs harass and prey upon them. Human interference with wild animals and their habitats often has serious consequences on wild animal populations.

Game Farms and Preserves

Privately owned game farms and ranches have been established in many parts of North America for a variety of purposes. Some of these are private efforts to preserve natural habitats for use by wild animals. However, most game farms and ranches must also earn a profit and be self sustaining, see Figure 19-16.

Some game farms raise wild species in captivity, and then sell the animals to hunting preserves. Birds such as pheasants and quail are probably the most common species raised for this pur-

Figure 19-16: Game farms raise wild species in captivity for use in commercial sport. Government agencies often regulate this kind of enterprise. *Photo courtesy of Leonard Lee Rue III.*

pose. Fish are also raised by private farms for commercial sport fishing operations.

Some large farms and ranches depend on the sale of private hunting permits for a major portion of their income. Many such ranches charge fees to hunters for the privilege of hunting on private land. In some instances large areas have been fenced to control the resident populations of big game animals, and habitats are carefully managed to assure that the herds are healthy.

Government agencies often regulate this kind of enterprise. Local fish and game laws must be followed, and permits may be required to operate commercial game farms and ranches. The success of a game farm or ranch depends on the ability of the owner to provide a quality hunting experience for his or her clients. Quality hunting is possible only when attention has been paid to the environmental needs of the wild animals.

Game preserves are of two general types. Some preserves are established to protect wild animals throughout the year, and they provide areas in which birds and animals are never hunted. Other game preserves are privately owned by an organization that sells memberships.

Members of game preserves are usually entitled to hunt within the boundaries of the preserve during regular hunting seasons, but wild birds and other animals are protected during the rest of the year. One priority on these preserves is to establish critical wildlife habitat. Mammal and bird populations on the preserve are sometimes supplemented during hunting seasons by purchases of animals from game farms.

Game preserves have been operated by landowners in Europe for a long time, but it is a fairly new concept in North America. Wild animals existed in great numbers when the first settlers came to America, and citizens have always been able to pursue their hunting interests. Only through the decline in some bird and animal

Figure 19-17: Wetlands provide habitats for a wide variety of species. *Photo courtesy of USDA.*

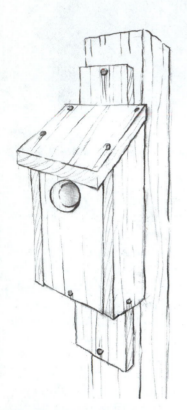

Figure 19-18: Nesting boxes are an innovative way to restore shelter in critical management situations.

populations was interest created in private game preserves. These are areas in which game animals are protected during critical periods in their life cycles. Members of such organizations are usually the only persons allowed to hunt on the property during the hunting season.

Some of the most common game preserves have been established by privately funded organizations such as "Ducks Unlimited." Funds are raised for the purpose of purchasing land that is converted into wetlands or other habitat for wildlife, see Figure 19-17. These attempts to restore wildlife habitats on private lands have greatly aided migratory waterfowl by providing safe places for them to rest during their annual migrations. Nesting areas have also been restored in critical areas to assure that safe nesting sites are available, see Figure 19-18.

ROLE OF SPECIAL INTEREST GROUPS

Public institutions are organizations whose membership is open to anyone who shares common interests and who desires to join. Some of these organizations have large memberships, and they are able to direct large sums of money to purposes that are priorities to their members. These organizations help support research, education, lobbying efforts, and court actions.

Environmental Organizations

The role of environmental organizations in North America is usually to create public awareness of problems or potential problems related to the environment. They also exert political pressure on lawmakers and resource managers through intense lobbying activities and through action in the courts.

Environmental organizations can contribute to improvements for wildlife by conducting unbiased research that leads to solutions

Figure 19-19: Environmental organizations that base their initiatives on good scientific data are valuable conservation partners. *Photo courtesy of USDA.*

for environmental problems. One challenge that environmental groups face is to assure that their activities are based on good science.

Research activities must be conducted by competent scientists who follow proven scientific procedures, see Figure 19-19, and study results made available to the scientific community for peer review. This process validates the research procedures and scientific processes that were used. Many environmental organizations spend their time and resources attempting to influence public opinion and to educate the public on issues that are important to their members. They do this by preparing printed materials and developing videos for use by television stations and private citizens. They visit schools and provide programs for public meetings. Environmental organizations have worked together to create a powerful lobby on environmental issues. These groups are beginning to have widespread influence with legislators at all levels of government. They have expanded their influence by bringing numerous court actions in high profile legal disputes.

It is sometimes difficult for organizations to make choices that are based on science because the results of a scientific study may not always agree with the position of the organizations members. Environmental organizations that base their initiatives on good scientific data are valuable conservation partners. Only organizations whose decisions and initiatives are based on emotions or flawed science can damage the cause of wildlife and natural resource conservation.

Recreational Interests

Recreation is an important part of life in America. Many people spend part of their leisure time in outdoor recreation, such as hiking, boating, horseback riding, hunting, and fishing, see Figure

Figure 19-20: Horseback riding is a popular outdoor recreation.
Photo courtesy of Michael Dzaman.

19-20. All of these activities take people into environments that are inhabited by wild animals.

It is the responsibility of all persons who use the outdoors to do so without abusing it. Trash that is left in recreation areas detracts from the appeal of the site, and some materials such as plastics and metals can be dangerous to wildlife. People should be willing to clean up garbage that has been left by less responsible individuals.

Some forms of outdoor recreation are capable of damaging the environment when they are pursued excessively or under certain conditions. One of these is driving vehicles in erosion sensitive areas, see Figure 19-21. Ruts in the surface of the terrain sometimes become channels for runoff water, and heavy rainfall can turn a rut into an eroded gully.

Snowmobiles make it possible for people to get into wintering areas that are used by big game animals such as deer and elk. Getting too close to these animals when they are stressed by harsh winter conditions can be fatal to them. They need all the energy they are able to consume just to keep warm during the cold weather. Chasing or causing undue stress to these animals not only depletes their limited energy reserves, but it is inhumane and illegal.

Outdoor sports take many people into the living environments of wild animals. Illegal poaching is a temptation to some people, and has become a serious problem in some areas. It can be controlled when responsible people who have observed suspicious activities report them to local law enforcement agencies. Hotlines have been established in many regions for this purpose, and they can be effective in bringing violators to justice.

ROLE OF GOVERNMENT

The role of government in the conservation and management of wildlife and natural resources is to manage public lands in a man-

Figure 19-21: Rut damage caused by offroad vehicles. *Photo courtesy of USDA.*

ner that is consistent with scientifically validated conservation practices. Government also plays a role in the management of private lands by providing expertise and financial support to natural resource conservation projects. Still another role of government is to enforce the laws that govern the uses of land, water, wildlife, and other natural resources.

Government Agencies

A number of government agencies have been charged with the responsibility of managing wildlife and other natural resources. Federal agencies have been established to resolve resource problems of many kinds. Some of them also function as law enforcement agencies with responsibility for ensuring that laws concerning wildlife and natural resources are obeyed.

The U.S. Fish and Wildlife Service is responsible for maintaining wildlife populations and for operating the system of National Wildlife Refuges. One of the duties of this agency is to enforce the provisions of the Endangered Species Act. It is the primary purpose of the agency to protect and nurture wildlife species. The professionals who work for this agency are expected to study declining animal populations, and take steps to preserve them.

The U.S. Forest Service is part of a large agency called the U.S. Department of Agriculture. The Forest Service is responsible for managing much of the forestlands and some of the rangelands that are owned by the federal government. They make decisions concerning timber sales and grazing allotments for livestock. The decisions of these managers sometimes have major impacts on wildlife populations.

The U.S. Bureau of Land Management is responsible for managing large tracts of arid lands in the western United States. Much of this land is used for grazing by cattle and sheep, see Figure 19-22. These areas also provide food and shelter for many birds and

Figure 19-22: The U.S. Bureau of Land Management is responsible for grazing land in the western United States. *Photo courtesy of Michael Dzaman.*

Figure 19-23: Old Faithful at Yellowstone National Park.
Photo courtesy of Leonard Lee Rue III.

other animals. This agency manages federal lands based on the concept of **multiple use.** This management strategy attempts to utilize natural resources in such a way that considers the varying needs of different groups of users.

The U.S. Soil Conservation Service is a federal agency charged with the responsibility for developing plans to reclaim damaged soils and to prevent soil erosion. This agency works with private landowners and government agencies to classify soils and to develop management plans. All of these activities affect the habitats of wild birds and other animals.

Many other government agencies have responsibilities for **oversight** or supervision of public and private lands and natural resources. The legislative branch of government is responsible to see that laws and regulations are enacted. The executive branch implements legislation. The judicial branch of government is responsible to ensure that the laws are fairly implemented.

National Parks and Monuments

National parks and monuments are areas that are set aside to preserve natural sites that are of scientific or historic interest. These areas are often chosen because they are areas of scenic beauty or unusual geological features. Wildlife habitats and populations are protected within the boundaries of national parks and monuments.

The National Park Service is part of the U.S. Department of the Interior and it was established by an act of Congress in 1916. It is responsible for conserving natural scenery, wildlife, and historic sites. Many different sites have been set aside in locations all over the United States. A strong effort is made within the national park system to preserve wildlife species in their natural habitats in such a manner that they are preserved for the enjoyment of future generations.

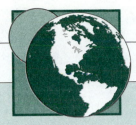

Ecology Profile

YELLOWSTONE NATIONAL PARK

Yellowstone National Park is one of the great treasures of North America. It was established in 1872 by the U.S. Congress, and it is the oldest and largest national park, see Figure 19-23. The park boundary encompasses 2,221,773 acres of land, and it is filled with unusual geological features such as geysers, hot water pools, petrified forests and waterfalls.

The land and water resources in Yellowstone are managed in such a manner that the environment is disturbed as little as possible. Hunting is not allowed inside the park, and rules governing fishing are strictly enforced. The park is filled with many wild animals, and they are often visible from the roads and trails. It supports one of the few remaining grizzly bear populations in the lower 48 states. It is one of the most popular vacation sites in North America.

Figure 19-24: Fish and game agencies strictly enforce hunting and fishing rules and regulations. *Photo courtesy of USDA.*

National and International Law

Laws and regulations are important tools in protecting wild animals and the environments in which they live. Without them, some populations of wild animals would soon be reduced beyond their ability to recover. This is because there are people in every society who will abuse the right to harvest wild game populations. Some of them abuse hunting and fishing privileges by ignoring laws that restrict these activities. Illegally taking or killing animals that are protected by law is known as **poaching.**

State and provincial governments have passed laws that set the rules for hunting and fishing. They have also established agencies to enforce these laws, see Figure 19-24. Fish and game agencies determine rules such as the number of fish, birds, or game animals that may be taken, the dates of the hunting and fishing seasons, and the number of permits that will be sold each year.

Federal laws that are enacted to protect wild animals and their environments are often targeted at specific abuses, such as point sources of pollution of streams and lakes. Such laws have resulted in less raw sewage and toxic wastes being dumped into waterways than in years past. The results are encouraging, and some of our most polluted waters are beginning to recover in response to federal environmental laws.

The Endangered Species Act is a federal law that identifies declining or nearly extinct populations of specific plants and animals, and requires changes in human behaviors that impact the survival of these animals. Laws of this kind may have wide-reaching effects on human activities because they may require changes in the way we live and work, see Figure 19-25. Many families living in the Pacific Northwest will have to seek new employment as

Figure 19-25: The logging industry has suffered from many reductions due to laws set forth by the Endangered Species Act. *Photo courtesy of Michael Dzaman.*

a result of logging reductions that are designed to protect the endangered spotted owl.

International regulations have been negotiated among nations to restrict harvests of whales, seals, sea turtles, dolphins, and other species of animals that live in international waters. Similar agreements protect some migrating species that regularly cross the borders between nations.

International laws have been implemented to restrict trade involving endangered species and their products, but black markets continue to offer such items for sale. For example, the high value of ivory as an international trade item continues to encourage poaching of endangered wild elephant populations. Trade in rare exotic birds is another activity that has a negative impact on the ability of endangered populations to increase. International laws restricting these and similar activities have been difficult to pass. They are even more difficult to enforce, but without these laws, there would be no legal restrictions on trade practices involving endangered species of animals and birds. Such laws need to be strengthened.

MULTIPLE USE CONCEPT OF MANAGEMENT

The multiple-use concept is a management strategy that has been used for many years on public lands. It implies that land and other natural resources can be managed in such a way that people with different interests and needs can use the same resources without depleting them. To do this, all users must assume full responsibility for understanding and using the resources prudently.

Multiple use of resources is the best management strategy when all users exercise good judgment in managing the resource. Examples of proper use of resources are evident in many management units. Sheep that are grazed in units where tree seedlings

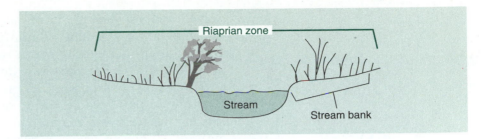

Figure 19-26: The riparian zone is the land adjacent to the bank of a stream.

have been planted can reduce the competition from weeds and other vegetation without damage to the trees. Recreational water sports are completely compatible with the need to store irrigation water in reservoirs or man-made lakes. Migrating waterfowl are also benefitted by such activities. Nitrates and phosphates are removed from surface waters as they flow through restored wetlands and marshes.

On the opposite end of the spectrum, examples can be cited in which untimely use of a resource becomes abuse of the resource. A motorcycle enthusiast who rides his or her bike across dirt trails when they are wet is likely to contribute to erosion. Heavy fishing in prime spawning areas during the spawning season may significantly reduce wild fish populations. Hunting game animals illegally out-of-season may reduce breeding populations to dangerously low levels. These same activities conducted at other times may have little effect on either the environment or the wild animal populations.

The **riparian zone** is the land adjacent to the bank of a stream, river, or other waterway, see Figure 19-26. A rancher who allows his cattle to repeatedly overgraze the riparian zone may contribute to severe damage of the natural plant cover in the area. Loss of plant cover accelerates the loss of topsoil, and increases the **silt load** of streams and rivers. Silt destroys spawning areas for fish, and it can kill young fish when the water is muddy for extended periods of time. Silt also fills in lakes and reservoirs as it settles out and sinks to the bottom.

A single abuser of a natural resource is likely to damage the trust relationship that is necessary for the multiple-use concept of management to work effectively. Blame for abuse to a natural resource is often assigned to all of the people who are known to use the resource. In some cases they are considered to be abusers even when they can prove that they are responsible managers.

LOOKING BACK

Conservation of wildlife resources is important because all of the members of the animal kingdom are linked together in a huge food web. They are either food animals that provide food for other animals, or they are predatory animals that eat meat. The rise or decline of a particular animal population affects all of the other animals that interact closely with it.

Humans are the only living organisms that can choose to exercise control over animal populations by modifying their environments, or by subjecting them to or protecting them from their natural enemies. Farmers, ranchers, government agencies and special interest groups should make reasonable efforts to work together to preserve declining species. This can be done by conserving habitats and managing wildlife resources in a responsible manner. Unbiased scientific research should guide all conservation efforts. Management decisions should also be balanced with the needs of humans to use natural resources.

REVIEW QUESTIONS

1. Describe some ways that private landowners can contribute to responsible management of wildlife resources.
2. List some positive and negative effects that modern farming practices have on wildlife habitats.
3. Suggest ways that farmers and ranchers might improve wildlife environments.
4. Describe ways that modern industries have affected wildlife environments, and suggest ways that such problems might be corrected.
5. Identify some sources of pollution that come from cities and towns, and describe the general effects of urban development on wildlife environments and populations.
6. Explain how restoration and conservation of wildlife habitats affects the profit potential of game farms, ranches, and preserves.
7. Define the roles of environmental organizations in conservation of wildlife populations and habitats.
8. Explain how some recreational activities contribute to damaged environments and cause stress to wildlife populations.
9. Suggest some practices that government agencies might use in managing and protecting wildlife habitats and populations.
10. Describe how national parks, monuments, and preserves contribute to the preservation of wildlife populations and habitats.
11. Propose ways in which national and international laws affect the preservation of wild animals and the environments in which they live.
12. Explain the multiple-use concept of management for public lands.
13. Discuss the effects of soil erosion on wildlife habitats and populations.

LEARNING ACTIVITIES

1. Invite a resource specialist to instruct the class on the correct procedure for conducting an environmental impact study. (Names of such people can be obtained from government agencies such as the Environmental Protection Agency, Bureau of Land Management, Forest Service, etc.). Choose an area near your school on which the class can conduct a limited study of the environmental impacts that might be expected if a subdivision or other development were to be constructed there. Use this exercise to demonstrate to the students that developing land affects the suitability of the area as wildlife habitat.
2. Contact the local Soil Conservation Service district office and request the help of their professionals in locating an area where severe soil erosion has occurred. Take a field trip to the area and observe the damaged site. Measure the depth of the soil layers and compare your findings to a soil map of the area. Discuss ways in which the soil might be managed to prevent further topsoil losses.

20 Conservation of Natural Resources

CONSERVATION is the practice of protecting natural resources against waste. It involves using less of a resource than is available, so that future generations may also benefit from its use. Conservation of resources does not necessarily mean that the resources are not used, but that they are not used in a wasteful or careless manner.

OBJECTIVES

After completing this chapter, you should be able to

- define conservation and provide examples of conservation practices affecting natural resources
- distinguish between renewable and nonrenewable resources
- identify the major destructive forces that contribute to soil erosion
- explain the relationship between soil erosion and water pollution
- describe some serious consequences that soil erosion imposes on wildlife resources
- suggest some conservation practices that are known to reduce soil losses due to erosion
- name the most common pollutants of water supplies, and describe their effects on wildlife
- suggest some ways that surface water can be protected against pollution
- discuss the effects of air pollution on wild animals and the environments in which they live
- list the sources that contribute most to air pollution
- discuss ways that air pollution can be reduced or eliminated
- describe some conservation practices that are used to preserve and restore wildlife populations and habitats
- appraise the use of biotechnology as a tool in reclaiming damaged and polluted resources.

Figure 20-1: Renewable resources include water, forests, and wildlife.
Photo courtesy of Wendy Troeger.

Natural resources fit into two broad categories. Resources that can be replaced by natural ecological cycles or by sound management practices are known as **renewable resources.** They include such resources as forests, water, and wildlife, see Figure 20-1. Resources that are depleted by use are called **nonrenewable resources,** see Figure 20-2. These resources include such things as soil, oil, coal, and mineral deposits.

Conservation practices for renewable resources should restrict their use to ensure that they are used no faster than they are regenerated. This rate of use should sustain a constant supply of such resources for as long as they may be needed.

Conservation of nonrenewable resources is accomplished by reducing the rate at which these resources are used to make them last longer. This can involve recycling, for instance of metals or even plastics (derived from petroleum) to reduce the necessity for extracting new materials. Conservation of these resources must ensure that they are not used up before we learn to replace them with other resources.

Natural Resources	
Renewable	**Non-renewable**
Forests	Coal
Water	Oil
Plants	Soil
Wildlife	Minerals

Figure 20-2: Renewable vs nonrenewable natural resources.

CONSERVING THE SOIL

Soil that is protected from damage is important in maintaining wildlife habitats. Without good soil, it is impossible for the plants to grow that are required by wild animals for food and shelter.

Soil conservation was defined in Chapter 19 as the practice of protecting soil from the destructive forces of wind and water. Erosion of soils was also cited as the greatest source of water pollution known to man. Many tons of soil are lost from the land each year through the combined effects of erosion by wind and by flowing water, see Figure 20-3.

Both wind and water carry soil particles to new locations. Massive amounts of **silt** consisting of tiny soil particles become suspended in water as it flows over exposed soil surfaces. When streams enter lakes and ponds, the rate of flow is reduced and the particles settle to the bottom forming large deposits of silt that eventually fill the lake or pond. Some ancient civilizations benefitted from upstream erosion because the rivers flooded their farms each spring depositing new silt. This kept their soils fertile and productive. Our problem today is that some land management practices result in damaging levels of silt being carried and deposited downstream beyond the natural flood plains.

Erosion of soil is not a new problem. It is a natural process, as evidenced by the river channels and canyons that have been cut through the surface of the land. Erosion is also evident in the formation of **alluvial fans** where streams enter from a gorge into a plain or where a tributary stream joins with the main stream. Alluvial fans are composed of rocks, gravel, sand, and silt that have been carried to the area by streams of rapidly flowing water. They are huge soil deposits formed from materials that eroded from the soil surface further up the stream. These materials are deposited where the flow rates of the streams and rivers slowed down. These events occur over long periods of time.

Figure 20-3: When the wind is particularly strong at the soil level, it can be a real problem. *Photo courtesy of USDA.*

Figure 20-4: Topsoil contains most of the growing nutrients and must be maintained to ensure a good crop. *Photo courtesy of USDA.*

Erosion of soils reduces their capacity to produce crops. The topsoil is the most affected layer, and it contains most of the nutrients that are needed by plants, see Figure 20-4. Soils are formed over long periods of time, and should be treated as nonrenewable resources. We must conserve them by reducing or eliminating erosion losses.

Erosion is a very destructive force that has serious consequences for crop production, and for fish and other wildlife. Fish spawning grounds that are filled with silt prevent developing fish eggs from getting the oxygen that they require to sustain life. Eggs become coated with silt particles and soon die. Young fish are also sometimes injured by water that is polluted with silt.

Erosion becomes more intense when the plant cover is removed on steep slopes, see Figure 20-5. As the North American continent was colonized, most of the land considered suitable for tillage was developed for the production of crops. Some of this farmland is located on slopes that are vulnerable to erosion. Tillage practices that remove plant cover during the winter season leave the soil exposed to the forces of erosion from wind, heavy seasonal rains, and snowmelt.

Serious soil erosion often follows range, forest, or grassland fires, see Figure 20-6. This is due partly to the loss of the plant cover that protects the soil surface. Fire also breaks down the soil structure, and causes it to be more easily damaged. Special conservation measures are necessary following fires that cover large areas. One important practice is to reseed the area with grasses and other cover plants to stabilize the soil as quickly as possible.

Forest lands are protected following timber harvests by digging holes in the forest floor to trap runoff water. In this manner, it is held on the surface until it is absorbed into the ground. This practice is effective in preventing excess water from flowing across the

Figure 20-5: Erosion on a steep slope. *Photo courtesy of USDA.*

soil surface where it might cause erosion. Properly constructed logging roads prevent water from running down the road surfaces in large streams. Water is channeled off the road and into areas that have stable ground cover.

Soil conservation practices for farms include many practical ways of protecting soil surfaces and slowing the movement of water or wind across the soil. Examples of such practices are planting windbreaks, creating dikes along the contours of fields and hills, adapting no-tillage or minimum tillage farming practices, planting grass waterways, and planting high risk fields to permanent cover crops, see Figure 20-7.

Figure 20-6: Erosion in San Bernardino, California as a result of firestorms. *Photo courtesy of USDA.*

Figure 20-7: Windbreaks are created to restrict water and wind erosion. *Photo courtesy USDA.*

MAINTAINING A PURE WATER SUPPLY

Pure water is one of the most important resources required by fish and wildlife, see Figure 20-8. It provides drinking water and living environments to many of the animals that make up natural ecosystems. Pollution of aquatic resources can have serious consequences for fish and wildlife.

Water pollution occurs when any foreign substance is dissolved or suspended in water. Suspended silt particles are the most frequent pollutants of water supplies, but many other substances pollute water. Some of the most common pollutants are industrial wastes, fertilizers, pesticides, oil, chemical spills, and sewage. All of these substances can be found in water supplies of North America, and all of them damage wildlife populations.

Figure 20-8: Clean water is necessary for the preservation of fish and wildlife. *Photo courtesy of Wendy Troeger.*

Dissolved Pollutants

Problem:
- High dissolved nitrates/phosphates
- Algae in surface water

Effects:
- Low oxygen levels
- Toxin or poison production
- Dead fish

Solution:
- Eliminate nitrate & phosphate from surface water

Figure 20-9: The problems, effects, and solutions of dissolved pollutants.

Clean water can be maintained only when everyone acts in a responsible manner to protect it. Legislation has been enacted to penalize people and institutions who are negligent in protecting water resources, but it is sometimes difficult to locate the exact source of water pollution. However, laws have helped to reduce industrial pollution of lakes and rivers in recent years.

Pollutants that sometimes occur in surface water are nitrogen and phosphorous compounds. The most common nitrogen pollutants exist in the form of **nitrates.** Phosphorous pollutants usually form compounds called **phosphates.** These compounds are found in fertilizers and in organic matter. They are also found in residues from detergents. They occur in water supplies when inadequately treated sewage is dumped into lakes and rivers, or when fertilizers are dissolved in runoff water from fields, lawns, and gardens.

High phosphate or nitrate content in water promotes the growth of tiny water plants called filamentous **algae.** Large growths of algae in a water supply uses up the dissolved oxygen required by fish and other aquatic organisms. Severe algae blooms sometimes release toxins or poisons into the water supply that can kill animals that drink it or live in it, see Figure 20-9.

Water pollution can be reduced in several ways. All industrial wastes and raw sewage must be treated using proven methods to remove pollutants before they are dumped into surface waters. Water runoff from fields must be trapped, when possible, and recycled back to the fields to prevent it from carrying dissolved pesticides and fertilizers into streams and lakes. Buffer zones should be created between human activities and streams to prevent erosion and accidental contamination of surface waters.

Runoff water from streets in cities and towns must also be treated before it is released back into the environment, see Figure 20-10. Toxic chemicals and their containers should be properly disposed of to prevent contamination of water resources. Garbage must no longer be dumped into oceans and waterways. Greater care has to be taken to prevent oil and chemical spills from ships, railcars, and trucks, and better methods are needed for cleaning up the spills that do occur.

Individual citizens must learn to appreciate the importance of preserving water resources and maintaining water purity. Until this happens, we will always be faced with pollution problems caused by the careless or irresponsible actions of people.

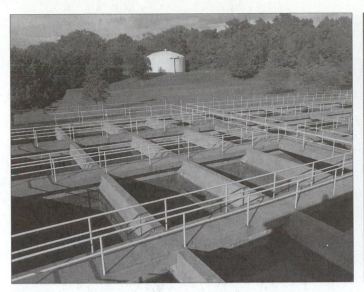
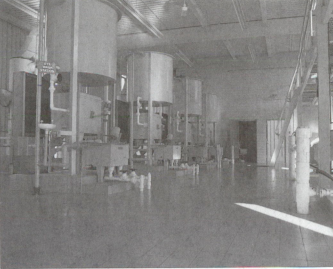

Figure 20-10: (A) The exterior and (B) interior of a water treatment plant. *Photos courtesy of Michael Dzaman.*

PRESERVING AIR QUALITY

Air is a natural resource upon which plants and animals depend for survival. The greatest hazard to wild animals and fish, however, comes from the effects of acid rain on the surface waters and plants that provide them food and shelter.

The weak acids that are formed when rain combines with pollutants in the atmosphere are capable of killing forests and destroying living organisms in streams and lakes. These problems are evident in North America and in many other industrialized nations.

The exhaust gases from cars and trucks are the greatest sources of atmospheric pollution, see Figure 20-11, but factories and electrical power plants that burn coal and petroleum products also emit large amounts of polluted gases into the atmosphere. Ultraviolet light from the sun reacts with atmospheric pollutants adding to the atmospheric haze. The result of this pollution to the atmosphere is a great cloud of polluted air called **smog.** Part of the

Career Option

ENVIRONMENTAL ANALYST

An environmental analyst uses science and engineering principles to find ways to remove pollutants from the environment. This career requires good research skills and an ability to gather data and analyze it properly. Data is gathered from samples of air, soil, water, plants, animals, and other sources. A college degree is required that combines the disciplines of science, engineering, and mathematics (statistics).

Figure 20-11: Exhaust gases from cars and trucks are the greatest sources of air pollution. *Photo courtesy of USDA.*

problem has eased since the development of vehicles that burn unleaded gas, but exhaust gases continue to pose a threat to the environment near large population centers because people drive more cars and longer distances every year.

One serious aspect of air pollution is that pollutants are carried by the wind to other areas. This creates damage to the environments of wild animals and fish, particularly those that live near large cities. Some regions along the eastern and western coasts of North America have sustained considerable amounts of damage from acid rain.

The solution to these problems is to remove as much of these pollutants from emission gases as possible before they are released to the atmosphere. We would also be wise to cut back on the amount of gases that are produced. The best possibility we have for doing this is to create effective mass transit systems in our cities that reduce dependence on personal cars. We should also research new industrial processes that require less energy.

As people begin to experience the effects of pollution to the atmosphere in the form of lung and skin diseases, there will be greater motivation to solve the problems that are created by harmful atmospheric gases. It is not likely that humans will do much to improve air quality for wild animals, but wild creatures will benefit when humans improve air quality for themselves.

PRESERVING AND RESTORING WILDLIFE POPULATIONS AND HABITATS

The settlement and colonization of North America by European immigrants opened new frontiers to the nations of Europe. They saw an abundance of natural resources and raw materials in America that eventually attracted new industries to the New World. Forests were cut to provide homes and fuel, land was cleared and swamps were drained to produce crops, and wild animals were harvested to supply food and clothing.

This pattern continued as settlements moved west. Industries were established to process raw materials, and vast acreages of land were converted from wildlife habitat to farms, see Figure 20-12. Many wildlife populations were reduced, and some animals became extinct as land was converted to new uses. Some species were completely eliminated from their historic ranges due to the establishment of settlements.

Wildlife habitats can never be restored as they were when the Native Americans were the only people who occupied the land. Herds of bison will never roam the Great Plains as they once did, and the rivers will probably never have the abundance of fish that once existed there. All of this is in our past, but some wildlife habitats and populations can be restored, and those that still exist can be conserved and enhanced.

Several proven methods are available to restore fish and wildlife to suitable habitats. They include transplanting birds or other animals from areas where they are abundant to areas where populations have been depleted or no longer exist. Some of the most successful of these programs have moved species such as elk, bighorn sheep, and wild turkeys to locations where they were no longer found or where they had never existed.

Many different species of fish are raised in hatcheries for the purpose of transplanting them to streams, rivers, and lakes, see Figure 20-13. This is done to supplement wild fish populations, and to provide adequate numbers of fish for sport fishing. Large numbers of game birds are also raised each year for the purpose of releasing them into the wild for hunting. Birds and fish that are raised in captivity are released with the intent of harvesting them. This is because they are likely to have a hard time surviving without the survival skills that wild fish and birds have learned.

Figure 20-12: As the industries need for raw materials increased, so did the conversion of wildlife habitat to farms. *Photo courtesy of Michael Dzaman.*

Figure 20-13: Trout being raised in a hatchery. *Photo courtesy of USDA.*

Some endangered species of birds are raised in captivity with the intention of adding them to wild flocks. The people who care for young birds must take steps to prevent them from **imprinting** or learning to mimic behaviors of humans. It is important that they learn to recognize and be attracted to birds and not to humans. They must learn to act like the birds of their species, and not like the people who raised them if they are to survive in the wild.

Ecology Profile

CALIFORNIA CONDOR RECOVERY EFFORT

One of the most interesting recovery efforts ever undertaken by scientists is the attempt to save the California Condor from extinction, see Figure 20-14. Condor eggs have been removed from the nests of wild breeding pairs, and the young birds raised in captivity. They are fed by a puppet-like foster parent from the time of hatching until they learn to gather their own food.

This practice was adopted to keep them from imprinting on human parents.

Several birds have been returned to the wild with mixed results. Some of the birds have been recaptured because they were not adapting well to life in the wild. Others appear to be adjusting, but long term survival may prove to be difficult for them.

Figure 20-14: The endangered California condor. *Photo courtesy of U.S. Fish and Wildlife Service.*

RECLAIMING DAMAGED OR POLLUTED RESOURCES

Pollution of a resource is usually quite difficult to overcome. It requires finding the origin of the pollutants and reducing or eliminating their release at the source. It also may be necessary to remove the pollutants from the contaminated area, see Figure 20-15. If this cannot be done, then ways must be found to dilute the pollutant or to break it down into nontoxic substances.

The science of **biotechnology** is relatively new, but it has important environmental applications. Scientists in this field are altering the genetic composition of living organisms. They have, for example, found ways to modify bacteria to allow them to eat and detoxify pollutants. Through a biotechnology practice known as **genetic engineering,** the genes of bacteria are modified to allow them to ingest and break down pesticides and other chemicals. This is an important scientific advancement that may help us to reclaim damaged resources.

Most of the pollution to our water resources comes from untreated sewage and industrial waste. Since the Clean Water Act was passed by Congress in 1972, billions of dollars have been spent to build treatment plants to reduce waste discharges into waterways. This law, which was strengthened in 1987, is administered by the Environmental Protection Agency. The goal of the Clean Water Act is to clean wastewater well enough that we will be able to swim and fish in it. It is now illegal for cities to dump sewage that has not been treated, and industries are required to stop pollution using the best practicable technology that is available, see Figure 20-16.

Water that contains nitrates and phosphates can be cleaned by bacteria and/or plants before it is released back into streams. This

Figure 20-15: Clean up of an oil spill is a long process. *Photo courtesy of U.S. Fish and Wildlife Service.*

Figure 20-16: The goal of the Clean Water Act is to clean wastewater well enough that we will be able to swim and fish in it. *Photo courtesy of USDA.*

process is known to occur naturally as water passes through marshes and swampy areas, see Figure 20-17. It is also possible to create man-made marshes that are capable of performing the same function.

A lagoon is a man-made pond where wastewater is stored for three or four weeks. During this period, bacteria and algae metabolize dissolved nutrients and many of the solids. Several lagoons that are linked together in a series are capable of cleaning water sufficiently to allow it to be released into streams and rivers.

As with water, the first step in reclaiming polluted air is to eliminate the source of the pollution. The Clean Air Act was passed by Congress in 1970. This act requires that air pollutants such as nitrogen oxides, hydrocarbons, sulphur oxides, and carbon monox-

Figure 20-17: Marshes and swamps act as natural water purifiers.

ide be reduced. The responsibility for setting clean air standards was assigned to The Environmental Protection Agency (EPA).

EPA standards require newer vehicles to be fitted with emission control equipment that removes significant amounts of pollutants from exhaust fumes. A **catalytic converter** is installed to convert dangerous emissions to harmless materials. Engines have also been designed to operate smoothly while burning unleaded gasoline. This reduces lead contamination in the air, and protects the catalytic converter in the exhaust system. Lead forms a coating on the catalyst metal when it is present in gasoline. This coating prevents the catalyst from reacting with exhaust gases and thereby reducing exhaust emissions of pollutants, especially nitrogen oxides and hydrocarbons.

Smoke and gases from factories are cleaned in several ways. Catalytic converters decrease nitrogen oxide and hydrocarbon emissions into the atmosphere. A **wet scrubber** is a device that sprays water into a chamber through which polluted gases are passed. It washes particles from the gases and absorbs water soluble materials such as sulphur dioxide. A new technology known as fluidized-bed combustion allows conversion of sulphur dioxide to gypsum by mixing limestone with coal in a controlled combustion chamber.

Particles can also be removed from gases by passing the gases over two electrically charged fields of opposite charges. The first field charges the particles in the gases, and the second attracts and holds them, see Figure 20-18. These devices are called **electrostatic precipitators.**

The 1970 Clean Air Act has helped to reduce air pollution. Tall smoke stacks that disperse harmful gases over larger areas instead of eliminating their release are no longer legal for pollution control.

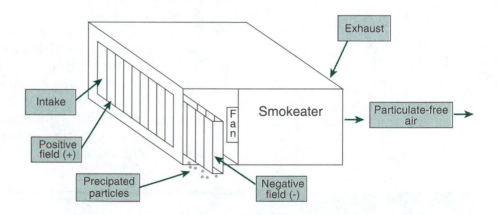

Figure 20-18: Electrostatic precipitators.

LOOKING BACK

Conservation of natural resources is the practice of using resources carefully to avoid waste. Renewable resources are regenerated, and wise use of them will ensure a constant supply of these resources for the future. Nonrenewable resources are those that cannot be replaced. Conservation of these resources involves using them in limited amounts to slow the rate at which they are depleted. Wildlife resources are usually considered to be renewable.

Soil resources are eroded by the forces of wind and flowing water. Water pollution occurs when silt and other pollutants contaminate water supplies. Air becomes polluted by waste gases and particles that are produced from the combustion of wood, coal, and petroleum products. All of these resources can be protected, and attempts to reclaim damaged or polluted resources are being made using the technologies available through modern science.

REVIEW QUESTIONS

1. Define conservation as it relates to wildlife and natural resources.
2. Distinguish between renewable and nonrenewable resources, and give examples of each.
3. Identify the major destructive forces that contribute to soil erosion, and describe how these forces erode soil.
4. Explain the relationship between soil erosion and water pollution.
5. Describe some of the damaging effects of soil erosion on wildlife resources.
6. List some conservation practices that are known to reduce erosion, and explain why the practices are effective.
7. Name the most common pollutants of water supplies, and describe ways that they may affect wildlife.
8. Suggest some ways of protecting surface water against pollution.
9. Discuss the effects of acid precipitation on wild creatures and the environments in which they live.
10. Name the most significant sources of air pollution.
11. Suggest ways that air pollution can be reduced or eliminated.
12. Describe some conservation practices that are used to preserve and restore wildlife populations and improve habitats.
13. List ways that genetic engineering is used to reclaim damaged and polluted resources.

LEARNING ACTIVITIES

1. Prepare a demonstration of the effect of slope on erosion. Fill several trays with soil of the same type and texture. Raise one end of each tray to a different height to represent variations in slopes. Release a measured amount of water over the surface of each tray, making sure that the water is released at the same rate in each tray. This might be done by pouring the water

through a gallon can with holes poked in the bottom. Collect the water that runs off the end of each tray using a plastic bag or other device. Filter the silt out of each water sample, and weigh the dried filter to determine the amount of soil that was eroded from each tray. Create a graph that summarizes the amount of erosion that occurred at each slope.

2. Invite a soil technician or scientist from the U.S. Soil Conservation Service to discuss the soil conservation practices recommended for your region. A field trip might be arranged to view some of the practices at the sites where they have been implemented.

3. Working in student teams, identify the major air pollution sources in your community or region. Research how the pollution sources have changed over the past ten or twenty years, and what the relative contributions of industry, motor vehicles, incineration of trash, etc. have been to the pollution problem. Repeat exercise for water pollution.

21

The Human Connection to Wildlife and Natural Resources

HUMANS have always used natural resources and wildlife for their own purposes. We developed our food plants from wild plant varieties, and our domestic breeds of livestock and poultry came from wild animals and birds. Therefore, our very existence is possible because of wild plants and animals since most of our food and clothing can be directly linked to them. Wildlife and other natural resources are also valuable to humans for recreational and for other purposes. It is difficult to imagine a world where the songs of birds did not exist, or where wild creatures were gone from the landscapes. Wild animals bring pleasure to humans, and they add to our own enjoyment of the world we share.

OBJECTIVES

After completing this chapter, you should be able to

- explain how the dependence of humans on wild animals and plants became the basis for modern agriculture
- describe how medical science has benefited from plant materials
- appraise the importance of tropical forests as sources of new medicines for the relief of human ailments
- explain how a watershed functions to reduce flooding and to supply a constant flow of water in rivers and streams
- analyze the importance of natural cycles in renewing and cleansing the environment
- identify some recreational activities that are associated with wild animals and their environments
- relate the importance of wildlife and outdoor sports to commerce
- explain the principle of stewardship as it relates to outdoor environments and wild animals
- speculate on the importance of human ethics in maintaining natural resources.

Figure 21-1: A biota consists of all species of plants and animals occurring in a specified area.

The total population of living organisms that is naturally found in an area is known as its **biota.** The biota of North America includes all the forms of life that are found there, see Figure 21-1. Many different products and life-sustaining materials, such as medicines, are derived from the biota of North America.

FOUNDATION FOR AGRICULTURE

Agriculture would not exist without wildlife, plants, and other natural resources. The first **domestic** animals were obtained from wild animal populations. They were tamed by our human ancestors, and used to provide dependable sources of food and clothing. Later, they were also used to pull farming implements.

Wild plants were also gathered by our early ancestors. Some of these plants were eventually raised near the homes of the people who used them. The concentration of food plants in the area made it easier to harvest them. The domesticated plants and animals that are used by farmers and ranchers today are quite different from their early ancestors, see Figure 21-2. Meat animals were selectively bred to bring about the genetic changes that are required to produce desirable meat in a short time. Dairy goats and cattle were selected for their ability to produce large amounts of milk, and poultry species were selected to produce eggs and meat. Less attention was paid to vigor because domestic animals were protected from severe conditions. The result of selective breeding is that domestic animals no longer have the vigor or the resistance to parasites that is evident in their wild relatives.

Wild animals and plants are selected mostly for their ability to survive in the environment in which they live. As a result, wild animals and birds do not produce as much meat, eggs, or milk as domestic breeds, but they have the genetic ability to withstand severe conditions in their natural environments. Wild animals that do not have these survival traits do not survive to reproduce large numbers of offspring.

Figure 21-2: Through selective breeding, the shape and genetic makeup of meat animals is quite different from their early ancestors. *Photo courtesy of Michael Dzaman.*

Figure 21-3: Recombinant DNA technology makes it possible to produce an entire plant from a single modified plant cell. *Photo courtesy of Utah Agricultural Experiment Station.*

Scientists are examining the genetic makeup of many wild species of animals and plants in their search for natural resistance to diseases and parasites. We would be wise to investigate the potential genetic value of wild species to our domestic species of plants and animals. Genetic engineers are likely to find many important genetic traits among wild species of animals and plants that can be transferred to domesticated species, see Figure 21-3.

SOURCE OF MATERIALS FOR MEDICINES

Many of the medicines that we use today were originally derived from plant materials. Many of the old herbal remedies that have been used by native peoples around the world have been found to have medicinal value. Plant extracts contain many chemical compounds that are useful in treating human ailments.

One of the most common medicines for headache, arthritis, and even to prevent heart attacks is aspirin. This is a trade name for a natural plant material called salicylic acid. This medicine is now produced in chemical laboratories, but it is found in nature in the fruits, blossoms, and stems of many different plants, such as willows. It was the active ingredient in plant materials used by some "medicine men."

The insulin that is needed by people who suffer from the disease called sugar diabetes is obtained from the pancreas organs of meat animals. Other human drugs obtained from these animals include estrogen, epinephrine, heparin, thrombin and many more.

Scientists tell us that only a few of the plants that exist in tropical rain forests have been examined for their medicinal qualities. They also tell us that many of these plants are likely to become extinct before we know their value if we continue to destroy the forests in these regions. Consider the value to the human race that might be derived from the cure of a single human disease using one of these plants and animals.

MECHANISM FOR NATURAL CYCLES

Some of the most valuable resources in nature are marshes and swamps, see Figure 21-4. These areas are natural water treatment plants, and they remove a variety of contaminants from our surface waters. They require no supplemental energy sources, and they function without human interference. The cleansing agents are plants, bacteria, and other aquatic organisms.

Figure 21-4: Marshes and swamps are some of our most valuable resources.

Figure 21-5: A watershed is an area in which rainwater and melting snow is absorbed to emerge as springs of water or artesian wells at lower elevations. *Photo courtesy of Utah Agricultural Experiment Station.*

Vast expanses of forest and other vegetation are required to regulate a uniform flow of water in rivers and streams. Precipitation on land surfaces can cause severe soil erosion unless plants are available to slow the flow of water over the land. Slow-flowing water infiltrates into the soil and comes out of the earth purified from contaminants. An area where precipitation is absorbed in the soil to form groundwater is called a **watershed**, see Figure 21-5. Each watershed is separated from other watersheds by natural divides or geological formations, and each is drained to a particular stream or body of water. Watersheds are valuable because they act like huge sponges, soaking up water from precipitation and melting snow and releasing it slowly.

Water is also cleansed by plants as they take in contaminated water and release clean water vapor into the atmosphere through the process called **transpiration**, see Figure 21-6. Our air is cleansed as the water cycle operates. Contaminants are trapped by falling rain or snow, and they are carried to the ground. As the

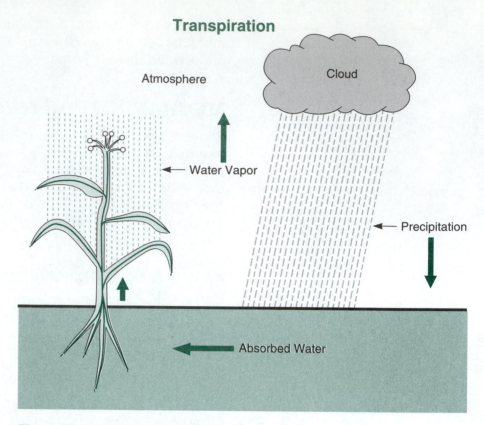

Transpiration

Atmosphere

Cloud

← Water Vapor

← Precipitation

← Absorbed Water

Figure 21-6: The process of transpiration.

water moves through the rest of the water cycle, the contaminants are left behind and clean water vapor enters the cycle.

The elemental cycles were discussed in Chapter 1. Elements tend to cycle from plants to animals through the food chain. When an animal dies, these elements may be transferred to a scavenger or released to the soil or to the atmosphere by bacteria and other organisms. Eventually the element is taken up again by plants, and the cycle begins again. How can we calculate the value of elemental cycles to the human population? The elemental cycles are of infinite value.

RECREATION

Wildlife and other natural resources have recreational value to humans. Much of our leisure time is spent in outdoor activities involving the use of natural resources and wildlife. Fishing and hunting are sports that focus on the use of wildlife resources, see Figure 21-7. Camping, boating, hiking, birding and other wildlife observation are all activities in which humans interact with the environment, see Figure 21-8. These uses are appropriate human activities. They can be engaged in with minimal effects on the environment as long as people are careful to exercise good stewardship.

Figure 21-7: Hiking is a sport enjoyed by many year round. *Photo courtesy of David Mosher.*

Figure 21-8: Fishing is an activity where humans interact with the environment. *Photo courtesy of Cathy Esperti.*

Intrinsic Value

The **intrinsic value** of fish and wildlife resources is the worth or value that exists within a resource. It is hard to measure, but it is very real. Millions of people visit outdoor recreation areas and national parks each year to enjoy nature. They place value on the natural resources.

Some wild animals are very difficult to view in their natural habitats. This is because many of them are wary of people and avoid contact with them. The only large wild animals that many people will ever see are the animals that live in zoos, see Figure 21-9. Some zoos attempt to place their animals in realistic outdoor set-

Figure 21-9: Most people will never see a porpoise in the wild, but many have enjoyed watching them perform at zoos. *Photo courtesy of Wendy Troeger.*

Figure 21-10: Gamefarms and petting zoos offer people the opportunity to interact with wildlife. *Photo courtesy of Wendy Troeger.*

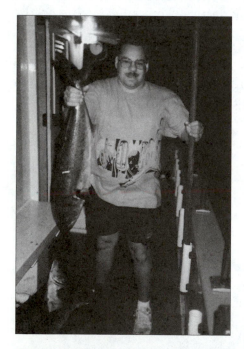

Figure 21-11: Hatcheries help stock popular fishing waters for recreational use. *Photo courtesy of Russ Berg.*

tings that resemble their natural habitats. This helps people to understand the habitat needs of animals. Most people do value wild animals or they would never pay to go see them in a zoo.

Bird-watching is an important leisure activity for many people. Birds are some of the most visible wild creatures, and they often adapt to living near people. Bird-watchers have formed societies since ancient times. Audubon Societies are modern organizations composed of people who appreciate birds and other wildlife. Birds have intrinsic value to many people.

Wild animals of many kinds draw crowds of people who simply want to watch them. The annual migrations of whales often bring them near enough to shore where people can observe them. Large commercial boats are often filled to capacity with people who want to see the whales from a better vantage point. People find excitement and fulfillment through observing wild animals, see Figure 21-10. Many visitors hike along trails in Glacier and Yellowstone National Parks hoping to see large wildlife such as elk, pronghorn deer, and bears.

Hunting and Fishing for Sport

Large numbers of people participate in the outdoor sports of fishing and hunting. North America has healthy populations of game species—mammals, birds, and fish—and this has contributed to the popularity of these sports. Fishing has long been considered to be a family activity, and large numbers of fish are raised in government hatcheries each year to supplement wild fish populations.

Hatchery stocks of fish are placed in popular fishing waters with the expectation that they will be harvested during the season, see Figure 21-11. Most fish and game agencies are funded by the sales of licenses, and it is in the best interest of these agencies to assure that adequate populations of game species are available for hunting and fishing.

Unrestricted hunting nearly destroyed the populations of many game animals in the early years of the twentieth century. The U.S. Congress passed the Pittman-Robertson Act in 1937 that taxed the sale of guns and ammunition. It also required that money from hunting licenses be used to restore wildlife and wildlife habitats. This has been accomplished in many areas. This act still provides the continuing funding base for state wildlife programs. Funds from this source are used to support research, land acquisition, construction, and maintenance of state wildlife areas, and management of wildlife programs.

One of the challenges of wildlife agencies is to accurately assess the carrying capacity of wildlife habitats. Sport hunting is a tool that is used to reduce game populations to the level of the food supply. This is necessary in areas that lack natural predators to prevent suffering and starvation of these animals during the winter. Hunting is a humane way to control game populations in comparison with the mass starvation that occurs when winter food supplies are inadequate. Hunting seasons and bag limits are regulated for the purposes of harvesting excess wildlife populations and estab-

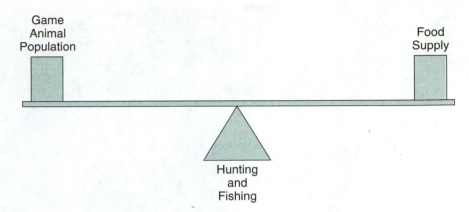

Figure 21-12: The balance between game animal populations and the food supply is carefully monitored.

lishing a balance between game animal populations and their food supplies, see Figure 21-12.

Hunting has become a more controversial issue in recent years. People who are against hunting cite different reasons for their opposition to this activity. Some people believe it is immoral to kill wild animals. Others believe that humans should not interfere with the laws of nature. They prefer that natural forces such as predators, starvation, and diseases control excess animal and bird populations.

Fishing and hunting privileges are sometimes abused by people who take fish and game animals illegally. Some of these people take more game than the law allows. Others take game when the hunting or fishing seasons are closed. These illegal activities are known as poaching. Irresponsible behaviors such as poaching and harassing game animals are difficult to control, but they cannot be ignored by law-abiding citizens.

Using Campgrounds, Trails, Roads and Waterways

There is such a great demand for outdoor recreation in North America that government agencies and private businesses have established camping areas, many equipped with restroom facilities and drinking water. A small user fee is charged for the use of government-owned camping areas to help defray the costs of maintaining these facilities. Commercial campgrounds are in business to make a profit.

A number of large tracts of wilderness have been set aside in North America. These areas are often remote from civilization, and designated wilderness areas are often restricted to backpacking and horse travel. Trails into these areas, see Figure 21-13, are maintained by government employees, and a strong effort is made to maintain wilderness areas in **pristine** or primitive conditions wherein resources exist in their pure, original form. People who use wilderness areas should take responsibility for carrying out their trash when they leave.

Figure 21-13: A strong effort is made to maintain wilderness trails in their pure, original form. *Photo courtesy of John Fisher.*

Many of the forests in North America are in areas where it is difficult to build and maintain roads. It has also been demonstrated that road surfaces are a major source of silt that pollutes streams and rivers flowing out of these regions. This problem is compounded when recreational vehicles create ruts in the roads by using them when conditions are too wet.

The agencies that manage government-owned land often close logging roads when they are no longer needed. This reduces the cost of maintaining roads in remote areas, and it helps to reduce erosion and silting problems in these areas. It is important for people to keep their vehicles off roads that have been closed to motor vehicles. Responsible use of backcountry roads is necessary to protect the environment and the wild animal populations that live in these areas. Irresponsible behavior by a few owners of recreational vehicles often leads to road closures in environmentally sensitive areas.

Many people who enjoy outdoor recreation gain access to natural environments by using boats. Numerous rivers, reservoirs, and lakes provide excellent boating opportunities. More people own boats today than at any time in our history, and large numbers of people make recreational use of our waterways each year, see Figure 21-14. Access to waters should also imply responsibility to respect the animal species that live there. Motorboats have become a serious threat to the survival of such diverse species as the loon, the manatee, and many other animals that live in water habitats.

BASIS FOR COMMERCE

The commercial fishing industry is completely dependent on healthy and abundant populations of fish. The fish that are harvested from the oceans are an important source of high protein food for people in many parts of the world. Commercial fishing fleets operate along both coasts of North America and this industry would cease to exist if our fish populations were lost.

Figure 21-14: More people own boats today than at any time in our history. It is important that boaters respect the animal species that live in the waters. *Photo courtesy of Carolyn Miller.*

Outdoor sports are the basis for a large segment of the tourism industry. Outdoor sporting goods stores meet the demand for outdoor sports equipment and supplies, see Figure 21-15. Among their products are hunting and fishing supplies and equipment, boats, recreational vehicles, camping gear, binoculars, backpacks, guidebooks, bikes, and motorcycles.

An entire industry has developed to support sport fishing and hunting. Professional guides and outfitters provide their equipment and services to fishing and hunting clients for a fee, see Figure 21-16. These business people earn their living as fishing and hunting consultants. They use their expert knowledge of an

Figure 21-15: Outdoor sporting goods stores meet the demand for necessary equipment and supplies.

Figure 21-16: Under the guidance of a professional outfitter, these rafters can enjoy a ride down the rapids. *Photo courtesy of Wendy Troeger.*

area and the habits of the game species that live there to help their clients achieve fishing and hunting successes.

THE PRINCIPLE OF STEWARDSHIP

A land **steward** is an administrator or supervisor who manages property for someone else. Federal, state, and provincial governments own large tracts of land in North America, and governments depend on professional managers to oversee these lands in a responsible manner, see Figure 21-17. Stewardship is responsible

Figure 21-17: Stewardship must be practiced to maintain habitat requirements of plant and animal communities. *Courtesy USDA Agricultural Research Service.*

Figure 21-18: Outfitters often use horses to assist them on their expeditions. *Photo courtesy of Leonard Lee Rue III.*

management of land, property, or resources and it should be practiced by all of the people who use the resources in addition to professional managers.

The principle of stewardship as it applies to natural resources implies that the agencies and people who have responsibilities to manage wildlife and other natural resources should perform their duties with a long-term view and commitment to the resources being managed. The ability to exercise good stewardship requires knowledge of proven management practices, and the wisdom and skills to implement them before habitats or resources are damaged. It also requires a knowledge of the ecology and habitat requirements of the plant and animal communities in the area.

Wise stewardship occurs when managers of natural resources make management decisions based on dependable information. This will require them to distinguish facts from fantasies and dependable research from biased research. It will also require that researchers know what questions to ask to get the research information that they need to make sound management decisions. Wise managers will adjust their management practices to protect natural resources against uses that cause damage to them. The principle of stewardship requires resource managers to use resources in ways that will preserve them for their children and grandchildren.

Career Option

HUNTING/FISHING OUTFITTER

Outfitters for hunting and fishing expeditions are business people who provide equipment and professional guide services to hunters and fishermen. Hunting outfitters often provide horses and vehicles such as trucks, boats, and airplanes to transport nonresident hunters and game to and from base camps or hunting lodges. Outfitters are usually excellent horsemen who are adept at packing supplies and game animals with pack animals, see Figure 21-18. They spend a lot of time scouting out areas where big game animals are located.

The clients of outfitters are usually unfamiliar with the area where they plan to hunt. They depend on the outfitter/guide to furnish food, supplies, transportation, and expert consulting services. It is the duty of the outfitter to get his/her client in position to legally collect a game animal or fish for which the client has a permit. The outfitter also accepts the responsibility of properly dressing and transporting wild game.

ETHICS, PRIVILEGES AND RESPONSIBILITIES

Ethics is defined as the code of morals of a particular person or group of people. It is important that people who use our wild environments and natural resources should adopt a "code of ethics" that governs their personal behavior. An example of such a code has been provided by the author.

Figure 21-19: Rules and regulations must be followed to protect our natural resources. *Photo courtesy of Wendy Troeger.*

The Outdoor Sportsman's Code of Ethics

I am a sportsman with a love for outdoor environments and wild creatures. I treasure the privilege of using the trails and roads, of camping and hunting in the deserts, woods, and forests, and of fishing and boating on the rivers, streams, and lakes. I appreciate the beauty of a sunrise on a high mountain peak, and the glory of a sunset over a restless sea.

I will respect each wild creature and protect the land that supports life. I will do more than my share to keep my world clean, and I will encourage others to do likewise. I will teach my children to enjoy and use our natural resources as responsible citizens. I will protect my personal privilege of using natural resources by becoming an informed and responsible steward of the environment. I am a sportsman with a love for outdoor environments and wild creatures.

Ethical behavior depends on the will of individual citizens to act responsibly, even when other people do not. The opportunity to use public lands and natural resources is a privilege, not a right. It is a privilege that must be earned by demonstrating responsible and ethical behavior. There is no statute or constitutional rule that gives anyone the right to abuse public lands, natural resources, habitats, or wildlife.

North America is endowed with an abundance of wildlife and natural resources. Citizens are allowed to use public lands for many purposes. Those who claim the privilege of using public resources must be willing to obey the regulations that have been established to protect these resources, see Figure 21-19. Anyone who cannot be trusted to act in an ethical manner should lose this privilege. We cannot sustain our natural resource base over an extended period of time, unless individual citizens become wise and fair stewards who care for the land, its plants and wildlife, and the environment.

LOOKING BACK

Humans have always depended on wildlife and natural resources for their own existence. We continue to do so today. While we no longer use the skins of wild animals for clothing, and few of us depend on wild animals for food, we do depend on domesticated plants and animals for these necessities of life. Modern agriculture has evolved from man's dependence on wild plants and animals.

We continue to obtain medicines from plant and animal products. The natural elemental cycles that function in nature continue to clean our world and replenish our supply of clean water and air. We use outdoor environments and observe wildlife for recreation, and we fish and hunt for sport. Many people earn their living through the sale of goods and services related to outdoor sports. It is important that we demonstrate ethical behavior, and exercise wise stewardship in our relationships with wild plants and animals and the environments in which they live.

REVIEW QUESTIONS

1. Explain how domestic livestock and crops evolved from the dependence of early human societies on wild animals and plants for food, clothing, and shelter.
2. Name some medicines that are obtained from plant and animal materials to treat human diseases and ailments.
3. Explain why preserving tropical forests is important to medical science.
4. Describe what a watershed is and explain why it is important to humans.
5. Analyze the importance of natural cycles such as the water cycle, nitrogen cycle, and carbon cycle in renewing and cleansing the environment.
6. List several recreational activities that are associated with wild animals and outdoor environments.
7. Identify some ways that human interests in wildlife and outdoor sports affect commerce.
8. Explain the principle of stewardship as it relates to outdoor environments and wild animals.
9. Speculate on the importance of human ethics in preventing abuses to wildlife and other natural resources.

LEARNING ACTIVITIES

1. Assign each class member to write a code of ethics for some form of outdoor sport. Limit their choices to no more than two or three separate sports. Examples might include "A Code of Ethics for Deer Hunters," or "A Code of Ethics for Off-Road Vehicles." A single paragraph of three to four sentences from each student will be adequate. Combine the papers together for each sport, and use the ideas from the students' papers to write a single code of ethics for each outdoor sport that was assigned.

2. Divide the class into two groups. One group will play the part of the U.S. Senate. The other group will play the role of the U.S. House of Representatives. Assign each of the groups to write a bill that will regulate the use of motorized vehicles on land that is managed by the U.S. Forest Service. When each group has passed a bill, assign two or three students from each group to create a compromise bill that is acceptable to both houses of Congress. Print a final copy of this bill for each class member.

Glossary

A

Accipiter   A Woodland hawk belonging to the genus Accipiter.

Acid Precipitation   Precipitation that has become polluted with sulphur or nitrogen compounds that form weak acids when they are dissolved in water, and that may damage plants and animals.

Adipose fin   A fin on some fishes that stores fat.

Aerated water   Water that contains high levels of dissolved oxygen.

Aerie   The nest of a raptor located in a high location such as a cliff or mountain top.

Algae filamentous   Tiny plants that are suspended in fresh water and which multiply rapidly when high levels of nitrates and/or phosphates are present, depleting the supply of dissolved oxygen in surface water.

Alien species   A species that is not native to an environment and that competes with native species for food and shelter.

Allantois   A membrane found in a developing egg that is filled with blood vessels, and through which respiration gases and wastes are exchanged.

Alluvial fan   A geological formation of gravel, clay, sand, and silt that has been deposited by water, and that is often seen near a location where a stream slows down as it enters a plain or where a tributary joins with a main stream.

Amnion   The innermost protective membrane in an egg that surrounds the embryo and which contains the fluid in which the embryo is immersed.

Amniotie egg   An egg with an inner membrane and a porous shell that surrounds a developing embryo and its food supply.

Amniotic fluid   A salty liquid in which a developing embryo of a bird, reptile or mammal is suspended as it matures.

Amphibian   A cold-blooded vertebrate with a moist skin that changes in its body form from a gilled, aquatic larva to an air-breathing adult through the process of metamorphosis. Their eggs must be laid in water.

Anadromous fish   A fish that migrates up a river from the sea to spawn.

Anal fin   A fin that is located on the underside of a fish between the anus and the tail.

Anatomy   The physical structure of an organism.

Antenna   A sensing organ or feeler (usually paired) on the heads of organisms such as insects and crustaceans.

Antler   Paired, bony, branching horn-like structures found on male and some female members of the deer family.

Aquaculturist   A person whose career is raising fish or shellfish for human consumption or use.

Aquatic species   Organisms that live in water habitats.

Auklet   A small penguin-like bird of the North Pacific coast that is capable of flight.

Avian   Of, relating to, or derived from birds.

Avocet A large, long-legged shorebird whose long bill curves upwards.

B

Baleen whale A whale that has whalebone in its mouth instead of teeth, and that strains plankton and other small organisms from the water for its food.

Barbel A whisker or feeler that is located on the lips of certain fishes such as catfishes.

Barging A practice in which migrating smolts are collected above a dam, loaded in barges and transported around the dam in an effort to help them avoid being killed in the turbines that are used to generate electricity.

Beard An unusual tuft of feathers that dangles from the breast of a bird, such as a wild turkey.

Billfish A fish, such as a marlin, spearfish, or sailfish, having a long upper jaw shaped like a sword that is used to strike and kill its prey.

Billy Goat A male goat.

Biodegradable Capable of being broken down by living organisms to form nontoxic products.

Biological succession Changes that occur as living organisms replace other lower order organisms in an environment.

Biologist A person whose career involves the scientific study of the characteristics, life processes, needs, and habits of plants and animals.

Biome A group of ecosystems within a region that have similar types of vegetation and similar climatic conditions.

Biosphere Consists of all of the ecosystems of the earth capable of supporting life.

Biota The community of living organisms that is naturally found in an environment or area.

Biotechnology A scientific field that applies principles of science to organisms.

Biotic Potential A measure of the ability of an organism to reproduce sufficient numbers of offspring to maintain a stable population.

Bison A large member of the cattle family, commonly known as a buffalo, that is native to the plains region.

Bivalve An organism that has a two-part hinged shell, such as a clam, that strains food particles from water sucked through its body.

Blowhole A nostril located on the top of the head of a whale.

Blubber A thick layer of fat that is found in the bodies of whales and other sea mammals.

Boar A mature male pig or bear.

Brood A family of young birds.

Buck A male deer, pronghorn, goat, or rabbit.

Bull An adult male elk, moose, whale, walrus, bison, or musk ox.

Bullhead Any of several common freshwater catfishes found in North America.

C

Caiman A species of alligator whose range extends from southern mexico and southern Florida to South America.

Calf The young of various large animals such as the elephant, whale, and bison.

Camouflage The use of protective coloration which blends in with an animal's surroundings to avoid detection by enemies.

Cannibalistic The tendency of an animal to eat other members of its own kind.

Canopy The highest level of vegetation in a forest, consisting of the branches and foliage of the tallest trees.

Carapace A protective outer coat of bony or chitinous armor, such as the external covering of a turtle, crab, or armadillo.

Carnivore An animal that eats meat.

Carp One of the most common fishes in the world that has been widely found in North America, and is often included in the human diet in other parts of the world.

Carrion The rotting flesh of a dead animal.

Carrying capacity A measure of the maximum number of animals that a habitat can support based on the amount of food and shelter that is available.

Cartilaginous fish A group of fishes including sharks and rays whose skeletons are composed of hard cartilage instead of bone.

Catadromous fish A freshwater fish that migrates the ocean to spawn.

Catalytic converter A device that removes significant amounts of pollutants from exhaust gases of motor vehicles or wood stoves.

Catkin The flowering parts of some trees that are used as food by grouse and other birds during the winter season.

Caviar The eggs of sturgeon often considered a delicacy food by humans.

Cephalopoda A class of marine mollusks with 8–10 arms equipped with suction organs that are used to capture prey. The Cephalopoda includes the squids, octopuses, and cuttlefishes. These animals have sacs containing ink which is released to hide them from their enemies.

Cetacean Any members of the order cetacea, that it includes the dolphins, porpoises and related forms that have one or two nares opening at the top of their heads and a horizontally flattened tail for swimming.

Char Brook trout and other closely related fishes.

Chemical energy Energy that is stored in plant tissues as sugars, oils and starches during the process of photosynthesis.

Chorion A protective membrane that grows out of the embryo, surrounding it and the other three membranes.

Class An intermediate division in the taxonomy of living organisms into which organisms of the same phylum are divided. It ranks above the order and below the phylum or division.

Cloven-hoofed A condition in which the hoof is divided into two parts, as found in sheep.

Clutch A nest of eggs or a brood of chicks.

Community All living organisms within a defined area such as a log, a woodland, or a marsh. It can be as small as a rotting log or as big as an entire forest.

Competitive advantage A condition that exists when one organism is better able to survive in an environment than another.

Competitive exclusion principle A principle of ecology that states that no two species of plants or animals can occupy the same niche in the environment indefinitely.

Compound eye A complex eye, as found in insects, that is made up of many simple eyes functioning together.

Conical bill A shape of a bird's bill that is particularly adapted for shelling grass and weed seeds.

Conifer A tree that produces seeds in cones.

Coniferous forest biome A group of evergreen forest ecosystems made up of a larch, pine, spruce, and other cone-bearing trees located in the northern regions of North America.

Conservation The practice of protecting natural resources against waste.

Constrictor A snake that squeezes its prey to death in its coils.

Continental shelf Land submerged beneath the surface of the ocean that slopes gradually away from the shore toward deeper water.

Cormorant A long-necked seabird that dives for fish from the surface of the ocean, and pursues them underwater by swimming with both its wings and its feet.

Covey A small flock of quail or partridges often members of the same family group.

Cow An adult female of various large animals such as an elk, moose, whale, walrus, bison, or musk ox.

Crop An organ located in the digestive tracts of birds and some other organisms in which food is stored before it is digested.

Crustacean A shellfish such as a lobster, crab, or shrimp that has a hard outer shell on its body. These are arthropods in the class crustacea.

Cud A small portion of plant material from the rumen of an animal that is regurgitated, chewed thoroughly, and swallowed again.

Cuttlefish A squid-like cephalopod mollusk with a hard internal shell and ten arms equipped with suction devices.

Cygnet A young swan.

D

Dabbling duck A duck that feeds on or near the surface of the water by submerging its head as it feeds.

Darter A small, bottom-dwelling food fish related to the perch, upon which many larger fish prey.

Deciduous forest biome A group of ecosystems in which broadleaf trees are abundant, and annual precipitation exceeds 30 inches per year. It includes only those broadleaf trees that shed their leaves each year.

Decompose Bacteria, fungi, insects, worms, etc. that are capable of breaking down complex substances to form simple elemental components, making them available to plants.

Delayed gestation A condition in which a fertilized embryo does not attach to the inner surface of the mother's uterus until conditions favor survival of the offspring.

Demersal spawner A fish, such as the herring, that attaches its eggs to weeds or other materials on the ocean floor.

Denitrification A natural process through which nitrates are broken down by bacteria to release nitrogen gas to the atmosphere.

Desert biome A terrestrial environment in which annual precipitation is less than 10 inches, inhabited by drought-tolerant species of plants and animals.

Dextral fish A flatfish whose eyes are always located on the right side of its head.

Diadromous fish A fish that migrates in either direction between freshwater and marine habitats.

Dipper A small, wrenlike bird with a habit of bobbing up and down continuously. It eats aquatic insects that it catches by walking beneath the water on the bottoms of swiftly moving streams.

Diurnal The tendency of an animal or bird to be active only during periods of daylight.

Doe A female deer, pronghorn, goat, or rabbit.

Dog Any member of the dog family, or a male fox, coyote, or wolf.

Dolphin A small, predatory, toothed whale (family Delphinidae) with a beaklike snout that is found in warm ocean waters.

Domestic A condition in which a plant or animal is raised in captivity or controlled conditions.

Dorsal fin A large prominent fin located on the backs of most fish.

Dove A small pigeon.

Down The soft, fluffy feathers beneath the outer feathers of a bird that insulate it from heat and cold.

Duckling A young duck.

E

Eaglet A young eagle.

Ecdysis Molting or shedding an outer layer of skin.

Ecologist A scientist who studies relationships between living organisms and their environments.

Ecology The branch of biology that describes relationships between living organisms and the environments in which they live.

Ecosphere Same as biosphere. All the ecosystems on the earth capable of supporting life.

Ecosystem A community of living organisms plus all of the nonliving features of the environment such as water, air, sunlight, and soil.

Ectotherm A cold-blooded animal that depends on the surrounding environment for body heat.

Eel A long snakelike fish with a long dorsal fin that extends around its tail.

Eft A land-based stage in the metamorphosis of a newt that occurs between the aquatic larval and adult stages, during which the newt changes in color to red or orange.

Egg case A pouchlike container filled with fertilized eggs that is expelled from the body of an oviparous female shark.

Elapid A poisonous snake having fangs that are permanently fixed in place.

Electrical energy Energy that is in the form of electricity.

Electrostatic precipitator A device that removes particles from smoke by charging the particles as they pass through an electrical field composed of small wires, and then attracting them to a similar electrical field having the opposite charge.

Elemental cycle The recurring circular flow of elements from living organisms to nonliving materials.

Elver A young freshwater eel that has just emerged from the larval stage.

Embryonic membrane Any of four different membranes that develop around an embryo as it matures.

Endangered species A legal designation assigned to a species or subspecies that is in immediate danger of becoming extinct due to small numbers of survivors in the population. This classification is assigned for the purpose of providing protection to such organisms.

Endotherm A warm-blooded animal that obtains body heat from the metabolism of its food.

Energy The ability to do work or to cause changes to occur.

Erosion The loss of topsoil from a region due to the forces of flowing water or strong winds.

Estivation A state in which an animal spends the hot summer in a dormant condition similar to hibernation.

Estuary An aquatic environment in which freshwater and salt water mix in areas where rivers and streams flow into the oceans.

Ethics The code of morals that governs the behavior of a person or group of people.

Evolution A process in which physical changes occur in organisms over long periods of time, during which those physical traits that help the organism survive in the environment become dominant and are expressed more frequently.

Ewe A female sheep.

Extinct No longer existing due to the deaths of all the members of a population.

Eyrie See Aerie.

F

Family An intermediate division in the taxonomy of living organisms into which organisms of the same order are divided. Family ranks above the genus and below the order.

Fauna The animals that are found in a particular region or environment.

Fawn A young deer or pronghorn.

Fecundity A fertile, prolific or productive trait in an animal.

Finfeet Another name for seals that acknowledges the development of fins for locomotion instead of feet.

Fingerling A young fish that is less than a year old.

First law of energy A law of science stating that energy cannot be created or destroyed, but it can be converted from one form of energy to another (e.g., light to heat).

Flatfish A flat fish such as a flounder that lives on the ocean floor, and whose eyes are both located on the same side of its head.

Fledge To grow enough feathers for a young bird to be capable of flying.

Flora The plants that are found in a particular region or environment.

Food chain A series of steps through which energy from the sun is transferred to living organisms. Members of the food chain feed on lower ranking members of the community.

Food web A group of interwoven food chains.

Forest floor The layer of decaying plant materials on the soil surface; it acts as a mulch which preserves moisture.

Fossil fuel A fuel that comes from deposits of natural gas, coal, and crude oil that are formed in the earth from plant or animal remains.

Freshwater Water that is not high in salt content.

Freshwater biome A set of similar ecological communities found in or near water that is not salty.

Frontal shield A red or white fleshy growth that extends from the top of the bill to the foreheads of some birds, and that is used by strong males to intimidate their rivals.

Fry   A tiny fish that has recently hatched from an egg.

G

Gallinaceous   Of an order of birds (galliformes) such as pheasants and grouse that build nests on the ground. Females of this order lay large clutches of eggs in comparison with other birds.

Gastropoda   A class of mollusks such as snails and slugs that move about on a muscular foot.

Gecko   A lizard belonging to either of two families in North America that has suction pads on its feet and that is capable of making loud calls.

Genetic engineering   Human modification of the genetic makeup of organisms.

Genus   An intermediate division in the taxonomy of living organisms into which organisms of the same family are divided. The genus ranks above the species and below the family.

Gestation   The period of pregnancy during which the young are carried in the uterus of the female.

Gizzard   A muscular organ in the digestive tract of birds, reptiles and other organisms that uses small rocks and pebbles to grind food into small particles.

Gosling   A young goose.

Grassland biome   A terrestrial environment sometimes, called a prairie, that is located in the middle of the continent and that lacks tree cover. The dominant plants are grass and broadleafed herbs.

Grebe   An aquatic diving bird with lobes on its toes to facilitate swimming that eats insects and small aquatic animals.

Gregarious   The tendency of some animals to prefer the company of their own kind, and to spend their time together in flocks, colonies, or herds.

Grit   Small rocks and pebbles that are swallowed whole by birds for use in grinding seeds and other food materials in the gizzard.

Groundwater   Water that is located under the surface of the earth in underground streams and reservoirs.

H

Habitat   An environment in which a plant or animal lives.

Harrier   A genus of hawks that are excellent hunters in heavy vegetation such as tall grass or marsh reeds, and that nest in groups of several pairs instead of defending individual territories.

Hatchery fish   A fish that has been spawned and raised in the artificial environment of a fish hatchery.

Hazardous material   Chemicals or poisonous materials that are dangerous to living organisms.

Herbicide   A pesticide that is used to kill unwanted plants.

Herbivore   An animal or other organism which eats plants.

Herb layer   The bottom layer of vegetation in a forest consisting of ferns, grasses, and other low plants that grow on the forest floor beneath the shrub layer.

Hermaphrodite   An organism which possesses both male and female sex organs at the same time.

Herpetologist   A scientist who specializes in the study of reptiles and amphibians.

Hibernate   To spend the cold winter season in a resting state in which the body temperature is reduced, body processes slow down and nutrition is derived from stored body fat.

Horn   Hard, bony projections on the heads of some hoofed animals that are used for fighting with rivals for protection against predators.

Hybrid   The offspring of a mating between two different species.

I

Ichthyologist   A scientist who studies fishes.

Ichthyology   The branch of zoology that is concerned with the classification, structure, and life cycles of fishes.

Ideal environment A habitat in which all of the living conditions are compatible with the needs of a particular organism.

Iguana Any of several tropical American lizards with a serrated dorsal crest that eat insects or plants.

Imprinting A learning process whereby young animals learn to mimic the behavior of a parent or trusted caregiver to establish a behavior pattern, such as recognition of and attraction to its own kind.

Incubation A process by which fertilized eggs are warmed by the body of a parent to create a constant temperature until they hatch.

Industrial waste Harmful chemicals, poisonous metal compounds, acids and other caustic materials that are left over from manufacturing processes.

Insecticide A pesticide that is used to kill insects.

Insectivore An animal whose diet consists of insects.

Intertidal zone An area near the shore of the ocean that is covered with water during high tide, and that is exposed above the water level during low tide.

Intrinsic value Appreciation for or essential nature of the wild creatures and the environments in which they live.

Isospondylous fish A fish whose vertebrae are essentially the same near its head as they are in its tail.

J

Jacobson's organ A sensing organ located in the mouth of a snake that is used for smelling.

Jaguar The largest predatory cat in North America with a range extending from Texas to Argentina, and prey consisting mostly of deer, peccaries, and domestic livestock.

Jaguarundi A predatory cat that prefers dense undergrowth, and that ranges from the southwestern United States to South America.

Javelina Same as peccary. A cloven-hoofed mammal, also known as a peccary, that is incorrectly called a wild pig but belonging to a different family.

K

Keel A ridge on the carapace of a turtle.

Kestrel An American falcon that is able to hover in a fixed position above the ground by flying into the wind.

Kid A young goat.

Kinetic energy Energy that is associated with motion and movement in animals.

Kingdom The highest division in the taxonomy of living organisms ranking above the phylum.

Kite One of several species of hawks with a forked tail and long, pointed wings that is found in warm climates.

Krill Tiny, shrimplike crustaceans that make up much of the diet of baleen whales.

L

Lamb A young sheep.

Lamellae Small, comblike projections on the bills of some ducks with which they strain small food particles out of the water.

Larva Immature, forms of some fishes, insects, salamanders, frogs, and animals.

Laterally compressed A description of a fish that is deeper from its back to its belly than it is from side to side.

Law of conservation of matter A basic law of physics that states that matter can be changed from one form to another, but it cannot be created or destroyed by ordinary physical or chemical processes.

Lemming A small arctic rodent that resembles a mouse and has a short tail and fur-covered feet.

Lentic habitat An aquatic environment characterized by still water, such as a marsh, swamp, pond, or lake.

Leptocephali   The marine larval stage of development of various eels.

Limnologist   A person who studies freshwater habitats.

Lobe   A fleshy growth on the sides of the toes of some birds that enables them to swim.

Loon   A solitary diving bird with webbed feet and a sharp bill found in northern habitats that captures and feeds on fish.

Lotic habitat   An aquatic environment characterized by actively moving water, such as a stream or river, where the flowing water restricts plant growth and food for fish and other aquatic animals is transported from distant source.

M

Mammal   Warm-blooded animals that have bony skeletons, protective hair coats, and mammary glands that produce milk to nourish their offspring.

Mammary gland   An mammalian organ which in the female gender secretes milk.

Manatee   A large, aquatic herbaceous mammal that ranges in warm coastal waters along the coast of Florida.

Margay   A predatory cat that resembles the ocelot, and that ranges from southernmost Texas to South America.

Marine biologist   A person who studies the aquatic organisms that live in oceans.

Marine biome   An aquatic environment that makes up the world's largest biome, and consists of the oceans where the salt concentration ranges from 3 to 3.7 percent.

Marine mammal   A mammal whose body is adapted to living in the ocean.

Marsupial   An unusual order of mammals, of which the opossum is the only North American example, that gives birth to offspring before they are fully developed and rears them in an external abdominal pouch.

Marsupium   The external pouch of a female marsupials in which the mammary glands are located, and in which the offspringare reared.

Martin   Any of several birds that are members of the swallow family.

Metabolism   The process by which food is digested and used by the cells of the body to release energy.

Metamorphosis   A process during which immature organisms such as insects and amphibians change in their physical structure and appearance to become more like adult members of their species.

Milt   The sperm-containing fluid of a male fish.

Mollusk   A shellfish of the phylum Molluseca, such as a clam, oyster, mussel, or snail, with a soft, unsegmented body enclosed totally or in part by a hard outer shell.

Molt   A process by which an animal loses it old outer covering such feathers, skins, shells, etc. and replaces them with new ones.

Monogamous   A mating behavior in which a male and a female bond as a pair, and mate.

Multiple use   A management strategy for natural resources that considers the needs of the different groups of people who use or desire to use the resources.

Musk   A foul smelling fluid that is secreted from specialized glands in some mammals as in the case of the weasel or musk ox.

N

Nares   Nostrils or nasal passages of vertebrates.

Narwhal   An unusual whale found in the northern seas, with a long, hollow, twisted tusk in the left upper jaw of the male.

Naturalist   A person who studies nature by observing plants and animals.

Neotenic   A tendency of newts and other organisms to metamorphose directly to the sexually mature adult stage from the larval stage, without entering a land-based stage of development.

Neritic zone   The area of the ocean beyond the intertidal zone that extends to the outer edge of the continental shelf.

Nestling   A young immature bird that has not yet left the nest.

Niche A specific role or function within a habitat that is performed by an organism allowing different organisms to occupy different niches in the same habitat.

Nightjar A nocturnal bird with a harsh call and a large mouth lined with bristles that aid in capturing insects during flight.

Nitrate An important plant nutrient that becomes a pollutant when it is present in groundwater or in surface water in excessive concentrations.

Nitrogen cycle The circular flow of nitrogen from free nitrogen gas in the air to ammonia and nitrates in soil, water, and organisms, and back to atmospheric nitrogen.

Nitrogen fixation A process by which nitrogen gas from the atmosphere is converted to ammonia by soil microorganisms.

Nitrogen-fixing bacteria Bacteria that live in soil, water, or in nodules or colonies on the roots of certain plants. These bacteria are capable of changing nitrogen gas to ammonia.

Nocturnal A tendency of some mammals or birds to sleep during the day and forage for food during the night.

Nonadaptive behavior Failure of an organism to adapt to a changing environment.

Nonbiodegradable The resistance of a substance or material to being broken down by living organisms.

Nonpoint source pollution Pollution that comes from several sources, and that cannot be traced back to a single point of origin.

Nonrenewable resource A resource such as minerals or oil that cannot be replaced when it is lost due to excessive use or abuse.

O

Oceanic zone An area beginning at the outer edge of the continental shelf that includes the deep ocean region extending to the continental shelf of the opposite shore.

Oceanography The scientific study of ocean environments including the chemistry and physics of water, plants, animals, reefs and other oceanic features.

Oceanologist A scientist who studies ocean environments.

Ocelot A spotted predatory cat that ranges from Texas to South America.

Omnivore An animal that eats both plants and other animals.

Order An intermediate division in the taxonomy of living organisms into which organisms of the same class are divided. The order ranks above the family and below the class.

Organism An individual plant, animal, or other life form with organs and parts that function together.

Ornithologist A scientist who studies birds.

Ornithology The branch of zoology that deals with the study of birds.

Overgrazing A condition in which domestic or wild animals destroy the vegetational cover in an area by harvesting or trampling the plants beyond their ability to recover. This occurs when the carrying capacity of the area is exceeded.

Oversight Responsible care and management public and private lands, and natural resources.

Oviparous Reproducing by laying eggs that develop and hatch outside the body of the female.

Ovoviviparous Reproducing from eggs that remain inside the female until they have hatched.

Ovum An mature female egg that is capable of developing into a new member of the species after it has been fertilized.

P

Parasitic bird A bird that lays its eggs in the nests of other birds.

Peccary Same as javelina. Sometimes incorrectly called a wild pig, but belonging to a different family.

Pectoral fin A fin that fish use to control their direction of travel, and which represents the forelimbs in most fish.

Pelagic Inhabiting the surface waters of the open ocean as in some fishes, birds and other organisms.

Pelvic fin   A fin that occurs in pairs on the lower rear of a fish, and which corresponds to hind legs in vertebrates.

Pesticide   A chemical that is used to kill insects, weeds, rodents, fungi, or other pests. Pesticides include herbicides, insecticides, rodenticides, and fungicides.

Petroleum   An oily, flammable liquid that occurs naturally in underground deposits.

Phalarope   A bird that is unusual in that females are brightly colored, and the drab-colored males incubate the eggs.

Phosphate   An important plant nutrient that becomes a pollutant in groundwater or in surface water when it is present in excess amounts.

Photoperiod   The recurring cycle of daylight and darkness that influences favorable for an organism to mature or develop physical functions of organisms such as sexual maturity.

Phylum   An intermediate division in the taxonomy of living organisms into which organisms of the same kingdom are divided.   The phylum is above the class and below the kingdom.

Phytoplankton   Microscopic plants that float on the surface of water.

Pigeon milk   A food that is composed of cells secreted in the crops of both parents, having the appearance and food value of cottage cheese.

Pika   A small, short-eared rodent having two sets of upper incisor teeth, and found in the rocky uplands of western North America

Pinniped   Carnivorous aquatic animals, such as seals and walruses. The limbs of these animals are modified flippers.

Placenta   A vascular organ consisting of a fluid-filled sac that is connected to the blood supply of the mother's uterus. It encloses the babies of most mammals protecting them from injury, eliminating wastes, and nourishing them inside the body of the mother until they are born.

Placental mammal   A mammal whose offspring develop inside the body of the mother inside the uterus surrounded by the placenta.

Plankton   Microscopic plants and animals that float in surface waters, and that provide food for fish and other aquatic animals.

Plastron   The lower shell of a turtle or tortoise.

Plover   A family of small- to medium-sized precocial shore-birds that feeds on insects and small aquatic animals.

Plumage   All the feathers of a bird.

Plume   A showy feather that develops upward and curves forward on the head of a quail or other plumed bird.

Poaching   To take anything, especially wild game, by illegal methods.

Pod   A group of whales that live together.

Point source pollution   Pollution of the environment that can be traced back to its point of origin.

Pollutant   A waste material or harmful chemical that is discharged into the environment.

Pollywog   The same as tadpole. The larval stage of development in the metamorphosis of a frog or toad that is sometimes called a tadpole.

Polyandry   A mating behavior in which a female mates with more than one male.

Polygamous   A mating behavior in which an animal has more than one mate at a time.

Polygynous   A mating behavior in which a male mates with more than one female.

Population   A group of similar organisms that is found in the same area.

Precocial   Of a type of bird whose young are covered with down, and which becomes active at the time of hatching, and that are capable of a somewhat independent life from birth.

Predator   An animal that kills and eats other animals.

Prehensile   Adapted for grasping as in the tail of the opposum.

Primary consumer   An animal that eats plants.

Primary succession   The development of an ecological community in an area where living organisms were not previously found, such as on a newly formed volcanic island.

Pristine   Of or constituting a pure or undamaged resource.

Producer   A green plant that converts solar energy to starches and sugars.

Prolific Having the capacity to produce large numbers of offspring.

Promiscuity A mating behavior wherein animals mate with numerous members of the opposite sex, and they do not form pair bonds.

Pronghorn A member of the deer family having hollow, two-pronged horns that resembles an antelope, and that lives in deserts or plains of western North America. Also called a pronghorn antelope.

Puma A large predatory cat that is also known as a cougar, panther, or mountain lion.

R

Race A division in the taxonomy of living organisms into which organisms of the same species are divided.

Radiant energy Energy that comes from the sun.

Rail A group of small to medium wading birds of which some are flightless that lives in the reeds that line the shores of wetlands. They have long toes that enable them to walk on soft mud.

Ram A male sheep.

Range of tolerance The limits in environmental conditions within which an organism can survive and function.

Raptor A bird of prey such as a hawk, owl, or eagle.

Renewable resource A resource such as a forest or wild animal that is capable of being replaced following its use.

Reptile A cold-blooded animal having scales, plates, or shields on its body and claws on its feet. Most reptiles lay eggs with hard leathery shells.

Riparian zone The land adjacent to the bank of a stream, river, or other waterway.

Rodent A small, gnawing animal that is identified by the four large incisor teeth located in the front of its mouth.

Rodenticide A chemical poison that is used to kill rodents such as mice or rats.

Roe Clusters of fresh fish eggs enclosed in ovarian membranes.

Rumen A large stomach compartment in a ruminant animal in which high-fiber plant materials are soaked, warmed, and digested with the aid of bacteria.

Ruminant An animal such as a deer or bison that has a series of four stomach compartments that are capable of digesting food that is high in fiber.

Rut The breeding period for some of the large North American ruminants during which males become aggressive in their behavior.

S

Salinity A measure of the salt concentration in water affecting the survival of organisms living in aquatic environments.

Salmonid A family of freshwater and marine fishes including salmon, trout, and whitefishes, that migrates upstream to spawn.

Scavenger A bird or mammal that eats any kind of food that it can find, including the carcasses of dead animals and garbage.

School A large group of fish that lives together in the same habitat.

Scrape A shallow depression that has been scratched out of the ground, and that is sometimes lined with dry vegetation for a nest.

Scute A large bony plate that is found on the head, back, or sides of an animal such as the outer covering on a sturgeon.

Secondary consumer A carnivorous animal that obtains its nutrition by eating primary consumers and other carnivores.

Secondary succession The gradual change in species of plants that live in an area during the time that a damaged ecosystem is returning to its original stage of ecological development.

Second law of energy A scientific law stating that each time energy is converted from one form to another, some energy is lost in the form of heat.

Shad Anadromous fishes that are members of the herring family.

Shellfish An animal with a soft body and hard outer shell, such as a mollusk or crustacean.

Shrike   Black and white birds with shrill voices that gather insects and store them for later meals.

Shrub layer   Vegetation consisting of short woody plants that occupies the stratum between the herb layer and the understory of a forest.

Silt   Tiny soil particles that are easily eroded by becoming suspended in flowing water or blown as dust in the wind.

Silt load   The amount of eroded soil that is carried in the flowing waters of streams and rivers.

Sinistral fish   A flatfish whose eyes are always on the left side of its head.

Skink   Any of several lizards having shiny bodies, short legs, and smooth scales.

Smelt   A small, slender food fish found in large schools in most coastal regions, and upon which many larger fishes prey.

Smog   Pollution of the atmosphere due to a poisonous mixture of fog and smoke.

Smolt   A young fish whose body is changing to allow it to leave its freshwater habitat and enter the saltwater environment of the ocean.

Soil conservation   The practice of protecting soil from erosion caused by strong winds or flowing water.

Solid waste   Includes most of the waste materials that are thrown away as garbage, but does not include liquid waste materials or gaseous effluents.

Sow   An adult female pig or bear.

Spawning   The process of reproduction in fish in which females deposit eggs on the bed of a stream, pond, or lake and males discharge sperm on the surface of the eggs.

Species   A division in the taxonomy of living organisms into which organisms of the same genus are divided. The species is below the genus, and is comprised of organisms or populations whose members can interbreed and who share a common gene pool.

Squab   A young pigeon or dove.

Steelhead   A race of large anadromous rainbow trout.

Steward, land   An administrator or supervisor who manages resources and/or property.

Stewardship, land   Exercising responsible and careful management over resources and/or property entrusted to one's care.

Stoop   A fast dive by a bird to capture a prey in flight.

Stratum   Consists of one of several levels or layers of plant growth in an ecosystem such as a forest.

Surface water   Water that is located on the earth's surface in rivers, streams, ponds, and lakes.

Swift   A migratory bird resembling a swallow with long wings, a slightly forked tail and weak legs and feet that feeds on insects. It remains in flight almost continuously except for the nesting season.

Swimmeret   A small swimming appendage occurring in pairs that is located on the abdomen of some crustaceans, and that function in reproduction and locomotion.

Symbiosis   A relationship between two organisms in which each organism receives benefits from its association with the other organism.

T

Tadpole   The larval stage of development in the metamorphosis of a frog or toad that is characterized by a long, rounded body and external gills. It is also called a pollywog.

Talon   The toe and sharp claw of a bird of prey with which its grasps and kills its prey.

Tanager   A brightly colored songbird that prefers habitats along streams near willow and cottonwood trees.

Tapetum lucidum   A special tissue located in the eyes of walleye and other fish that gives them night vision.

Taxonomist   A scientist who classifies living organisms and defines their relationships with other organisms.

Taxonomy   The field of science that classifies living organisms and defines their relationships with other organisms.

Temperate Forest Biome   Same as deciduous forest biome   A group of ecosystems in which

Broadleaf trees are abundant and annual precipitation exceeds 30 inches per year.

Terrapin Any of several turtles that live in freshwater habitats or saltwater marshes.

Terrestrial biome A large community of plants, animals, and other living organisms that live on land.

Terrestrial To live on land.

Territorial A behavior in which birds and other animals establish and defend living and hunting areas against other competing members of their species.

Thermal Energy Energy that is released as heat when fuels or nutrients are burned or digested.

Thermal Stratification Differences in water temperatures at various depths with deep water being colder than water near the surface.

Threatened species Species that are at risk due to declining numbers in their population, and that can reasonably be expected to survive if immediate steps are taken to protect the remaining populations and their habitats. It is a legal status that is established through The Endangered Species Act to provide extra protection and management.

Toothed whale A whale with a large mouth and sharp teeth, with which it attacks large prey and bites it into pieces that can be swallowed. They include dolphins, porpoises, and killer whales.

Torpedo A Pacific Electric Ray.

Torpid To be in a dormant condition due to cold temperatures.

Torpor A physical state similar to hibernation that occurs in some animals during periods of cold weather in which their metabolism slows down and they require little or no food.

Tortoise A turtle that is adapted for living on land.

Toxic waste Waste products that are poisonous to living organisms.

Transpiration The loss of water to the atmosphere from the leaves of plants.

Tree creeper A small, brown bird usually observed climbing tree trunks probing for insects with its long curved bill.

Tundra biome A group of ecosystems located in the frozen northern regions of the continent where evaporation rates are low, precipitation is minimal and swamplike conditions exist during the summer because water cannot penetrate the frozen soil.

Turbid Muddy or cloudy water conditions.

Turtle A reptile with a toothless beak and a soft body inside a two-part shell, into which it can withdraw its head, legs, and tail.

Tympanum The vibrating membrane in the middle ear, or the round external eardrum of a frog or toad.

U

Understory Short trees in a forest that fill an intermediate stratum of vegetation beneath the canopy created by the branches and foliage of the tallest trees.

Ungulate A mammal with hooves.

Uterus An internal female organ to which the placenta becomes attached, and which encloses the growing fetus as it develops prior to birth.

V

Velvet The soft skin containing blood vessels that nourishes and protects the developing antlers of deer.

Vertebrate An animal with a segmented backbone that surrounds the spinal cord.

Viper A poisonous snake having hinged fangs in the front of its mouth that fold back out of the way when they are not in use.

Viviparous To give birth to live offspring.

Vixen A female fox.

Vole A small mouselike rodent with a stout body and short tail.

W

Water cycle   The movement of water in the form of vapor from the oceans to the clouds to the earth as precipitation, and back to the oceans through rivers and streams.

Waterfowl   Swimming game birds that live in water habitats.

Watershed   An area bounded by geographic features where precipitation is absorbed in the soil to form groundwater that eventually emerges to become surface water and that ultimately drains to a particular water course or body of water.

Wattles   Red tissue on the face of a bird such as a turkey that swells with blood and takes on a bright red color during the mating display.

Wetlands   Land areas that are flooded during all or part of the year.

Wet scrubber   A device that sprays water through a chamber to remove particles and water-soluble materials from polluted gases and smoke fumes before they are released into the atmosphere.

Whalebone   A comblike bony structure in the mouths of some whales that is used to strain food such as plankton and other small organisms from ocean water. (aka baleen)

Whitecoat   A baby harp seal that has a white fur coat at birth.

Wild fish   A strain of fish that reproduces naturally in the wild.

Y

Yolk   A yellow substance found in eggs containing protein and fat for the nourishment of embryos and newly hatched fish, birds, and reptiles.

Yolk sac   A pouchlike organ found in young fish, reptiles, and birds containing a highly nutritious material called yolk that nourishes the young before and after hatching.

Z

Zooplankton   Microscopic animals that live at or near the surface of a body of water.

Index

NOTE: Page numbers in *italics* reference non-text material.

F